Human–Computer Interaction and Technology Integration in Modern Society

Hakikur Rahman
Institute of Computer Management and Science, Bangladesh

A volume in the Advances in Human and Social Aspects of Technology (AHSAT) Book Series

Published in the United States of America by
IGI Global
Engineering Science Reference (an imprint of IGI Global)
701 E. Chocolate Avenue
Hershey PA, USA 17033
Tel: 717-533-8845
Fax: 717-533-8661
E-mail: cust@igi-global.com
Web site: http://www.igi-global.com

Library of Congress Cataloging-in-Publication Data

Names: Rahman, Hakikur, 1957- editor.
Title: Human-computer interaction and technology integration in modern society / Hakikur Rahman, editor.
Description: Hershey, PA : Information Science Reference, [2021] | Includes bibliographical references and index. | Summary: "This book will serve as an authoritative reference foundation for the latest scholarly research on the widespread incorporation of technological innovations around the globe by investigating how applications of ubiquitous computing technologies affects various aspects of human lives"-- Provided by publisher.
Identifiers: LCCN 2020019408 (print) | LCCN 2020019409 (ebook) | ISBN 9781799858492 (h/c) | ISBN 9781799858508 (s/c) | ISBN 9781799858515 (ebook)
Subjects: LCSH: Telecommunication--Social aspects. | Information technolgy--Social aspects. | Information society. | Human-computer interaction.
Classification: LCC HM851 .H853 2021 (print) | LCC HM851 (ebook) | DDC 303.48/33--dc23
LC record available at https://lccn.loc.gov/2020019408
LC ebook record available at https://lccn.loc.gov/2020019409

This book is published in the IGI Global book series Advances in Human and Social Aspects of Technology (AHSAT) (ISSN: 2328-1316; eISSN: 2328-1324)

British Cataloguing in Publication Data
A Cataloguing in Publication record for this book is available from the British Library.

Advances in Human and Social Aspects of Technology (AHSAT) Book Series

Ashish Dwivedi
The University of Hull, UK

ISSN:2328-1316
EISSN:2328-1324

Mission

In recent years, the societal impact of technology has been noted as we become increasingly more connected and are presented with more digital tools and devices. With the popularity of digital devices such as cell phones and tablets, it is crucial to consider the implications of our digital dependence and the presence of technology in our everyday lives.

The **Advances in Human and Social Aspects of Technology (AHSAT) Book Series** seeks to explore the ways in which society and human beings have been affected by technology and how the technological revolution has changed the way we conduct our lives as well as our behavior. The AHSAT book series aims to publish the most cutting-edge research on human behavior and interaction with technology and the ways in which the digital age is changing society.

Coverage

- Technology Dependence
- Digital Identity
- Philosophy of technology
- Human Rights and Digitization
- Computer-Mediated Communication
- Technology Adoption
- ICTs and social change
- Cyber Bullying
- Technology and Social Change
- Cultural Influence of ICTs

IGI Global is currently accepting manuscripts for publication within this series. To submit a proposal for a volume in this series, please contact our Acquisition Editors at Acquisitions@igi-global.com or visit: http://www.igi-global.com/publish/.

The Advances in Human and Social Aspects of Technology (AHSAT) Book Series (ISSN 2328-1316) is published by IGI Global, 701 E. Chocolate Avenue, Hershey, PA 17033-1240, USA, www.igi-global.com. This series is composed of titles available for purchase individually; each title is edited to be contextually exclusive from any other title within the series. For pricing and ordering information please visit http://www.igi-global.com/book-series/advances-human-social-aspects-technology/37145. Postmaster: Send all address changes to above address.

Titles in this Series

For a list of additional titles in this series, please visit: www.igi-global.com/book-series
www.igi-global.com/book-series/advances-human-social-aspects-technology/37145

Present and Future Paradigms of Cyberculture in the 21st Century
Simber Atay (Dokuz Eylül Üniversity, Turkey) Gülsün Kurubacak-Meriç (Anadolu University, Turkey) and Serap Sisman-Uğur (Anadolu University, Turkey)
Information Science Reference • ©2021 • 280pp • H/C (ISBN: 9781522580249) • US $185.00

Machine Law, Ethics, and Morality in the Age of Artificial Intelligence
Steven John Thompson (University of California, Davis, USA & University of Maryland Global Campus, USA)
Engineering Science Reference • ©2021 • 266pp • H/C (ISBN: 9781799848943) • US $295.00

Responsible AI and Ethical Issues for Businesses and Governments
Bistra Vassileva (University of Economics, Varna, Bulgaria) and Moti Zwilling (Ariel University, Israel)
Engineering Science Reference • ©2021 • 259pp • H/C (ISBN: 9781799842859) • US $225.00

Bridging the Gap Between AI, Cognitive Science, and Narratology With Narrative Generation
Takashi Ogata (Iwate Prefectural University, Japan) and Jumpei Ono (Aomori University, Japan)
Information Science Reference • ©2021 • 409pp • H/C (ISBN: 9781799848646) • US $190.00

Reductive Model of the Conscious Mind
Wieslaw Galus (Independent Researcher, Poland) and Janusz Starzyk (Ohio University, USA)
Information Science Reference • ©2021 • 296pp • H/C (ISBN: 9781799856535) • US $195.00

Dyslexia and Accessibility in the Modern Era Emerging Research and Opportunities
Kamila Balharová (Pedagogical and Psychological Counseling Center, Brno, Czech Republic) Jakub Balhar (Gisat s.r.o., Czech Republic) and Věra Vojtová (Masaryk University, Czech Republic)
Information Science Reference • ©2021 • 279pp • H/C (ISBN: 9781799842675) • US $165.00

Understanding the Role of Artificial Intelligence and Its Future Social Impact
Salim Sheikh (Saïd Business School, University of Oxford, UK)
Engineering Science Reference • ©2021 • 284pp • H/C (ISBN: 9781799846079) • US $200.00

Overcoming Barriers for Women of Color in STEM Fields Emerging Research and Opportunities
Pamela M. Leggett-Robinson (PLR Consulting, USA) and Brandi Campbell Villa (Belay Consulting, USA)
Information Science Reference • ©2020 • 199pp • H/C (ISBN: 9781799848585) • US $145.00

701 East Chocolate Avenue, Hershey, PA 17033, USA
Tel: 717-533-8845 x100 • Fax: 717-533-8661
E-Mail: cust@igi-global.com • www.igi-global.com

Table of Contents

Detailed Table of Contents

Information and communication technologies (ICT) are influencing every discipline under the sun including education. It is affecting every aspect of education from teaching-learning to assessment and evaluation. It improves the effectiveness of education. It aids literacy movements. It enhances scope of education by facilitating mobile learning and inclusive education. It facilitates research and scholarly communication. Impact of ICT and its potential for the education field is manifold. It positively affects all the stakeholders of the education field. The chapter discusses the same along with the various projects and schemes adopted by the country to spread ICT literacy and competency.

Online/web-based collaborative tools enable teachers and students to perform a wide range of tasks, such as interactive discussions, online collaboration activities, sharing and accessing electronic learning resources, and many more others. It not only promotes critical thinking and reflection in students but also encourages them to develop a sense of community, thus enabling the creation of an environment in which further collaborative work can take place. The author has categorised various tools into 11 types that deal with idea generation and brainstorming, live conference, robotics and coding tools, mapping, design, online group work and document collaboration, and online communication and content development. The chapter explains the online collaboration with its features, preparation required by institution and role of teacher presence in online learning. It also emphasizes that library consultations (i.e., librarians) directly boost student learning, so the active collaboration of librarians is a must.

While talking about successful entrepreneurship and value addition within an enterprise through innovation, one could comprehend that the innovation paradigm has been shifted from simple introduction of new thoughts and products to accumulation of diversified actions, actors, and agents along the process. Furthermore, when the innovation process is not being constrained within the closed nature of it, the process takes many forms during its evolution. Innovations have been seen as closed innovation or open innovation, depending on its nature of action, but contemporary world may have seen many forms of innovation, such as technological innovation, products/service innovation, process/production innovation, operational/management/organizational innovation, business model innovation, or disruptive innovation, though often they are robustly interrelated.

Innovation has been treated as a recognized driver of economic prosperity of a country through the sustained growth of its entrepreneurships. Moreover, the recently coined term open innovation is increasingly taking the lead in enterprise management in terms of value addition and knowledge building. Foci of academics, researchers, and practitioners nowadays are revolving around various innovation models, comprising innovation techniques, processes, and strategies. This chapter seeks to find out open innovation research and practices that are being carried out circumscribing development of entrepreneurships, particularly the sector belonging to the small and medium enterprises (SMEs) through a longitudinal study.

Despite the increased acceptance by corporate business houses around the globe, the adaptation of strategies and concepts belonging to the newly evolved dimension of entrepreneurships, the open innovation (OI), countries in the East, West, or South are yet to adapt appropriate strategies in their business practices, especially in order to reach out to the grass roots and marginal communities. By far, firms belonging to the small and medium sized enterprises (SMEs) sector, irrespective of their numbers and contributions towards their national economies, are lagging behind in accepting open innovation strategies for their business improvements. While talking about this newly emerged business dimension, it comprises of complex and dynamically developed concepts like, management of intellectual property aspects, administration of patents, copyright and trademark issues, or supervision of market trend for minute details related to knowledge attainment.

Ethics has always been a main domain of philosophical study. Hypertextual discursive expression has never been considered as the technology that can revolutionise the way we interpret ethical events and actions. This chapter intends to show that hypertextual discursive expression is not only useful, but also vital in any undertaking that seeks to reframe our ethical deliberations and judgements. The final objective is to convey the message that technology is becoming ever more essential in transforming philosophical thinking.

Despite their recent emergence, WhatsApp neighbourhood crime prevention (WNCP) groups are an already pervasive phenomenon in the Netherlands. This study draws on interviews and focus groups to provide an in-depth analysis of the watchfulness and surveillance activities within these groups. The conceptualisation of WNCP through the lens of practice theory shows that the use of ICTs in the form of WhatsApp amplified all three dimensions of neighbourhood watchfulness practices. It examines how friction at the intersections of materialities, competencies, and meanings affect neighbourhood dynamics as well as the personal lives and experiences of people involved. While voluntary citizen participation in crime prevention leads to an increase in social support, feelings of safety, and active prevention of break-ins, it also defaults to forms of lateral surveillance which transcend digital monitoring practices. Pressing issues related to social media use, participatory policing, surveillance, and the normalisation of distrust and intolerance have an impact beyond its localised Dutch context.

The introduction of digital media into university writing courses, while leading to innovative ideas on multimedia as a rhetorical enhancement means, has also resulted in profound changes in writing pedagogy at almost all levels of its theory and practice. Because traditional approaches to examining and discussing assigned texts in the classroom were developed to help students analyze different genres of print-based texts, many university educators find these methods prohibitively deficient when applied to digital reading environments. Even strategies in reading and text annotation need to be reconsidered methodologically in order to manage effectively the ongoing shift from print to digital or electronic media formats within first year composition. The current study proposes one of the first and most extensive attempts to analyze fully how students engage with digital modes of reading to demonstrate if and how students may benefit from reading digital texts using computer-assisted text analysis (CATA) software.

This chapter focuses on one element in the digitalisation of work: the forms and conditions of working in the so-called 'sharing economy' (SE). Based on an analysis of 67 SE platforms, it distinguishes three segments, each of which constitutes a distinctive institutional sphere within the sharing economy: these are an 'exchange and gift economy', a 'niche and sideline economy', and the 'platform economy'. In a further step, the study then identified and compared five dimensions of work within these segments: the type of activity, the form of compensation or recompense involved (monetary or non-monetary), skills and competencies required, the role of technology, and control mechanisms. Each segment is associated with a particular pattern of these dimensions. The chapter then discusses the shift in the

Preface

In recent years, incorporating a human computer interaction (HCI) perspective into the systems development life cycle (SDLC) is particularly necessary to information systems (IS) success and, in turn, to the success of not only entrepreneurships, but also for the development of human society. However, based on various studies, it is evident that the modern SDLC models are based more on organizational needs than human needs. Furthermore, the human interaction aspect of an information system is usually considered far too little (only the screen interface) and far too late in the IS development process (only at the design stage). Thus, there exists a gap between satisfying organizational needs and supporting and enriching common human users. (Zhang, P., Carey, J., Te'eni, D., & Tremaine, M., 2005).

As human computer interaction (HCI) and interactive systems design have developed a sense of people living with and through the ubiquitous usage of technologies, the main concerns have broadened from usability to include wider qualities of people's experiences with technology. Evidently, the user experience has become a focal interest in leading HCI textbooks, e.g. monographs, edited collections, researches and even the websites of leading entrepreneurs. There have also been specific technical proposals for enhancing people's experience with the interactive usage of computers. (McCarthy, J., Wright, P., Wallace, J., & Dearden, A., 2006).

Hence, the increased interest in issues such as engagement, care, and meaning, contemporary HCI has increasingly come to examine the nature of interactions between artifacts, humans, and environments through concepts such as user experience and implication. Researches indicate that, during the transition from usability metrics to user experience, what appears lacking is a more explicit characterization of what it is HCI now strives for as a discipline, for example, what constitutes a 'good' user experience? (Fallman, D., 2011). Therefore, this book has tried to incorporate issues, studies and researches similar to them.

Along these perspectives, this has been observed that in most of the countries demographic developments tend towards more and more people living in the marginal contexts. Improving the quality of life for those people at the grass roots is an emerging issue within the information society. It has been found that the good user interfaces have tremendous implications for appropriate accessibility. Though, it has to be stated that the user interfaces should not only be easily accessible, they should also be useful, usable and most of all enjoyable and a benefit for people (Holzinger, A., Ziefle, M., & Röcker, C., 2010). The book has tried to incorporate these issues at the extended states.

Furthermore, the interactive spaces are ubiquitous computing environments for a computer supported collaboration that builds on and enhances the preexisting motor, spatial, social, and cognitive skills of users (Jetter, H. C., Reiterer, H., & Geyer, F., 2014). Hence, it can be stated that the HCI is an emerging discipline which aims at an established understanding and designing of different interfaces between humans and computers in a way that it defines systems that are enjoyable to use, are engaging and are accessible. However, while HCI leads to efficient user handling, the questions remain about increasing human dependency on computers and how it leads to a change such as, not being able to respond fully to the perspective of it, are ignored with development of technology integrations. Hence, with novel touch to the upcoming future, the HCI practitioners in the coming years should strive for a confliction free world where technology and quotidian life exists in synchronization (Gupta, R., 2012).

Though, the development of the smart systems presents new challenges to the application of human-computer interaction and user experience (HCI/UX), providing many possibilities for smart users to suggest additional practical necessities (Zhao, Y., Zhang, X., & Crabtree, J., 2016). Keeping these in the context, the book has incorporated several studies and researches along these perspectives and deserves further extended studies.

ORGANIZATION OF THE BOOK

The book *Human-Computer Interaction and Technology Integration in Modern Society* has incorporated 12 highly ranked studies and researches across the globe that ranges from education, learning, open innovation, entrepreneurship, ethics, citizens' participation, collaboration, sharing and policy implementation.

Chapter 1 discusses on issues of ICT Integration in Education and explores various perspectives in relation to this. Information and Communication Technologies (ICT) are extremely influencing every discipline under the sun including Education. It is affecting every aspect of education from teaching-learning to assessment and evaluation. It improves the effectiveness of education. It aids literacy movements. It enhances scope of education by facilitating mobile learning and inclusive education. The chapter discusses the same along with the various projects and schemes adopted by the Indian context to spread ICT literacy and competency.

Chapter 2 discusses on aspects of online collaborative learning tools and types and their key role in managing resources and classrooms. The research mentions that Online/ web based collaborative tools that enable teachers and students to perform a wide range of tasks, such as interactive discussions, online collaboration activities, sharing and accessing electronic learning resources, and many more others. It not only promotes critical thinking and reflection in students but also encourages them to develop a sense of community, thus enabling the creation of an environment in which further collaborative work can take place. The chapter explains the online collaboration with its features, preparation required by institution and role of teacher presence in online learning.

Chapter 3 explores issues related to reach out to the grass roots through small and medium enterprises through open innovation. While talking about successful entrepreneurship and value addition within an enterprise through innovation, one could comprehend that the innovation paradigm has been shifted from simple introduction of new thoughts and products to accumulation of diversified actions, actors and agents along the process. Furthermore, when the innovation process is not being constrained within the closed nature of it, the process takes many forms during its evolution.

Chapter 4 discusses on various research issues related to the introduction of open innovation in small and medium enterprises. The chapter mentions that innovation has been treated as a recognized driver of economic prosperity of a country through the sustained growth of its entrepreneurships. Moreover, recently coined term open innovation is increasingly taking the lead in enterprise management in terms of value addition and knowledge building. This chapter seeks to find out open innovation researches and practices that are being carried out circumscribing development of entrepreneurships, particularly the sector belonging to the small and medium enterprises (SMEs) through a longitudinal study.

Chapter 5 compares introduction of open innovation to small and medium enterprises among developing and developed nations. The chapter emphasizes that in spite of the increased acceptance by corporate business houses around the globe, the adaptation of strategies and concepts belonging to the newly evolved dimension of entrepreneurships, the open innovation (OI), countries in the East, West or South are yet to adapt appropriate strategies in their business practices, especially in order to reach out to the grass roots and marginal communities.

Chapter 6 talks about various issues on visualizing ethics through hypertext. The research mentions that ethics has always been a main domain of philosophical study. Hence, hyper textual discursive expression has never been considered as the technology that can revolutionize the way we interpret ethical events and actions. This chapter intends to show that hyper textual discursive expression is not only useful, but also vital in any undertaking that seeks to reframe our ethical deliberations and judgments.

Chapter 7 confers on issues related to citizen participation in community surveillance and mentions that mapping the dynamics of WhatsApp, neighbourhood crime prevention practices can be enhanced. As a country study, this research states that despite their recent emergence, WhatsApp Neighbourhood Crime Prevention (WNCP) groups are an already pervasive phenomenon in the Netherlands. This research draws on interviews and focus groups to provide an in-depth analysis of the watchfulness and surveillance activities within these groups. The conceptualization of WNCP through the lens of practice theory shows that the use of ICTs in the form of WhatsApp amplified all three dimensions of neighborhood watchfulness practices.

Chapter 8 focuses on developing digital reading practices in the humanities with computer assisted text analysis tools. Authors mention that the introduction of digital media into university writing courses, while leading to innovative ideas on multimedia as a rhetorical enhancement means, has also resulted in profound changes in writing pedagogy at almost all levels of its theory and practice. They further mention that due to the traditional approaches to examining and discussing assigned texts in the classroom were developed to help students analyze different genres of print-based texts, many university educators find these methods prohibitively deficient when applied to digital reading environments.

Chapter 9 focuses on one element in the digitalisation of work, such as: the forms and conditions of working in the so-called 'sharing economy' (SE). Based on an analysis of 67 SE platforms, it distinguishes three segments, each of which constitutes a distinctive institutional sphere within the sharing economy: these are an 'exchange and gift economy', a 'niche and sideline economy', and the 'platform economy'. In a further step, the study then identified and compares five dimensions of work within these segments: the type of activity, the form of compensation or recompense involved (monetary or non-monetary), skills and competencies required, the role of technology, and control mechanisms.

Chapter 10 discusses on the role of big data and business analytics in decision making. This research mentions that the amount of data in our world has been exploding and Big Data represents a fundamental shift in business decision-making. Analyzing such so-called Big Data is today a keystone of competition and the success of organizations depends on fast and well-founded decisions taken by relevant people in their specific area of responsibility. In this chapter, author describes and analyzes Business Intelligence and Analytics (BI&A) and four popular open source systems – BIRT, Jaspersoft, Pentaho, and SpagoBI.

Chapter 11 illustrates e-government policy implementation in Thailand as a case study. This research investigates the current e-government and ICT policy in Thailand from an actor-centered perspective. It reviews existing literature on e-government implementation, while looking into the interaction of government institutions and citizens. It seeks to answer questions, such as the following. What are the key actors in driving the implementation of e-government policies in Thailand? How do Thai citizens perceive e-government efforts and ICT policy implementation especially in the context of present military government power?

Finally, Chapter 12 explores on issues related to the evaluation of measuring the publicness level of interiors in public building design through the Visual Graph Analysis (VGA) approach. This chapter examines the publicness level of the interior spaces of public buildings. As a method, VGA (Visual Graph Analysis) is used for analyzing the early design phases of selected municipal service buildings. The municipal service buildings, which reflect public structure, identity and the society's periodic ideological stance, represent an important type in these public administration structures. Additionally, this examination reveals how public spaces perform in terms of their publicness level.

CONCLUSION

It has been observed that the local scientific resources are always poorly developed, nourished and disseminated in many countries. Small states are particularly disadvantaged by the absence of employment opportunities for local scientists and researchers, and remained heavily dependent on outside assistance. Hence, serious attention must be paid for their capacity building in scientific education; building relations between science and communities; and targeting limited scientific resources through effective policy and governance. This book can serve as a knowledge provider for the researchers and academics, including policy initiators at all levels for the benefit of common masses.

Throughout the book contemporary technology issues have been portrayed with lessons learned from different researches that complement and applied for economic, technical and social benefit, and it is expected that the content of this book will become an added asset to the traditional knowledge systems, especially in the developing countries.

Therefore, this may encourage all stakeholders of similar nature to undertake participatory reviews of their social perspectives of development in the aspect of incorporation of human computer interactions and technology integrations to ensure that the contexts, learning techniques, researches and achievements of basic development goals for parallel citizenship requirements. The outcome of researches should focus on making changes in both learning and researches to strengthen development techniques towards strengthening of civic consciousness and the revitalization of social resources. It is expected that this could be a long-term but ultimately effective approach to empower marginal communities at the grass roots.

Hakikur Rahman
Institute of Computer Management and Science, Bangladesh
October 2020

REFERENCES

Fallman, D. (2011, May). The new good: exploring the potential of philosophy of technology to contribute to human-computer interaction. In *Proceedings of the SIGCHI conference on human factors in computing systems* (pp. 1051-1060). 10.1145/1978942.1979099

Gupta, R. (2012). Human Computer Interaction–A Modern Overview. *International Journal Computer Technology Application*, *3*(5), 1736–1740.

Holzinger, A., Ziefle, M., & Röcker, C. (2010, July). Human-computer interaction and usability engineering for elderly (HCI4AGING): introduction to the special thematic session. In *International Conference on Computers for Handicapped Persons* (pp. 556-559). Springer. 10.1007/978-3-642-14100-3_83

Jetter, H. C., Reiterer, H., & Geyer, F. (2014). Blended Interaction: Understanding natural human–computer interaction in post-WIMP interactive spaces. *Personal and Ubiquitous Computing*, *18*(5), 1139–1158. doi:10.100700779-013-0725-4

McCarthy, J., Wright, P., Wallace, J., & Dearden, A. (2006). The experience of enchantment in human–computer interaction. *Personal and Ubiquitous Computing*, *10*(6), 369–378. doi:10.100700779-005-0055-2

Zhang, P., Carey, J., Te'eni, D., & Tremaine, M. (2005). Integrating human-computer interaction development into the systems development life cycle: A methodology. *Communications of the Association for Information Systems*, *15*(1), 29. doi:10.17705/1CAIS.01529

Zhao, Y., Zhang, X., & Crabtree, J. (2016). Human-computer interaction and user experience in smart home research: A critical analysis. *Issues in Information Systems*, *17*(3).

Acknowledgment

The editor would like to acknowledge the help of all involved in the accumulation and review process of the book, without whose support the project could not have been satisfactorily completed. Thanks go to all the authors of chapters included in this book whose relentless cooperation and devotion have made the successful completion of the book. Special thanks go my close friends and colleagues for their encouragement during the entire process.

Special thanks also go to the publishing team at IGI Global. Particularly to Carlee Nilphai, Jordan Tepper, Maria Rohde and Eric Whalen for their continuous suggestions and supports via e-mail for keeping the project on schedule, and to Mehdi Khosrow-Pour and Jan Travers for their continuing professional supports. Finally, I want to thank my family members for their love and support throughout this period.

Hakikur Rahman
Institute of Computer Management and Science, Bangladesh
October 2020

Chapter 1
ICT Integration in Education: Indian Scenario

Madhuri Tikam
H. R. College of Commerce and Economics, Mumbai, India

ABSTRACT

Information and communication technologies (ICT) are influencing every discipline under the sun including education. It is affecting every aspect of education from teaching-learning to assessment and evaluation. It improves the effectiveness of education. It aids literacy movements. It enhances scope of education by facilitating mobile learning and inclusive education. It facilitates research and scholarly communication. Impact of ICT and its potential for the education field is manifold. It positively affects all the stakeholders of the education field. The chapter discusses the same along with the various projects and schemes adopted by the country to spread ICT literacy and competency.

INTRODUCTION

"Education for All" is a global movement led by UNESCO (United Nation educational, Scientific and Cultural Organization), aiming to meet the learning needs of all children, youth and adults by 2015.It faces three major challenges:

1. Providing access to all
2. Improving learning environment
3. Measuring learning outcomes.

To succeed in "Education for All" movement, access to education should be provided to all irrespective of gender, physical, geographical, language, economical, social or any other barrier. The poorest people, residents of remote areas, and the most disadvantaged populations - for example, girls and members of ethnic and religious minorities, physically challenged people are the main category of people to whom education should reach. To deal with varied needs of learners from different stratum is a complex issue to handle in a traditional mode of teaching learning. It demands for an open and flexible approach of

DOI: 10.4018/978-1-7998-5849-2.ch001

education and distance or virtual modes of learning. These advanced demands of education delivery cannot be met in the developed and developing world without the help of Information and Communication Technologies (ICT) (UNESCO, 2009). The impact of ICT on trainers, learners, researchers and the entire learned society is tremendous. It is changing the contours of the education delivery system in the world by enhancing access to information for all. It also ensures effective and inclusive education. ICT supports the concept of open learning where the thrust is upon enhanced student access and the development of student autonomy.

Roles of ICT in Education

ICT can play varied roles in developing an effective learning environment. It helps in offering access as well as enhances the learning environment. It acts as a teacher and explains core content concepts and addresses misconceptions. It acts as a stimulant and fosters analytical thinking and interdisciplinary studies. It networks a learner with the peers and experts and develops collaborative atmosphere. It plays the role of a guide and mentor by providing tailor made instructions to meet individual needs. Online learning facilitates learning through digital mode. With the help of multimedia, it enhances effectiveness of teaching-learning and hence proves crucial for early learners, slow learners and differently abled learners. Modern ICT tools not only deliver the content but also replicate formal learning experience via virtual learning. The intention of virtual classrooms is to extend the structure and services that accompany formal education programs from the campus to learners.

ICT also addresses the need of mobile learning. It offers independent space and flexibility that comes from working away from the learning institute or tutor. It makes education accessible to all, irrespective of geographical barriers or resource constraints. Learners from remote areas, working people who want to learn further and update their knowledge and differently-abled students who find travelling an issue of concern - benefit from the mobile learning mode. As per Scott Motlik's technical evaluation report on "Mobile Learning in Developing Nations"; by comparison, mobile phone technology is widespread, easy to use, and familiar to learners and instructors (Motlik, 2008). An exploratory study of unsupervised mobile learning in rural India conducted by Anuj Kumar and his colleagues showed a reasonable level of academic learning and motivation among rural children who were voluntarily engaged in mobile learning. (Kumar, 2010). Similarly a study by Douglas Mcconatha, Matt Praul, and Michael J. Lynch, revealed that the use of mobile learning can make a positive and significant difference in the outcome performance than traditional methods of class lectures, notes and reviews (Mcconatha, 2008). Dr. Fahad N. Al-Fahad's study about students' attitudes and perceptions towards the effectiveness of mobile learning in King Saud University, Saudi Arabia, also supported that m learning makes the learners truly engaged learners who are behaviorally, intellectually and emotionally involved in their learning task. (Al-Fahad, 2009).

A lot of educational resources are available in m-learning formats. Google books have worked with major publishers to bring pages, chapters and books in mobile readable format. It also offer mobile version of book search since February 2009. Other publishers like Amazon, EBL, BBC, etc. also provide mobile collection of e-books and/or audio books to their users. Databases like JSTOR, LexisNexis, EBSCO, etc. are available on mobile devices. Many libraries all over the world like California State University, Fullerton Pollak Library, National University of Singapore Libraries; Public Library Münster, Germany; University of Virginia Library; Washington State University Libraries; London School of Economics, United Kingdom; etc. offer mobile based library services. Many mobile service provider companies are providing educational programs via mobiles. E.g. Tata Docoma launched mobile sex education program – Sparsh, they also offers programs like English Seekho, Storytelling via mobile. Universities like S.N.D.T. Women's University, Indira Gandhi Open University, etc. are developing tie ups with mobile service provider companies to offer distance education programs with the help of mobile phones. Publishers like Tata McGraw Hill also initiated mobile based courses like TOPCAT. Thus, ICT offers possibilities of transforming the learning paradigm and bringing knowledge to those who have not earlier been able to participate in education.

ICT and Inclusive Education

Inclusion helps to involve the identification and minimizing of barriers to learning. It encourages participation and maximizing of resources to support learning. This participation is irrespective of the gender, age, ability, ethnicity or impairment of learners. ICT proves a boon in such circumstances. It unlocks the hidden potential of those who have communication difficulties. It enables students to demonstrate achievement in ways which might not be possible with traditional methods by developing tailor-made tasks to suit individual skills and abilities. It improves independent access to students to educate themselves at their own pace. Visually impaired students using the internet can access information alongside their sighted peers while students with profound and multiple learning difficulties can communicate more easily besides gaining interest in education, confidence and social credibility at school and in their communities (UNESCO, 2006). An array of Assistive technology is available for supporting inclusive education for students with special needs. They can be broadly grouped in four categories:

1. Assistive technologies for Magnification
2. Assistive technologies for Format Conversion
3. Assistive technologies for Reading
4. Assistive technologies for Learning

Assistive technologies for Magnification include computers having larger screen so that magnified images of text as well as photographs, charts, etc. can be provided to the partially sighted. Software like Optelec's ClearView+ Desktop Video Magnifier system magnifies anything under the viewfinder from 2 to 50 times its original size. It has a specially designed monitor and lighting design for optimum visual enhancement and the entire unit can be controlled with a single button, and can be customized to meet the exact needs. The advanced versions like Optelec Compact + Portable Magnifier, Bonita (Portable mouse Magnifier), etc. are more powerful, smaller, portable and user friendly devices which can be used anywhere and anytime. It has six viewing modes (full color, high contrast black/white, white/black, yellow/black, yellow/blue and full color without light - ideal for magnifying displays such as the screen of your mobile phone). An integrated writing function enables a person to write while viewing what is being written. It is best suited for persons with low vision for attending to their daily reading-writing tasks.

Assistive technologies for Format Conversion used to digitize the notes, question papers, etc. Software like EasyConverter, Zoom Ex instant reader, etc. can convert any printed text into multiple accessible formats such as speech, large print, sound, text files and Braille versions within seconds. The speech mode offers a choice of accents; one can even select Indian Accent for better understanding which meet the needs of dyslexic, visually impaired and learning disabled students.

Software like JAWS (Jobs Access with Speech), KURZWEIL, SARA (Scanning and Reading Appliance), Dolphin Easy Reader, etc. convert the computer into a talking machine by providing the user with access to the information displayed on the screen via text to speech mode. They provides users with document creation, editing as well as study skills capabilities for preparing notes, summarizing and outlining the text. This software are multilingual and interactive. They allow the users to make their talking book directly from plain text.

Assistive technologies for Learning include hardware and software which assists the users with special needs to create their own information resource and learn by their own. Software like Talking Typing Teacher, Touch Typing Tutor reads out the text as it is typed on the keyboard. Recorders like Angel player, Victor Stream Reader and Digital Voice recorder can be used to record information which can be played back according to the need of the individual. Specially designed keyboards like Easy Link 12, allows the Braille user to effortlessly control PDA, PC and smartphone. It can be connected to a smartphone to send / receive SMS. Software like Spell well software used to learn spellings and pronunciation of different words. With the advancement of ICT, mainstream education became available, accessible, affordable and appropriate for students with disabilities. ICT removes the hindrance in accessing higher education and motivates the disabled students to pursue higher education by making them self-reliant.

ICT AND EVALUATION

Education and Evaluation goes hand in hand. ICT aids and speeds the process while maintaining transparency and accuracy. ICT offers a chance to treat every student as an individual and offer a tailor made solution for each. E.g. Reading Assessment software like AceReader takes into account the existing skills of a student as an individual and develops advanced assessments based on previous results. As a result, the coursework per student differs and caters their personalized needs. The ICT based evaluation proves exceedingly useful for distance education students where lack of feedback may de-motivate the students. The use of software like 'MarkIt'- used widely in Australia, provides students information about their various performances consistently. It also offers details about performance of their peers, plus the capacity for markers to enter detailed and consistent feedback at all stages of the marking process (Don, 2000). Such software enhances the inclusiveness and give rise to healthy competitions.

Computer based concept mapping with automated scoring helps summative assessment of critical and creative thinking about complex relationships. A large number of students can be analyzed for different sets of skills simultaneously with the help of ICT. ICT based evaluation avoids time-consuming manual data transcription and recoding before statistical analysis. Built-in error control in ICT based evaluation methods reduces administrative errors. As a result, assessment becomes speedy and transparent. Data analysis programs facilitate the tutor and learner to monitor their progress and understand the different facets of performance. Visual representations and study of different relations of evaluations are made easy by ICT. It facilitates teachers in understanding the difficulties faced by students in a particular area and helps in developing suitable measures. Overall, ICT offers greater flexibility, improved reliability and enhanced effectiveness to the assessment process.

ICT and Research

ICT offers a range of products to aid research work at every stage. It has tools for searching and keeping updated. It has tools for collaboration and communication. The introduction of the Open Access Resource initiative by ICT makes the sharing of scholarly ideas at a global level easy. Budapest Open

Access Initiative of Open Society Institute (OSI) activated this initiative in December 2001. It makes peer-reviewed journal articles and un-reviewed preprints that [scholars] might wish to put online for comment or to alert colleagues to important research findings freely available on the public internet, permitting any user to read, download, copy, distribute, print, search, or link to the full texts of these articles, crawl them for indexing, pass them as data to software, or use them for any other lawful purpose, without financial, legal, or technical barriers other than those inseparable from gaining access to the internet itself. The only constraint on reproduction and distribution, and the only role for copyright in this domain, should be to give authors control over the integrity of their work and the right to be properly acknowledged and cited (BOAI, 2011). It impacts the education field a lot as more and more researchers have started publishing and the latest updates in a subject field become available to all across the globe in real time. Initiatives like DOAJ, Google Books reveals a large database of scholarly communication to the researchers irrespective of subject, language and geographical barrier. Search engines like Google, Yahoo aids to search the required information in fraction of seconds. Tools like RSS, wikis, bookmarks helps to remain updated in one's field. Programs like Freesummarizer or Greatsummary helps to Summarize any text online in just a few seconds. Mind mapping tools like Freemind, Mindmeister helps to manage the knowledge base as well as the project progress.

For developing and conducting survey, survey tools like Survey Monkey (www.surveymonkey.com), Zoomerang (www.zoomerang.com), Survey Gizmo (www.surveygizmo.com), PollDaddy(www.polldaddy.com) helps a lot. They offer extensive reporting, with a flexible cross-tabulation report and advanced logic features like question and answer piping, randomization, text analysis for open responses, and integration with statistical software like IBM's SPSS. To conduct complex data mining and analysis, statistical software like Analytica, Matlab +Statistics toolbox, SPSS, etc. are very useful. One can use social media tools like Linkendin, skype, slideshare, etc. to share and disseminate the ideas and views. There are a few all in one solution like Zotero [zoh-TAIR-oh] are available for researchers which free, help to collect, organize, cite, and share research sources. Once the final research paper is ready, one can use bibliographic tools like Easybib (www.easybib.com), noodletools (www.noodletools.com/), Bibme (www.bibme.org), Google Docs Bibliographic Template, etc. to build their bibliography in APA, MLA, AMA, and Chicago Style. ICT became an integral part of the research and scholarly communication. It contributes positively to the all stages of research and impacts the scholarly world remarkably.

ICT IN INDIAN EDUCATION SYSTEM

Indian Education System is incorporating ICT in every aspect of education for better connectivity and access. Many schemes are introduced and implemented from school up to higher education system for making education more effective. A few schemes are discussed below:

For Maintenance of Information and Administration

Schools Geo Portal: (https://schoolgis.nic.in)

It is developed for seamless visualization of school locations across the country (Spatial and Non-Spatial). GIS offers a powerful decision-making toolkit that can be used in educational administration, educational policy, and in instruction. It ensures universal access within a reasonable distance of any habitation and without any discrimination and better utilization of resources available under different schemes.

ShaGun: (http://seshagun.gov.in/)

ShaGun, which has been coined from the words 'Shala' meaning Schools and 'Gunvatta' meaning Quality, It is a dedicated web portal for the Sarva Shiksha Abhiyan developed by MHRD . 'ShaGun' aims to capture and showcase innovations and progress in Elementary Education sector of India It comprises of three parts. 1. ShaGun Repository – It documents state-level innovations, success stories and best in the form of videos, testimonials, case studies, and images, which will display that are driving improvements in performance under SSA. 2. Online monitoring module to measure state-level performance and progress against key educational indicators. 3. 'Toolkit for Master Trainers in Preparing Teachers for Inclusive Education for Children with Special Needs': It has five training modules for the teachers offering practical information on effective inclusion of CWSN and related pedagogical practices.

Shala Darpan: (e.g.: http://rajshaladarpan.nic.in/)

It is an e-Governance platform developed by NIC for all Government schools in the State. This School Automation Application includes school profile, report cards, curriculum tracking system, SMS alerts for parents / administrators on students and teacher attendance, employee information, student attendance and, leave management. It aims to improve quality of learning, efficiency of school administration, governance of schools & service delivery to students, parents, teachers, community and schools.

Shaala Kosh: (App)

It is used for registration of all government and private schools in the state-affiliated by all boards. Through this app based web portal all students, teachers, schools and parents are able to manage the data on a daily basis such as student performance report, information on online exam forms, etc. It has enabled data consolidation, analysis and usage and empowered the stakeholders at block, district, state and central level for use data in effective decision making.

Saransh: (https://saransh.digitallocker.gov.in/)

It is online interface started by CBSE. It is an instrument for self-audit and investigation for schools and guardians of CBSE affiliated schools. It maintains communication and flow of relevant information among schools, instructors and guardians. It empowers them to keep a monitor the performance of the students and take remedial measures in time.

For Enhancement of Computer Literacy

CLASS

Computer Literacy and Studies (CLASS) is a project launched in 1984 was a joint initiative of MHRD, Department of Electronics, and NCERT. IT IS revised time to time. It the aimed at providing computer literacy in 10,000 schools, computer-assisted learning in 1,000 schools and computer-based learning in 100 schools. These 100 schools were called smart schools, and were designed to be agents of change seeking to promote the extensive use of computers in the teaching-learning process.

Giri Pragna

Giri Pragna Project has been successful in providing learning opportunities to tribals in Khammam District, Andhra Pradesh. 50 school complexes covering classes 6-10 and 10000 children every year are exposed to computer-aided learning. The project also focuses on teacher training. The latest technology and syllabus and specially designed local language CDs are used for training purpose.

Microsoft's Project Shiksha

According to Microsoft Corporation India, Project Shiksha aims to accelerate computer literacy in India by providing a comprehensive programme that includes software solutions, comprehensive training for teachers and students, development of a world-class IT curriculum, and scholarships for teachers and students. Teachers are exposed to an IT literacy curriculum with the key objective that they will return to the classroom and use IT interventions in their teaching. It offers opportunity to strengthen IT proficiency of schoolteachers and students across government schools.

For Facilitation of Self Learning

E-pathshala: (epathshala.nic.in)

Central Institute of Educational Technology (CIET), National Council of Educational Research and Training (NCERT) has designed and developed 'E-Pathshala' to showcase and disseminate 'all educational e-resources including textbooks, audio, video, periodicals and a variety of other print and non-print materials through web portal and mobile app'. It is an effort to solve the challenges of digital divide and improve accessibility by offering quality content anytime anywhere.

National Repository of Open Educational Resources (NROER): (https://nroer.gov.in)

It is an initiative of MHRD which is a collaborative effort of CIET and NCERT to develop a repository of multimedia resources for teaching and learning purposes. Both online and offline platforms have been designed. Different formats of Resources such as Audio, Video, Image, Document, Interactive, Books including flip books, and multimedia are available on NROER platform.

National Digital Library (NDL): (ndl.gov.in)

The National Digital Library of India (NDL) is a project to develop a framework of virtual repository of learning resources with a single-window search facility. NDL India is designed to hold content of any language and provides interface support for leading Indian languages. .There are more than 153 Lakh digital books available through the NDL. It is being focused to support all academic levels including researchers and life-long learners, all disciplines, all popular form of access devices and differently-abled learners. It is being developed to help students to prepare for entrance and competitive examination, to enable people to learn and prepare from best practices from all over the world and to facilitate researchers to perform inter-linked exploration from multiple sources.

SWAYAM: (swayam.gov.in)

The 'Study Webs of Active Learning for Young Aspiring Minds' (SWAYAM) an integrated platform for online courses. It is intended to accomplish the goal of equity, access and quality education for all. It ensures that the every student in the country has access to the best quality higher education at affordable cost. It covers all higher education subjects and skill sector courses covering school (9th to 12th) to Post Graduate Level. It also offers online courses for students, teachers and teacher educators.

SWAYAM PRABHA: (TV Channels)

SWAYAM PRABHA DTH-TV has 32 National Channels for transmission of educational e-contents through for utilization of satellite communication technologies. It aims to provide learning opportunities for the stake holders, as per their convenience. .CIET-NCERT coordinates one DTH TV channel called Kishore Manch which broadcasts educational programmes 24x7. Besides, NIOS is running 5 channels for teachers, for secondary and senior secondary levels and for sign language.

Online Labs: (https://amrita.olabs.edu.in/)

The Online Labs is based on the idea that lab experiments can be taught using the Internet, more efficiently and less expensively. The labs can also be made available to students with no access to physical labs or where equipment is not available owing to being scarce or costly. This helps them compete with students in better equipped schools and bridges the digital divide and geographical distances. The experiments can be accessed anytime and anywhere, overcoming the constraints on time felt when having access to the physical lab for only a short period of time. The labs make use of cutting edge simulation technology to create real world lab environments

CONCLUSION

The impact of ICT on the education sector is tremendous. It improves the effectiveness and inclusiveness of education. It saves resources and enhances access. However, it poses many challenges. While for many ICT offers a promise of greater inclusion of previously marginalized groups (whether marginalized by gender, disability, distance, language, culture, race, age or economic status), their use also brings with it very real dangers of increasing the marginalization of such groups inside the education system. India is focusing on spreading the ICT countrywide by adopting various projects and schemes. Many projects are successfully being run to offer better access and bring equity.

REFERENCES

Al-Fahad, F. (2009). Students' attitudes and perceptions towards the effectiveness of mobile learning in king Saud University, Saudi Arabia. *The Turkish Online Journal of Educational Technology*, *8*(2). Retrieved November 18, 2012, from http://www.tojet.net/articles/8210.pdf

Anonymous. (1998). *Maximum security: a hacker's guide to protecting your internet site and network*. Techmedia.

Berthon, H., & Webb, C. (2000). *The Moving Frontier: Archiving, Preservation and Tomorrow's Digital Heritage*. Paper presented at VALA 2000 - 10th VALA Biennial Conference and Exhibition, Melbourne. Retrieved October 27, 2009, from http://www.nla.gov.au/nla/staffpapers/hberthon2.html

Björk, B.-C., Welling, P., Laakso, M., Majlender, P., & Hedlund, T. (2010). Open Access to the Scientific Journal Literature. *PLoS ONE, 5*(6). Retrieved January 14, 2012, from www.plosone.org/article/info:doi/10.1371/journal.pone.0011273

BOAI: Budapest Open Access Initiative. (2011). *Frequently Asked Questions*. Retrieved December 19, 2012, from http://www.earlham.edu/~peters/fos/boaifaq.htm

Crawford, W. (2011). *Open access: what you need to know now*. ALA Publishing.

Current Publications. (2001). *The copyright act 1957: with short notes*. Current Publications.

Day M. (1998). Electronic Access: Archives in the New Millennium. Reports on a conference held at the Public Record Office, Kew on 3-4 June 1998. *Ariadane,* (16).

Diwan, P., Suri, R. K., & Kaushik, S. (2000). *IT Encyclopaedia.com* (Vol. 1-10). Pentagon Press.

Don Dingsdag, D., Armstrong, B., & Neil, D. (2000). *Electronic Assessment Software for Distance Education Students*. Retrieved January 14, 2012, from http://www. ascilite.org.au/ conferences/coffs00/papers/don_dingsdag.pdf

Douglas, M., Praul, M., & Lynch, M. (2008). Mobile learning in higher education: An empirical assessment of a new educational tool. *The Turkish Online Journal of Educational Technology*, *7*(3). http://www.tojet.net/articles/732.pdf

Freedman, A. (1996). *The computer desktop encyclopedia*. AMACOM.

Government of India: Ministry of Human Resource Development: Press Information Bureau. (2018). *Several steps have been taken to promote e-Education in the country*. Retrieved August18, 2019, from https://pib.gov.in/newsite/PrintRelease.aspx?relid=186501

Government of India: Ministry of Human Resource Development: Press Information Bureau. (2018). *Union HRD Minister launches 'ShaGun' - a web-portal for Sarva Shiksha Abhiyan*. Retrieved August18, 2019, from https://pib.gov.in/newsite/PrintRelease.aspx?relid=157488

Hidreth, C. (2001). Accounting for user's inflated assessment of online catalogue search performance and usefulness an experimental study. *Information Research*, *6*(2). Retrieved January 14, 2012, from http://www.informationR.net/ir/6-2/paper101.html

International Telecommunication Union (ITU). (2013). *The world in 2013: ICT facts and figures*. Retrieved September 10, 2014, from https://www.itu.int/en/ITU-D/Statistics/.../facts/ ICTFactsFigures2014-e.pdf

Johnson, L., Smith, R., Willis, H., Levine, A., & Haywood, K. (2011). *The 2011 Horizon Report*. The New Media Consortium.

Kanugo, S. (1999). *Making IT work*. Sage Publications.

Khona, Z. (2000, September 18). Copyright: A safeguard against piracy. *Express Computer, 11*(28), 15.

Kole, E. S. (2001). *Internet Information for African Women's Empowerment*. Paper presented at the seminar 'Women, Internet and the South', organized by the Vereniging Informatie en International Ontwikkeling (Society for Information and International Development, VIIO), 18 January 2001, Amsterdam. Retrieved January 14, 2012, from https://www.xs4all.nl/~ekole/public/endrapafrinh.html

Kumar, A. (1999). *Mass Media*. Anmol.

Kumar, A. (2010). *An exploratory study of unsupervised mobile learning in rural India*. Retrieved November 18, 2012, from http://www.cs.cmu.edu/~anujk1/ CHI2010.pdf

Lesk, M. (1995). *Keynote address: preserving digital objects: recurrent needs and challenges*. Papers from the National Preservation Office Annual Conference - 1995 Multimedia Preservation: Capturing The Rainbow. Retrieved October 27, 2009, from http://community.bellcore.com/lesk/auspres/aus.html

Mahajan, S. L. (2002). Information Communication Technology in distance education in India: A challenge. *University News, 40*(19), 1–9.

Motlik, S. (2008). Mobile Learning in Developing Nations. *The International Review of Research in Open and Distance Learning, 9*(2). Retrieved November 18, 2012, from http://www.irrodl.orgindex.php/irrodl/article/view/564

Singh, A. (2019). ICT initiatives in School Education of India *Indian Journal of Educational Technology, 1*(1), 38-49.

Strigel, C. (2011). *ICT and the Early Grade Reading Assessment: From Testing to Teaching*. Retrieved January 14, 2012, from https://edutechdebate.org/reading-skills-in-primary-schools/ict-and-the-early-grade-reading-assessment-from-testing-to-teaching/

Tikam, M (2013). Impact of ICT on education. *International Journal of Information Communication Technologies and Human Development, 5*(4), 1-9.

Tikam, M. (2016). ICT Integration in Education Potential and Challenges. In Human Development and Interaction in the Age of Ubiquitous Technology: Advances in Human and Social Aspects of Technology. IGI Global.

Tikam, M., & Lobo, A. (2012). Special Information Center for Visually Impaired Persons – A Case Study. *International Journal of Information Research, 1*(3), 42–49.

Traxler, J. (2007). Defining, discussing and evaluating mobile learning: The moving finger writes and having writ . . .. *International Review of Research in Open and Distance Learning, 8*(2). Advance online publication. doi:10.19173/irrodl.v8i2.346

Trilokekar, N. P. (2000). *A practical guide to information technology act, 2000: Indian cyber law*. Snow White Publications.

UNESCO. (2006). *ICTs in education for people with special needs united nations educational, scientific and cultural organizationunesco institute for information technologies in education specialized training course*. Moscow: UNESCO. Retrieved December 19, 2012, from https://iite.unesco.org/pics/publications/en/files/3214644.pdf

UNESCO. (2009). *Guide to measuring Information and Communication Technologies (ICT) in education*. UNESCO Institute for Statistics.

Valk, J., Rashid, A., & Elder, L. (2010). Using Mobile Phones to Improve Educational Outcomes: An Analysis of Evidence from Asia. *International Review of Research in Open and Distance Learning*, *11*(1), 117. doi:10.19173/irrodl.v11i1.794

World Bank. (2002). *Information and Communication Technologies: a World Bank group strategy*. World Bank.

Yadava, J. S., & Mathur, P. (1998). *Mass communication: the basic concepts* (Vol. 1 & 2). Kanishka.

Chapter 2
Online Collaborative Learning Tools and Types:
Their Key Role in Managing Classrooms Without Walls

Sarika Sawant
SNDT Women's University, India

ABSTRACT

Online/web-based collaborative tools enable teachers and students to perform a wide range of tasks, such as interactive discussions, online collaboration activities, sharing and accessing electronic learning resources, and many more others. It not only promotes critical thinking and reflection in students but also encourages them to develop a sense of community, thus enabling the creation of an environment in which further collaborative work can take place. The author has categorised various tools into 11 types that deal with idea generation and brainstorming, live conference, robotics and coding tools, mapping, design, online group work and document collaboration, and online communication and content development. The chapter explains the online collaboration with its features, preparation required by institution and role of teacher presence in online learning. It also emphasizes that library consultations (i.e., librarians) directly boost student learning, so the active collaboration of librarians is a must.

A. COLLABORATIVE LEARNING

Collaborative learning is a situation where students are able to socially interact with other students, as well as instructors. In essence, learners work together in order to expand their knowledge of a particular subject or skill.

Collaborative learning is based upon the principle that students can enrich their learning experiences by interacting with others and benefiting from one another's strengths. In collaborative learning situations, students are responsible for one another's actions and tasks which encourages teamwork as well (eLearning 101 – concepts, trends, applications, 2014).

DOI: 10.4018/978-1-7998-5849-2.ch002

Collaborative learning engages learners in knowledge sharing, inspiring each other, depending upon each other, and applying active social interaction in a small group. Therefore, collaborative learning depends upon the art of social interaction among learners rather than a mechanical process (Tu, 2004). The idea of group work in learning finds its root in work from the Russian psychologist Vygotsky (1978) who explored the causal relationships that exists between social interaction and individual learning providing a foundation of the social constructivist theory of learning (Muuro,Wagacha, Kihoro & Oboko, 2014).

Collaborative learning is based on the view that knowledge is a social construct. Collaborative activities are most often based on four principles:

- The learner or student is the primary focus of instruction.
- Interaction and "doing" are of primary importance
- Working in groups is an important mode of learning.
- Structured approaches to developing solutions to real-world problems should be incorporated into learning (Chandra, 2015).

Some activities or assignments well suited for collaborative learning include:

- Case studies
- Discussions
- Student-moderated discussions
- Debates
- Collaborative writing
- Collaborative presentation
- Games
- Demonstrations

Benefits of collaborative learning include:

- Development of higher-level thinking, oral communication, self-management, and leadership skills.
- Improves analytical skills and critical thinking
- It boosts confidence
- Promotion of student-faculty interaction.
- Increase in student retention, self-esteem, and responsibility.
- Exposure to and an increase in understanding of diverse perspectives.
- Preparation for real life social and employment situations.

B. ONLINE COLLABORATIVE LEARNING

Collaborative learning can be conducted either offline or on the web, and can be done asynchronously or synchronously. It allows students to learn from the ideas, skill sets, and experience of others enrolled in the course. By engaging in a shared task (whether it be a project or lesson) students gain the opportunity to learn a variety of skills, such as group analysis and collaborative teamwork building skills.

In addition, even students who are unable to attend a live event online can participate in collaborative learning, using online forums, message boards, and other various posting sites that don't rely on real-time interaction (eLearning 101 – concepts, trends, applications, 2014).

Collaborative learning technologies range from communication tools that allow text, voice, or video chat to online spaces that facilitate brainstorming, document editing, and remote presentations of topics (Mallon & Bernsten, 2015). According to Bates (2016) there is an increasing range of digitally based tools that can enrich the quality and range of student assessment. Therefore, the choice of assessment methods, and their relevance to other components, are vital elements of any effective learning environment.

1. Preparation Required by the Instructor/Trainer/Teacher to Implement Collaborative Online Learning

The right plans will help instructors to conceive and create interesting, informative learning coursework, activities and exercises that will fuel students' hunger for learning more. Also this will encourage them to open up towards their instructors and other students in their group who can ably support them by offering thoughtful advice and tips. So implementing right plans and using proper tools for implementing them, is a must for facilitating learning through online group collaboration.

- Plan activity or activities that fit into the course.
- Whenever teacher think about whether to introduce a new tool, activity or method into a course, it is essential to consider its usefulness. Teacher should have a clear goal/objective for introducing the new tool and need to be able to articulate this to students. It is required to think about how it alters or adds to the teaching methods, how it will fit with teaching philosophy and style, and perhaps most importantly from a student's perspective, how it will affect the assessment methods.
- Teacher need to spend as much time in advance as possible thinking through the new activity to balance interactivity and instructor workload.
- The more time teacher able to spend in advance, less time they will have to spend making important decisions about the course while it is in session. Online instruction can often mean more work for the instructor, but good course design and planning can help reduce the workload while the course is in session and can help make the quality of interaction between the instructor and the students more rewarding. Online discussions should not be viewed as an "add-on"; rather, they should replace something else.
- Plan to prepare the students for using the new tool or activity.
- Students cannot be expected to "know" how to discuss effectively either online or inperson. Nor can one expect them to "know" how to work effectively in a group setting, particularly in a virtual group. The teacher need to prepare students for the work they will be doing. Also need to prepare them for working in groups by arranging workshop on group work to teach them how to work as a team in a face-to-face setting so groups can begin to understand the dynamics of their team and what their own role in the group will be.
- Clear definition of expectations and purpose
- It needs to be clearly defined how a specific learning activity is relevant to the students and why being part of a group and working together will be beneficial for them. They need to be conveyed what is expected to be in the syllabus. Instructors need to describe various requirements for participation in an activity, the requisite process for participation as well as the specific online tools

needed for facilitating communication within a group. As students get to know about tools, they would have ample time to get familiar in their use. This is going to help them in learning as they will be able to work with more confidence and get improved results.

- Providing clear instructions to students in a group
- Smooth working in a group gets obstructed if students are unclear about objectives of an activity and if they have not been provided with proper and clear instructions. Instructors need to explain the purpose of an activity, provide specific due dates and necessary instructions. It is highly advised to set a due date around such time when course is nearing completion as this would enable students to get adjusted, be familiar with other students and cultivate rich relationships with them.
- Emphasis on keeping groups small
- To ensure active participation from students, it is better to make small sized groups which can be handled well. By making bigger groups constituting increased number of students, students tend to become unresponsive and do not actively contribute.
- Close monitoring and support to be provided by instructors
- Students might need help from time to time and teacher should be readily available for answering their queries. Students, who are not faring well, need support from their instructors and they should be able to provide students the much needed support in a fast manner. An instructor can readily pass instructions to students in a group through synchronous video sessions.
- Defining etiquette guidelines for proper participation
- Guidelines need to be created for making students aware about how they can better participate in an online group. They have to be pretty clear about how they should collaborate. This would enable them to work as a well-knit unit and contribute to a common output.
- Devising activities relevant to the topic
- Teachers should conceive such activities which are relevant and specific to the topic. These should not be stuffed with general information which will make students lose their interest. Activities that encourage exploration, improve engagement and relate with real life examples will invoke a much better response from the students. Putting relevant links, images, quizzes, videos and other engaging, informative material will bring out the best from students and they will enjoy the learning process (University of Waterloo, n.d.).

2. Role of Teacher Presence in Facilitating an Effective Online Discussion

- Set-up expectations for the students engaged in the activity. Rules for posting in discussion such as grammatically correct language, informal expression and the use of slang and emoticons. Students may need a reminder of acceptable online behaviour; to be courteous and respectful in their communication style, content, and tone. Provide a rubric to help students understand how they are going to be assessed on in the discussion activity. Helping the students get started in their group activities online is an important first step in ensuring success. As they start to discuss online, drop into their discussions to provide focus to the discussion or to draw attention to particular concepts or information that is necessary to frame or pursue knowledge growth.
- Help students get started in their discussion. An online icebreaker activity can help students get to know each other online and reduce the awkwardness of discussing a topic with strangers. Alternatively arrange for your students to meet face-to-face as a group before they start their online discussion if that option is possible. Encourage students to draw on previous knowledge and

experiences and respond to others' comments directly as they think critically about the discussion questions. E-mail people who are not participating to find out if they are experiencing technical difficulties with the online forum.

- Teacher's presence motivate and encourage students. Perhaps one of the most important aspects for the instructor who uses online discussions is teacher presence. This happens by posting the discussion questions, directing the groups in the discussions, and by providing feedback on how the discussion is going. Strategies include the following: ask questions (these are called "trigger questions"); give and ask for examples; identify students who are good at making connections between posts; create "weaving" posts to link other's good ideas together to advance the discussion, (example "V and X make a good point,... What do others think?"). When instructors explicitly recognize and reward this level of learning they can also encourage further knowledge growth.
- Provide direct instruction to the students. Direct instruction and feedback to the groups is sometimes necessary to keep them on track with the discussion. The instructor's comments and questions to the groups can be invaluable and can serve as a model for how the discussion should unfold.
- Provide access to resources. The instructor can provide access to a wealth of resources which students can be referred to for further individual or group study. Hyperlinks to online resources can be especially helpful, as they are easy for students who are already online to access.
- Provide technical assistance. The teachers may be asked to provide direct instruction about technical issues related to accessing the conferencing system, manipulation of the conferencing software, operation of other tools or resources and the technical aspects of dealing with any of the subject related tools and techniques. Have a plan in place to handle these requests.
- Practical considerations for facilitating online. It can become overwhelming to read through a busy discussion forum with lots of posts and replies. Teacher should ask students to create new threads if new topics evolve in the discussion. "Subscribing" to receive e-mail alerts of new postings can help participants keep up with a conversation without checking back into the discussion forum repeatedly (University of Waterloo, n.d.).

C. ONLINE COLLABORATIVE LEARNING TOOLS

Learning is no longer limited to classrooms now. The introduction of new technological tools has made it possible for far away located students to collaborate with their instructors and peers for learning new skills and acquiring enhanced knowledge, probably gain degree. Distance and time is no longer a barrier for imbibing knowledge. But for ensuring that students remain motivated and focused in their learning and gain the most from their interaction with their fellow students and instructors, proper strategies for learning activities need to be formulated and implemented. Otherwise, students will feel neglected and disengaged. For effective implementation of learning strategies, right aids need to be used. These online collaboration tools will enable students to communicate and collaborate quickly and easily. This will sustain their interest, improve focus and they would be able to contribute in best possible way and get quality results in turn.

Choosing precise tools is important for improved online collaboration for group learning. These online collaboration tools should have right functionality and should be easy to use. Only then students will feel confident and comfortable in carrying out their tasks and assignments (Thomson, 2014).

The enormous and distributed landscape of tools, makes it difficulty of finding one or a set of online learning tools to meet the requirements and it can be time intensive too.

While choosing for best online collaboration tool one can expect following features:

- facilitate real-time and asynchronous text, voice, and video communication.
- assist in basic project management activities.
- support co-creation by enabling groups to modify output in real-time or asynchronously.
- facilitate consensus building through group discussions and polling.
- simplify and streamline resource management.
- enable local and remote presentation and archiving of completed projects (Carnegie Mellon University, 2009).

Online collaboration tools can be divided into types according to its functions are as follows:

- Collaboration Suites
- Course / Learning Management Systems
- Project Management Tools
- Wikis
- Real-Time Communications
- Collaborative Concept Mapping
- List/Task Management
- Presentation & Slide Sharing
- Collaborative Writing
- Creative/Design Collaboration
- Robotics and Coding

Collaboration Suites

An online collaboration suite provides an integrated set of web tools that span a range of collaboration needs. While not every collaboration suite includes the same capabilities, they often feature web tools for email, instant messaging, contact management, calendars, file sharing, document management, project management, portals, workspaces, web conferencing, and social media tools such as forums and wikis (McCabe, 2010).

Some examples are as follows.

Google http://www.google.com/intl/en/options/

It includes:

- Google Apps offers web-based real-time collaboration: document, spreadsheet, presentation editing and more. It is free.
- Google Drive is a file storage and synchronization service which enables user cloud storage, file sharing and collaborative editing. It is free.

- Google Sites is a structured wiki- and web page-creation tool offered by Google as part of the Google Apps productivity suite.

Zimbra https://www.zimbra.com/

Zimbra is an open source unified collaboration platform for messaging, social and sharing. There is a free trial for Email and Collaborate option including an open source email and collaboration solution, Zimbra mobile and customer support and also a free edition of Build a Social Network which is a private social networking and online community solution, with full suite of social applications and built-in social analytics. Other options are paid.

Zoho https://www.zoho.com/

Zoho provides a wide, integrated portfolio of rich online applications for businesses. With more than 20 different applications spanning Collaboration, Business and Productivity applications, Zoho helps businesses and organizations get work done. Our applications are delivered over the internet, requiring nothing but a browser. This means you can focus on your business and rely on us to maintain the servers and keep your data safe.

Course / Learning Management Systems

An LMS is an integrated software environment supporting all these functionalities along with some mechanism for student enrolment into courses, general user management (including administrator, teacher, and student) and some important system functions like backup and restore. Some examples are as follows

Blackboard https://www.blackboard.com/

It is a virtual learning environment and course management system developed byBlackboard Inc. It is Web-based server software which features course management, customizable open architecture, and scalable design that allows integration with student information systems and authentication protocols. It may be installed on local servers or hosted by Blackboard ASP Solutions. Its main purposes are to add online elements to courses traditionally delivered face-to-face and to develop completely online courses with few or no face-to-face meetings.

Moodle https://moodle.org/

Moodle (Modular Object-Oriented Dynamic Learning Environment) Moodle is a Course Management System (CMS), also known as a Learning Management System (LMS) or a Virtual Learning Environment (VLE). It is an Open Source e-learning web application that educators can use to create effective online learning sites.

There are a range of potential applications of Moodle technology in education and training, delivery of learning materials and collaboration.

The basic features of Moodle include tools for creating resources and activities. These in turn provide the teacher managing the course various useful options. The Resources tab offers the teacher a choice of creating labels which are simply headings for each topic or week, creating text pages or web pages with a combination of text, images and links. Creating links to files or web sites/pages which can link to podcasts, videos and other files, creating directories which are folders one creates with a multitude of different files to be accessed by students or staff.

Another useful and collaborative section is the Activities tab which includes: assignments, chat, choice (one question with a choice of answers – answers are logged so statistics can be deducted), database which is a table created by the teacher and which is filled in by the students creating a database. Forum where everyone can post in response to discussion threads, glossary is a type of dictionary created by the teacher with terms used and their meanings.

Glossaries can also be an enjoyable, collaborative activity as well as a teaching tool. Lessons offer the flexibility of a web page, the interactivity of a quiz and branching capabilities. Quiz enables the creation of various types of quizzes, survey is a questionnaire which gathers feedback from students, wiki is a web page edited collaboratively.

Edmodo https://www.edmodo.com

With an intuitive, user-friendly interface designed by teachers for teachers, Edmodo operates as a communication portal for students, teachers, and parents. No more questions about assignments, quizzes, and grades. Edmodo creates a space where everyone can communicate confidently without the pressure of a group setting or the inconvenience of scheduling face-to-face conversations.

Project Management Tools

Collaborative project management is based on the principle of actively involving all project members in the planning and control process and of networking them using information, communication, and collaboration modules. Management is not regarded as an activity reserved solely for managers but as an integral part of the project work of all team members.

ActiveCollab https://www.activecollab.com/

Active Collab offers features such as task management, collaboration, time tracking, and invoicing. It provides users with needed flexibility to manage virtual projects. Active Collab runs in the cloud like most browser apps today, but can also install it on server too.

Basecamp https://www.basecamphq.com/

Basecamp is an excellent online software package that makes project management and collaboration easy. Basecamp helps to manage multiple projects at a time with to-do lists, file sharing, chatting, messages, calendars and time tracking

Wikis

Wiki is a piece of server software that allows users to freely create and edit web page content using any web browser. Wiki supports hyperlinks and has a simple text syntax for creating new pages and crosslinks between internal pages on the fly.

Wiki is unusual among group communication mechanisms in that it allows the organization of contributions to be edited in addition to the content itself.

Like many simple concepts, "open editing" has some profound and subtle effects on Wiki usage. Allowing everyday users to create and edit any page in a Web site is exciting in that it encourages democratic use of the Web and promotes content composition by nontechnical users.

PBwiki http://pbwiki.com/

PBworks provides a broad set of collaboration products that help businesses work more efficiently and effectively. Products such as Agency Hub, Legal Hub, and Project Hub serve markets such as advertising and marketing agencies, law firms, and education, as well as the broader business market. Millions use PBworks each month for partner/client collaboration, new business development, project management, social intranets, and knowledge management.

PBworks allows multiple users to create and edit a website without any special software or web-design skills. The owner(s) of the wiki can track changes, moderate comments, and control who has access to the wiki. These features make PBworks a useful tool for collaborative writing projects. Also publish schedules, lectures, notes, and assignments. It enables collaborative group projects. Even it keeps parents involved and informed. Some more features include

- Create student accounts without requiring email addresses
- Automated notifications
- keep everyone up to date
- Edit and format wiki pages without learning how to code Grant access to people inside or outside your organization
- Store, discuss, search & share wiki pages, files, and documents
- Every wiki page or file, accessible by computer, smartphone, or tablet

Wikispaces https://www.wikispaces.com/

Wikispaces Classroom is a social writing platform for education. It is incredibly easy to create a classroom workspace where teacher and students can communicate and work on writing projects alone or in teams. Rich assessment tools gives the power to measure student contribution and engagement in real-time. Wikispaces Classroom works great on modern browsers, tablets, and phones. Some applications are as follows

Some Social applications

- Create a safe, private network for your students
- Connect and communicate using a familiar newsfeed
- Monitor complete history of student discussions, writing, and file uploads

Writing

- Collaboratively edit pages using our visual editor
- Embed content from around the web, including videos, images, polls, documents, and more
- Comment on sections of text or the entire page

Projects

- Create individual or group assignments in seconds
- Choose to set assignment start and end dates, or create long-running projects
- At the end of the assignment, automatically publish projects to the entire class or students, parents, or others in your community

Real-Time Formative Assessment

- Watch student engagement in real time, literally as they type, without changing how you work
- Report on contributions to pages, discussions, and comments over time
- Focus on particular students, projects, or view reports across your entire class

Real-Time Communications

The real time communication is the communication in which sender and receiver exchange their information and data over a channel without any delay. Generally Real time communication (RTC) is called "live communication"

Characteristics of Real time communication:

- Timeliness
- Fast
- Low loss rate
- Low end to end delay
- Delivery of acknowledgement
- Peer to peer
- Cost effective

Some application of Real time communication (RTC)

- Instant messaging
- Internet telephony and voip
- Live video conferencing
- Teleconference
- Multimedia multicast
- Internet relay chat(IR)
- Amateur radio (Singh, & Passi, 2014)

Citrix GoToMeeting http://www.gotomeeting.com

Citrix GoToMeeting makes it simple and cost-effective to meet online with colleagues and customers. Best of all, meeting participants can share their webcams in high definition, so you can enjoy more personal interactions – without needing a complicated setup. You can meet from anywhere on any device – no training needed. Start a meeting and share your screen, video and audio with just a click. Show your screen, share your webcam and speak your mind.

GoToMeeting integrates everything – VoIP, telephone, HD video – for a clear and professional web conference. It's the next best thing to sitting at the same table.

It hosts over 30 million meetings a year. Each one is secured by end-to-end 128-bit AES encryption and backed by multiple datacenters around the globe

Skype https://www.skype.com/

Skype is an application that specializes in providing video chat and voice calls. Users can also exchange text and video messages, files and images, as well as create conference calls.

Skype is available on Microsoft Windows, Mac, or Linux, as well as Android, Blackberry, iOS and Windows smartphones and tablets. Skype is based on a freemium model. Much of the service is free, but users require Skype Credit or a subscription to call landline or mobile numbers.

Skype allows users to communicate by voice using a microphone, video by using a webcam, and instant messaging over the Internet. Skype-to-Skype calls to other users are free of charge, while calls to landline telephones and mobile phones (overtraditional telephone networks) are charged via a debit-based user account system called Skype Credit.

Skype boosts online group learning. Instructors can conduct video meetings with groups. They can also conveniently discuss progress or concerns with individual students through video meetings.

These online group collaboration tools have enabled smooth and fast communication, collaboration among distantly located students and their instructors. Thus students can gain more from the experience of their instructors. These tools along with instructional design and implementation strategies devised by instructors has led to enhancement of knowledge and sharpening of students' skills. This has allowed educational benefits to reach to more and more students globally beyond the confines of classrooms

BigMarker

This web conferencing service facilitates communication among learning group members through webinars. Webinars can be flexibly conducted from any location at present or in the future. Members can discuss over matters in real time. Members/contacts can be invited to participate in a webinar through automatic email invitations. Webinars can be tracked through calendar. Its live video chat feature enables synchronous communication among group members. Presentations, audio, chat and webcams can be effectively recorded for later viewing and sharing. So members missing the live events can view their recorded version and gain from them. Participation and attendance of group members in events can be effectively tracked through members page (Thomson, 2014).

Collaborative Concept Mapping

Concept mapping is a technique where users externalise their conceptual and propositional knowledge of a domain in a way that can be readily understood by others. It is widely used in education, so that a learner's understanding is made available to their peers and to teachers. There is considerable potential educational benefit in collaborative concept mapping (Martinez, Yacef & Kay, 2010)

Bubbl.us https://bubbl.us/

is a free online mind mapping tool that allows users to create colorful mind maps and even collaborate on a document. The interface is easy to use and includes several basic options. Maps can also be printed, emailed, downloaded as a graphic, or embedded into a website. This activity allows students to hone their skills for organizing and mapping ideas and collaborating with others.

Powered by Flash, Bubbl-us makes it easy for anyone to quickly start planning and sorting out their ideas through the use of linked text bubbles.

MindMeister https://www.mindmeister.com

MindMeister an online mind mapping and collaboration platform that runs in both web browser and mobile devices. It can be used for live or remote online collaboration. It's a great organizational tool for note taking, live collaborative story or concept mapping, and maps can easily be transformed into stand-alone or live presentations.

Slatebox https://www.slatebox.com

Slatebox is a slick tool for collaboratively creating mind maps and organizational charts. Slatebox offers a variety of good-looking templates and intuitive tools for designing and editing mind maps and charts. Creating a mind map is a simple matter of selecting a template and using the visual editor to place text and images in boxes. Those boxes can be resized and rearranged using the drag and drop editor. If you need more text boxes, simply add more.

List/Task Management

Task management is the process of managing a task through its life cycle. It involves planning, testing, tracking and reporting. Task management can help either individuals achieve goals, or groups of individuals collaborate and share knowledge for the accomplishment of collective goals

Remember the Milk https://www.rememberthemilk.com

Remember the Milk is available as a Web app, but not as a desktop app. You'll need an Internet connection to use it on the Web, i.e., there is no offline functionality. It formerly had offline features running through Google Gears, but with that program long gone and dead, offline functionality for Remember the Milk is, too.

On mobile devices, Remember the Milk is available on iOS, Android, and BlackBerry. It can also integrate with Gmail, Google Calendar, Outlook, Evernote, and Twitter.

In the app, you can create a task, assign a due date, attach notes (as typed text only, not uploaded files), and add tags. I like that you can also add a time estimate to any task, which isn't a feature I've seen in many task management apps for personal use, though it's more common among project management platforms.

Tasks can be grouped into Lists, sometimes called projects by other apps. For example, you might have a list for personal to-dos, family to-dos, and work tasks. Remember the Milk lets you share both individual tasks and lists with others. When you share a task or list, the collaborators have both read and write privileges. In the Web app, it's easy to spot a tab to the right of any task that's prominently labeled Share, though this feature is only in the Web app, not the mobile apps.

Thoughtboxes https://www.thoughtbox.es/

Thoughtboxes makes it easy to get organized, complete tasks, and collaborate.

- All your lists on one page
- Web-based, access anywhere
- Drag-and-drop controls
- Sharing & Collaboration

It breaks down bigger tasks into actionable steps, and so one see everything at a glance. It is available free of cost.

It manages a project, keep track of to-dos, or brainstorm a new idea. One can make lists public, or share them privately and collaborate with friends, family, or co-workers. Thoughtboxes is web-based, so there is nothing to download, and need to log on to access lists from anywhere.

Todoist: https://todoist.com/

Todoist is a cloud-based service, so all your tasks and notes from one app automatically sync to all the other places where you have Todoist installed.

To manage and conquer your to-dos, Todoist lets you create projects and add tasks to them. Projects can be color-coded to help you visually differentiate between them. Tasks can have subtasks, as well as due dates, reminders, flags noting the tasks' urgency, and more.

You can schedule a task to be recurring, and one neat feature is that Todoist lets you use natural language to set it up. For example, if you open the due date option on any task and type "every two weeks" or "every other Wednesday," Todoist will figure out what you want and schedule the task to recur accordingly. The natural-language input works on the Web, in the desktop apps, and in the mobile apps.

It has custom filter and good reminder options. Email notifications are included, as are a few notifications that are specific to mobile devices: push notifications, SMS notifications, and location-based reminders.

One can invite collaborators to access the tasks and projects, but need to sign up for a Todoist account

Presentation and Slide Sharing

SlideShare https://www.slideshare.net/

SlideShare is a Web 2.0 based slide hosting service. Users can upload files privately or publicly in the following file formats: PowerPoint, PDF, Keynote or OpenDocument presentations. Slide decks can then be viewed on the site itself, on hand held devices or embedded on other sites. Although the website is primarily a slide hosting service, it also supports documents, PDFs, videos and webinars. SlideShare also provides users the ability to rate, comment on, and share the uploaded content.

The website gets an estimated 70 million unique visitors a month, and has about 38 million registered users. SlideShare was voted among the World's Top 10 tools for education & e-learning in 2010

SlideRocket www.sliderocket.com

Attractive and engaging presentations can be created with this web based presentation tool and can be accessed from anywhere. Members in a learning group can collaboratively work on one presentation document. Each document has got a specific URL which can be be submitted to an instructor for easy viewing. This application can be embedded within discussion forums of learning management system platforms or web pages. Themes, pictures, audio and others can be combined in a presentation. Presentations and slides can be shared and reused. Invites can be sent for sharing presentations. Data can be pulled in real time from Google Spreadsheets, Twitter live feeds and Yahoo. SlideRocket analytics helps to measure a presentation's effectiveness by giving information about who viewed it and what was their response (Thomson, 2014).

Pear Deck https://www.peardeck.com/

According to the developer, Pear Desk is a tool for Google Slide presentations and templates that allows you to transform "presentations into classroom conversations with an array of interactive and formative assessment questions (TeachThought Staff, 2019)

Collaborative Writing

Group writing assignments have traditionally been confined to a word processing program where one group member would attempt to record all of the group's ideas into one coherent product. However, having students write collaboratively online allows them to share ideas in real time, peer edit, contribute to the writing process from any computer with internet access and they can share their live edits with teachers/peers with a unique web address for their paper. Online, collaborative writing is an outstanding platform for a group writing assignment because it allows all group members to participate in the writing process. Whether students are working in a computer lab or from a computer in their home across globe, the writing process is extended beyond the classroom with online, collaborative writing. Along with group writing assignments, with online collaborative writing, students can share their work in real time and one can have the ability to read, critique, leave feedback and comments directly in their paper during the writing process.

Some of the key advantage of online, collaborative writing include:

- Live, real time feedback.
- Editable & accessible from any computer with an internet connection.
- Specific word processing programs are not necessary, allowing students without specific programs to access their work in multiple locations.
- Student work can be easily published and shared online.
- Revision history of writing and progress is easily accessible and viewable by both the student and teacher
- Accessibility: Online documents can be shared as viewable or editable (Collaborative Writing, n.d.)

Etherpad: https://etherpad.org/

Etherpad is an open-source writing platform, which has been forked for dozens of iterations, such as PiratePad. While not as robust as Google Docs, or as elegantly designed, the fact that Etherpad is open-source means that it can be customized to suit the needs of a specific organization or project.

Draft https://draftin.com/

A very simple tool for working with editors or getting feedback from peers on a piece of writing. Like Editorially, Draft has a clean interface that keeps the focus on the text itself. Includes robust version control, carefully tracking changes made to a document. Draft does not facilitate co-authoring, as much as it attempts to smooth out and enhance the relationship between writers and editors. And the best part: you can shut off your internal editor by using "Hemingway mode," which basically disables your "delete" button and allows only one way through to the completion of a piece of writing: forward.

Examples of Online Collaborative Writing in the Classroom

- The Great Immigration Debate is a Google Docs lesson designed to help students study a topic related to patterns in immigration history, while gathering and analyzing data using primary source materials.
- Islamic Architecture is a Google Document resource created by students.(Collaborative Writing, n.d.)

Creative/Design Collaboration

Creative/Design Collaboration tools makes it easier and faster for designers to get feedback and approve artwork in a professional manner, and they come in all sort of forms, from free Android apps to Chrome extensions. These online tools allow designers to take part in collaborational work in real time. There are collaboration tools, from concept drafting and brainstorming to working on mock-ups and live project (He & Brandweiner, 2014)

Makers Empire https://www.makersempire.com/

By focusing on 3D design, Makers Empire gives students the opportunity to develop critical thinking and problem-solving skills in an online collaborative environment. The designers of Makers Empire began with a desire to encourage STEM (Science, Technology, Engineering, and Mathematics) learning at the early stages of education. The popularity of this online tool suggests they accomplished their goal.

Popplet https://popplet.com/

Part mind mapping tool, part PowerPoint, Popplet gives students a clear, concise way to compile multiple ideas on a single topic and share them with each other. From brainstorming for a writing project to visualizing relationships between newly discovered images and information, this cooperative application lets users record their thoughts from a tablet or computer and display them for the rest of the class to see.

ConceptShare http://www.conceptshare.com/

ConceptShare is a creative operations platform used by enterprises of all sizes to share, communicate and collaborate on creative work. ConceptShare helps eliminate the clutter, chaos and inefficiency of paper and email-based review and approval processes. Through the web-based system, users can initiate, track and report on reviews and approvals, and easily communicate clear and actionable feedback on project assets.

Red Pen: https://redpen.io/

Specifically created for designers, Red Pen lets you drag and drop your designs into a dashboard and invite specific colleagues (or even clients) to let you know their thoughts in real-time as you roll out your latest updates to a project. One of its best features is that it keeps track of the numerous versions made so you can always reclaim that earlier design if you changed you mind. Pricing starts at $20/month for 5 projects.

Stixy http://www.stixyexperience.com/work/stixy

Stixy helps users organize their world on flexible, shareable web-based bulletin boards called Stixyboards. Unlike most personal productivity or project management software, Stixy doesn't dictate how users should organize their information. Users can create tasks, appointments, files, photos, notes, and bookmarks on their Stixyboards, organized in whatever way makes sense to them. Then they can share Stixyboards with friends, family, and colleagues.

Book Creator https://bookcreator.com/

Book Creator is the simple way to make ebooks using the Chrome App or iOS App. Book Creator has real-time collaboration and is ideal for making all kinds of books, portfolios, comic books, photo books, journals, textbooks and more.

Cospaces Edu https://cospaces.io/edu/

CoSpaces Edu is an application that allows students and teachers to easily build their own 3D creations, animate them with code and explore them in Virtual or Augmented Reality.

WeVideo https://www.wevideo.com/education

WeVideo is an online and collaborative video editing tool. Students can easily edit and collaborate on videos on desktops, iOS, and Android. And, it's Chromebook compatible!

Subject Specific Tools

Economics-games.com

When it comes to exciting subject matter, most teachers will tell you that economics rarely tops the list. However, with web-based games that let students simulate real-world scenarios, Economics-Games.com weaves in a little friendly competition to make the subject come alive. Since there are no apps or software to download, students and teachers can access the games wherever there's an internet connection.

CueThink https://www.cuethink.com/

With the tagline "Make Math Social," this engaging app uses classroom interaction to enhance problem-solving skills. When presented with math problems, students use CueThink to select a strategy and display their work. The app also gives classmates the ability to share positive critique and feedback. As students learn to solve problems, they also gain the ability to communicate their thought process to others.

Minecraft for Education https://education.minecraft.net/

Students can work together to explore ecosystems, solve problems through design, architecture, etc.

VideoAnt https://ant.umn.edu/

In our Zero to Infinity module, students carved up an hour-long documentary about mathematics titled "The Story of One." Educators annotate certain times with questions for reflection and short answers, while students annotate with different follow-up questions, and clarifying comments. Due to our students' geographic diversity, there can be large gaps in their understanding of mathematics. This tool helps educators gain more insight into the student math experience before starting the module (TeachThought Staff, 2019).

Robotics and Coding

The Robot Factory by Tinybop is an exploratory app as kids build their dream bots in this beautiful robot factory. Kids who love to create, design, and experience free play can mix and match 50+ unique parts to experiment with different designs. The Robot Factory's is best for builder-types who have the patience to explore and stick to a task without a lot of reinforcement.

codeSpark Academy is an award-winning app that teaches kids how to code. The game-like interface makes coding fun for kids and they don't even realize they're learning. Kids learn to code with lovable characters called The Foos. Each world explores a fundamental coding concept where kids use logical thinking and problem solving skills to help The Foos accomplish tasks. The app empowers young thinkers to become makers as kids learn to create their own stories and games.

CodaKid is a kids coding platform that teaches kids how to use real programming languages and professional tools while creating games, coding apps, programming drones, building websites, and more. CodaKid teaches JavaScript, Java, Lua, and the Unreal Blueprints scripting language - with more on the way soon. Their coding courses teach booleans, conditionals, loops, variables, methods, arrays, switch statements, functions, and more. CodaKid classes also help boost students' proficiency in mathematics, problem solving, and critical thinking.

ScratchJr is a free coding app for young children. With ScratchJr, young children (ages 5-7) can program their own interactive stories and games. In the process, they learn to solve problems, design projects, and express themselves creatively on the computer. Children snap together graphical programming blocks to make characters move, jump, dance, and sing. Children can modify characters in the paint editor, add their own voices and sounds, even insert photos of themselves -- then use the programming blocks to make their characters come to life (10 Best Coding Apps for Kids, 2019).

Miscellaneous Applications Useful in Online Collaborative Learning

Twiddla https://www.twiddla.com/

Twiddla is a real-time web-based meeting tool that lets you "mark up websites, graphics, and photos, or start brainstorming on a blank canvas." This is a great alternative for those sometimes deadly boring online meetings in Adobe Connect and the like. No plug-ins. No sign-up. Just click, invite and go.

Synaptop https://www.synaptop.com

Synaptop lets students watch videos or lectures together, edit and annotate documents together, give presentations online, and store and share files. Plus, there are loads of apps, music, movies, games, and more. Synaptop works similarly to Google Hangouts, but without the need for screensharing or downloads. Everything is done via the cloud. Classroom instructors and librarians can useSynaptop for Education for online presentations, webinars, or even online courses. And librarians can co-browse the web with students for remote reference help. This is a tool worth looking at. (Hovious, 2013).

Quick Screenshare screencast-o-matic.com

Screencast-o-matic is a screen capture software that can be used to create video from your screen (i.e. short lectures or course tours), and it doesn't require any downloading or installing

TogetherJS https://togetherjs.com/

TogetherJS is a service you add to an existing website to add real-time collaboration features. Using the tool two or more visitors on a website or web application can see each other's mouse/cursor position, clicks, track each other's browsing, edit forms together, watch videos together, and chat via audio and WebRTC

GooseChase EDU https://www.goosechase.com/edu/

This exciting educational tool puts the "active" in "interactive." While GooseChase features an extensive game library that teachers can use to teach everything from physical education to basic grammar, the app also features scavenger hunt-style lessons that can also add an extra element of fun to field trips. As a bonus, GooseChase developers also incorporated staff training and professional development exercises.

Yammer

Despite its potential pitfalls, social media is still a powerful tool for collaboration. But in a school setting, the open source platform of Facebook, Twitter, and Instagram can be distracting for students. With its familiar, interactive design, Yammer allows customized groups to share ideas, information, and feedback, making it an ideal communication tool for teachers, students, and parents.

Microsoft Translator

Translate languages via text, voice, or photograph to more easily communicate in other languages. Even Google Translate is also a good option.

WeTransfer: Transfer files of almost any size with WeTransfer. Google Drive and Microsoft OneDrive other option are also available.

FlipGrid https://info.flipgrid.com/

It is a video tool that is meant to encourage discussion and engagement. These short video-logs allow students to share ideas and opinions in a fun and hands-on way, as video submissions are often more enticing to students than a written response (this is especially true for teachers of ESL learners, who sometimes get anxiety about their written work.)

FlipGrid provides another approach for long-distance collaboration: The time limits are an added challenge to students who sometimes struggle with brevity. It's important to note that the free version offers limited features, while the paid version features offer full student collaboration and video conversation.

Diigo https://www.diigo.com/

This tool has become a crucial part of our school's approach to managing project-based learning resources. Since we are constantly on-the-move, a few books, let alone a physical library, are impossible for us to reasonably transport. Diigo eliminates that concern, and helps our students curate and build an ever-growing library of bookmarks for our modules year after year. Countries and specific place- and project-based modules have their own groups for students to contribute to and annotate resources. When a guest speaker visits, we can quickly create a research group to curate a list of resources so that our entire student body is informed and attentive before the speaker arrives (TeachThought Staff, 2019).

Classcraft https://www.classcraft.com/

Classcraft is a game-based approach to teaching and behavior management. It's designed to encourage participation, good behavior, and 21st-century skills like collaboration.

Soundtrap Edu https://www.soundtrap.com/edu/

Soundtrap Edu is an online music studio where students can create together by recording and using loops.

D. ACADEMIC LIBRARY: EVOLUTION OF ACADEMIC LIBRARIAN TO DIGITAL INSTRUCTOR

Academic librarians are equally important as teachers. The main aim of academic librarians is to help students and faculty to achieve academic success. But the situation has been changed where librarians can no longer in position themselves at service desks and wait for students and faculty to come to them but to proactively work to bring and provides best services to them. There is need to blend library and information services into the teaching and learning process by applying "design thinking," which involves, first and foremost, putting librarians in the place of the user in order to understand how the user can receive the "optimal learning experience." The blended librarian on today's campus seeks to meet the user on the user's terms (Sinclair, 2009). According to the Benton (2009) faculty members can work more effectively with librarians to design research projects and to develop collections that support the undergraduate curriculum,". He also suggested that incorporating "new technology while preserving the traditional culture of scholarship and books," can help the library reemerge as "the center of undergraduate education." He strongly believed that collaboration between university/college/school librarians, teachers and students leads to grand success.

The blended librarian is focused on course goals and learning objectives outside of the library and across the curriculum. Books, articles, and reserve readings (both electronic and print) may meet the needs of many faculty and students. But for instructors who seek to use new forms of multimedia—streaming video, podcasts, digitized images, 3-D animations, screencasts, etc.—to engage students and enhance the learning experience, the blended librarian is there to provide guidance and expertise, as well. Perhaps the learning objectives are more collaborative in nature and would benefit from social software in the form of wikis, blogs, video sharing, discussion forums, and other tools offered through a learning management system. The blended librarian is versed in both print and online tools and can help faculty meet course goals, regardless of the medium or technology.

Essentially, this is a new call to outreach. The blended librarian seeks to build new collaborations with students, faculty, staff, and other information and instructional technology professionals both in and outside of the classroom—in physical spaces and virtual environments—in order to match learners and teachers with the information tools they need (Sinclair, 2009). Librarians are often more likely to be able to help teachers find the digital resources and websites that can best support instruction. They can also aid students by teaching lessons about research skills and digital citizenship and can help facilitate personalized learning and the building of literacy skills can be called as a 'digital instructor' (Harper, 2018).

There are a wide range of potential tools available, which has prompted some libraries to create guides to quality collaboration tools for their campus communities.

These guides, including the University of Queensland Library's Research Collaboration Tools and Harvard Law School Library's Collaboration Tools, suggest tools for a variety of applications.

1. Use of Collaborative Learning Tools in Providing Library Services

Library Instruction

In library instruction, online mind and concept mapping tools can be used to generate keywords for searches, to narrow down research topics, and for students in groups to give feedback on each other's topics. Fuchs (2014) describes a variety of possible uses of Padlet, including having students post terms they would use to search for their topic at the beginning of an instruction session. She points out that in addition to serving as a formative assessment, this activity engages students in peer learning and helps them to assess their own skills. The activity could be taken further by having students suggest additional terms for their peers and adding terms to the Padlet after the library instruction session.

Singapore Management University Library produces video tutorials and other online learning objects which promote self-paced learning in the use of library resources and aspects of the research process (Lee Yen, Yuyun, Sandra, Shameen, & Rajen, 2013)

Citation Tool https://www.citethisforme.com/

Cite This For Me - It is an automated citation machine that turns any of users sources into citations in just a click. It helps students to integrate referencing into their research and writing routine; turning a time-consuming ordeal into a simple task. A citation machine is essentially a works cited generator that accesses information from across the web, drawing the relevant information into a fully-formatted bibliography that clearly presents all of the sources that have contributed in the work. This tool supports APA, MLA, Chicago, ASA, IEEE and AMA * styles.

EasyBib https://www.easybib.com/

t is an intuitive information literacy platform that provides citation, note taking, and research tools that are easy-to-use and educational. EasyBib is not only accurate, fast, and comprehensive, but helps educators teach and students learn how to become effective and organized researchers. EasyBib cite according to the 8th and 7th ed. of MLA, 6th ed. of APA, and 16th and 17th ed. of Chicago (9th ed. Turabian). Many of their styles are powered by CSL, the Citation Styles Language from CitationStyles.org, which are licensed under a CC-BY-SA license (https://www.easybib.com/).

Online Group Work and Collaboration

Document collaboration tools are widely used among librarians to work together on presentations and instructional materials, and they can be used in one-shot sessions as well. Bobish (2011) provides creative examples of how Web 2.0 tools can be used in library instruction. Many of his ideas, such as creating a research timeline or a collaborative bibliography or wiki, could easily be adapted for use with Google Docs or Padlet. Bilby (2014) reports on a collaborative student research project developed with the theology librarian at the University of San Diego that required students to create and share annotated bibliographies throughout the semester using Google Drive. Librarians can also use these tools to facilitate small group discussions about research strategies. For example, groups of students could be assigned websites to evaluate and could then post their evaluations, and review other group's evaluations.

Library Space, Facilities and Services

In a flipped classroom and blended learning environment, students spend more time in collaborative learning spaces. The Singapore Management University Library provides such physical spaces especially for individual students and for collaborative study to the entire library community 24/7 open for them.

Bringing Real-Time Help to Patrons

Embedding live chat in specific web pages, to provide just-in-time services. The widget would also be mobile friendly and taps into the Singapore Management University library's extensive FAQ and LibGuides knowledge base, to provide a unified patron support service. Also uses a webinar platform to deliver research skills workshop (Lee Yen, Yuyun, Sandra, Shameen, & Rajen, 2013).

E. PROS AND CONS OF ONLINE COLLABORATIVE LEARNING

Pros are as follows

- Many collaborations online tools are free to use & available 27/7 for everyone
- Students can actively exchange, debate and negotiate ideas within their groups increases students' interest in learning.
- Students may feel good to learn virtually in group instead of learning virtually alone which can be boring
- It can help to build a professional network
- Planning, working & submission of project can be done in time as everyone in group has to complete task in time
- Learning from other group members can be possible

Cons are as follows

- Online collaborative tools are free tools but often disappear or become paid tools.
- Sometimes it is difficult to divide students into groups.
- Those tools offer anonymous participation, can lead to students starting conversations that are offensive or completely irrelevant to the topic
- Low or no participation of other group members
- Differences in skill/knowledge level of group members
- Delays in communication by some group members
- Some kind of hindrance due to not knowing group member personally

F. CONCLUSION

There is great potential for using online collaboration tools to engage students/researchers, provide an outlet for creative exploration of ideas. The variety of tools available means that there is a technology available for almost any classroom activity, right from idea generation and discussion at the beginning of a research project/assignment to its working, sharing to peer-review of research papers. For the effective collaborative learning, Muuro, Wagacha, Kihoro & Oboko (2014) has recommended that institutions should ensure their instructors do engage students in collaborative activities in their online courses and instructor's role is more emphasized during collaborative learning. Also find ways of motivating the students as well as teachers in order to increase their level of participation in collaborative learning

Collaboration can be effectively used to improve the quality and quantity of education in online learning environments. There are numerous tools and methods that can be used to facilitate and stimulate collaboration in online education. The author has made an effort to list and define the most important of those tools and methods. These tools have evolved very recently and will continue to evolve as the time progresses. The online collaborative learning methodology is likely to evolve and make significant benefits to education, and probably to post educational business collaboration as well (Lark, n.d.)

REFERENCES

Advanced Online Collaboration. (n.d.). *Where Creative Minds Meld.* Retrieved from http://www.yorku.ca/dzwick/what_is_octopz.htm

Bates, A. W. (2016). *Teaching in a digital age: guidelines for designing teaching and learning*. Opentextbook.

Bell, K. (2018). *15 Collaborative Tools for Your Classroom That Are NOT Google*. Retrieved from https://shakeuplearning.com/blog/15-collaborative-tools-for-your-classroom-that-are-not-google/

Bell, S., & Shank, J. (2007). *Academic Librarianship by Design: A Blended Librarian's Guide to the Tools and Techniques*. ALA.

Benton, T. H. (2009). A Laboratory of Collaborative Learning. *The Chronicle of Higher Education, 2009*(August). https://www.chronicle.com/article/A-Laboratory-of-Collaborative/47518/

Best Coding Apps for Kids. (2019). https://www.educationalappstore.com/best-apps/10-best-coding-apps-for-kids

Bilby, M. (2014). *Collaborative Student Research with Google Drive: Advantages and Challenges*. Presentation at the annual conference of the American Theological Library Association, New Orleans, LA. Retrieved October 27, 2015, from http://lanyrd.com/sctzbt

Bobish, G. (2011). Participation and Pedagogy: Connecting the Social Web to ACRL Learning Outcomes. *Journal of Academic Librarianship*, *37*(1), 54–63. doi:10.1016/j.acalib.2010.10.007

Center for Teaching Excellence. Carnegie Mellon University. (2009). *Collaboration tools*. Teaching with Technology White Paper. Carnegie Mellon university. Retrieved October 27, 2015, from https://www.cmu.edu/teaching/technology/whitepapers/CollaborationTools_Jan09.pdf

Chandra, R. (2015). Collaborative Learning for Educational Achievement. *IOSR Journal of Research & Method in Education, 5*(3), 4-7. Retrieved October 3, 2015, from http://www.iosrjournals.org/iosr-jrme/papers/Vol-5%20Issue-3/Version-1/B05310407.pdf

Collaborative writing. (n.d.). *EdTechTeacher*. Retrieved October 27, 2015, from http://tewt.org/collaborative-writing/

Cooper, B. B. (2015, November 18). *The Science of Collaboration: How to Optimize Working Together* [Web log post]. Retrieved October 27, 2015, from https://thenextweb.com/entrepreneur/2014/07/15/science-collaboration-optimize-working-together/

eLearning 101 – concepts, trends, applications. (2014). San Francisco: Epignosis. Retrieved October 27, 2015, from LLC http://www.talentlms.com/elearning/

Fuchs, B. (2014). The Writing is on the Wall: Using Padlet for Whole-Class Engagement. *LOEX Quarterly, 40*(4), 7-9. Retrieved October 27, 2015, from http://uknowledge.uky.edu /libraries_facpub/240

Green, L. S. (2019). Online Learning Is Here to Stay: Librarians transform into digital instructors. *American Libraries Magazine*. https://americanlibrariesmagazine.org/2019/06/03/online-learning-is-here-to-stay/

Harper, A. (2018). *The benefits of collaborating with school librarians*. https://www.educationdive.com/news/the-benefits-of-collaborating-with-school-librarians/533247/

He, J., & Brandweiner, N. (2014). *The 20 best tools for online collaboration*. Retrieved October 27, 2015, from http://www.creativebloq.com/design/online-collaboration-tools-912855

Hovious, A. (2013). *5 Free Real-Time Collaboration Tools*. Retrieved October 21, 2015, from https://designerlibrarian.wordpress.com/2013/09/15/5-free-real-time-collaboration-tools/

Krongard, S., & McCormick, J. (2013). *Real Time Visual Analytics to Evaluate Online Collaboration*. Presentation at the annual conference of NERCOMP, Providence, RI, Retrieved October 20, 2015, from http://www.educause.edu /nercomp-conference/2013/2013/real-time-visual-analytics-evaluate-online -collaboration.

Lark, J. (n.d.). Collaboration Tools in Online Learning Environments. *ALN Magazine*. Retrieved October 27, 2015, fromhttp://www.nspnvt.org/jim/aln-colab.pdf

Mallon, M., & Bernsten, S. (2015). *Collaborative Learning Technologies. Tips and Trends*. ACRL Instruction Section, Instructional Technologies Committee, Winter. Retrieved October 02, 2015, from http://bit.ly/tipsandtrendswi15

Martinez, R., Yacef, K., & Kay, J. (2010). *Collaborative concept mapping at the tabletop*. Technical Report 657. University of Sydney. Retrieved October 27, 2015, from https://sydney.edu.au/engineering/it/research/tr/tr657.pdf

McCabe, L. (2010). *What's a Collaboration Suite & Why Should You Care?* Retrieved October 27, 2015, from https://www.smallbusinesscomputing.com/biztools/article.php/3890601/Whats-a-Collaboration-Suite--Why-Should-You-Care.htm

Murugan, S. (2013). User Education: Academic Libraries. *International Journal of Information Technology and Library Science Research, 1*(1), 1-6. Retrieved October 27, 2015, from http://acascipub.com/Journals.php

Muuro, M., Wagacha, W., Kihoro, R., & Oboko, J. (2014). Students' Perceived Challenges in an Online Collaborative Learning Environment: A Case of Higher Learning Institutions in Nairobi, Kenya. *The International Review of Research in Open and Distributed Learning, 15*(6). Advance online publication. doi:10.19173/irrodl.v15i6.1768

Nine Tools for Collaboratively Creating Mind Maps. (2010). Retrieved from http://www.freetech4teachers.com/2010/03/nine-tools-for-collaboratively-creating.html#.VubuNPl97IV

Sinclair, B. (2009). The blended librarian in the learning commons: New skills for the blended library. *College & Research Libraries News, 70*(9), 504–516. doi:10.5860/crln.70.9.8250

Singh, S. P., & Passi, A. (2014). Real Time Communication. *International Journal of Recent Development in Engineering and Technology, 2*(3). Retrieved October 27, 2015, from http://www.ijrdet.com/files/Volume2Issue3/IJRDET_0314_23.pdf

Skyep. (n.d.). Retrieved from Wikipedia: https://en.wikipedia.org/wiki/Skype

Slideshare. (n.d.). Retrieved from Wikipedia: https://en.wikipedia.org/wiki/SlideShare

Smith, B. L., & MacGregor, J. T. (1992). What is collaborative learning? In A. S. Goodsell, M. R. Maher, & V. Tinto (Eds.), *Collaborative Learning: A Sourcebook for Higher Education. National Center on Postsecondary Teaching, Learning, & Assessment*. Syracuse University.

Task Management. (n.d.). Retrieved from Wikipedia: https://en.wikipedia.org/wiki/Task_management

TeachThought Staff. (2019). *30 Of The Best Digital Collaboration Tools For Students*. Retrieved from https://www.teachthought.com/technology/12-tech-tools-for-student-to-student-digital-collaboration/

These 10 Collaboration Tools for Education are Boosting Student Engagement and Bringing Interactive Fun to the Classroom. (n.d.). Retrieved from https://www.getcleartouch.com/what-are-the-best-online-collaboration-tools-for-students/

Thomson, S. (2014). *6 Online Collaboration Tools and Strategies for Boosting Learning*. Retrieved October 27, 2015, from https://elearningindustry.com/6-online-collaboration-tools-and-strategies-boosting-learning

Todoist. (2015). Retrieved from http://www.pcmag.com/article2/0,2817,2408574,00.asp

Tu, C.-H. (2004). *Online Collaborative Learning Communities: Twenty-One Designs to Building an Online Collaborative Learning Community*. Libraries Unlimited Inc.

University of Waterloo. (n.d.). *Collaborative online learning: fostering effective discussions*. Retrieved March 21, 2014, from https://uwaterloo.ca/centre-for-teaching-excellence/teaching-resources/teaching-tips/alternatives-lecturing/discussions/collaborative-online-learning

Vygotsky, L. S. (1978). *Mind in society: The development of higher psychological processes*. Harvard University Press.

Wiki. (n.d.). Retrieved from Wikipedia: http://wiki.org/wiki.cgi?WhatIsWiki

Yen, Yuyun, Shameen, & Rajen. (2013). *SMU Libraries' Role in Supporting SMU's Blended Learning Initiatives.* Retrieved October 27, 2015, from https://library.smu.edu.sg/sites/default/files/library/pdf/Librarys_Role_in_Blended_Learning.pdf

ADDITIONAL READING

Academic Training Group, University of Kentucky. (2015). Adobe Connect Best Practices. Retrieved on 30th July 2014 from http://www.uky.edu /acadtrain/connect/bestpractices

An, H., Kim, S., & Kim, B. (2008). Teacher perspectives on online collaborative learning: Factors perceived as facilitating and impeding successful online group work. *Contemporary Issues in Technology & Teacher Education*, *8*(1). https://www.citejournal.org/vol8/iss4/general/article1.cfm

Association of College and Research Libraries. (2017). *Academic Library Impact on Student Learning and Success: Findings from Assessment in Action Team Projects. Prepared by Karen Brown with contributions by Kara J. Malenfant.* Association of College and Research Libraries.

Barkely, E. F., Cross, K. P., & Howell Major, C. (2005). *Collaborative learning techniques: A handbook for college faculty*. Jossey-Bass.

Black, A. (2005). The use of asynchronous discussion: Creating a text of talk. Contemporary *Issues in Technology and Teacher Education, 5*(1), 5-24. Retrieved on 20th August 2014, fromhttp://www.citejournal.org/articles/v5i1languagearts1.pdf

Brindley, J., Blaschke, L. M., & Walti, C. (2009). Creating effective collaborative learning groups in an online environment. *International Review of Research in Open and Distance Learning*, *10*(3). Advance online publication. Retrieved January 5th, 2014, from http://www.irrodl.org/index.php/irrodl/article/view/675/1271. doi:10.19173/irrodl.v10i3.675

Bruffee, K. A. (1998). *Collaborative learning: Higher education, interdependence, and the authority of knowledge*. The Johns Hopkins University Press.

Byrne, R. (2014, November 19). Seven Free Online Whiteboard Tools for Teachers and Students. Free Technology for Teachers (Web log post), Retrieved on 30th July 2015 from http://www. freetech4teachers.com/2014/01/seven -free-online-whiteboard-tools-for.html

Capdeferro, N., & Romero, M. (2012). Are online learners frustrated with collaborative learning experiences? *International Review of Research in Open and Distance Learning*, *13*(2), 26–44. Retrieved January 5th, 2014, from http://www.irrodl.org/index.php/irrodl/article/view/1127. doi:10.19173/irrodl.v13i2.1127

Cavalier, R. (2008). Campus conversations: Modeling a diverse democracy through deliberative polling. *Diversity and Democracy*, *11*(1), 16–17.

Cavalier, R., & Bridges, M. (2007). Polling for an Educated Citizenry. *The Chronicle of Higher Education*, *53*(20).

Center for Research on Learning and Teaching, University of Michigan. (2015). Teaching with Online Collaboration Tools: U-M Faculty Examples. Retrieved on 30th July 2015 from http:// www.crlt.umich.edu/oct

Chiong, R., & Jovanovic, J. (2012). Collaborative Learning in Online Study Groups: An Evolutionary Game Theory Perspective. *Journal of Information Technology Education, 11*, 081–101. http://jite.informingscience.org/documents/Vol11/JITEv11p081-101Chiong1104.pdf. doi:10.28945/1574

Constructing Knowledge Together. (2008). *(21-45). Extract from Telecollaborative Language Learning. A guidebook to moderating intercultural collaboration online* (M. Dooly, Ed.). Peter Lang.

Curtis, D. D., & Lawson, M. J. (2001). Exploring collaborative online learning. *Journal of Asynchronous Learning Networks, 5*(1), 21–34. Retrieved August 20th, 2014, from https://wikieducator.org/images/6/60/ALN_Collaborative_Learning.pdf

Davis, B. G. (2009). *Tools for teaching* (2nd ed.). Jossey-Bass.

Desanctis, G., & Gallupe, B. R. (1987). A foundation for the study of group decision support systems. Retrieved on 30th July 2015 from *Management Science, 33* (5), 589-609. URL http://www.jstor.org/stable/2632288

EDUCAUSE Learning Initiative. (2014). Cloud Storage and Collaboration. 7 Things You Should Know About…. EDUCAUSE, Retrieved on 30th July 2015 from https://net.educause.edu/ir/library/pdf / ELI7108.pdf

Felder, R. M., Felder, G. N., & Dietz, E. J. (1998). A longitudinal study of engineering student performance and retention. V. Comparisons with traditionally-taught students. *Journal of Engineering Education, 87*(4), 469–480. doi:10.1002/j.2168-9830.1998.tb00381.x

Floyd, J. (2015, January 18). How to Use Google+ Hangouts in Higher Education: Distance Learning with Social Media. Jeremy Floyd (Web log post), Retrieved on 30th July 2015 fromhttp://www.jeremyfloyd.com/2013/05 /how-to-use-google-hangouts-in -higher-education-distance-learning -with-social-media/

Forsyth, D. R. (2009). *Group dynamics*. Wadsworth Cengage Learning.

Harasim, L., Hiltz, S. R., Teles, L., & Turoff, M. (1998). *Learning networks: A field guide to teaching and learning online*. The MIT Press.

Hassanien, A. (2007). A qualitative student evaluation of group learning in higher education. *Higher Education in Europe, 32*(2-3), 135-150. Retrieved on 30th July 2014 fromhttp://www.tandfonline.com/doi/pdf/10.1080/03797720701840633

Hershock, C., & LaVaque-Manty, M. (2012). Teaching in the Cloud: Leveraging Online Collaboration Tools to Enhance Student Engagement. CRLT Occasional Papers, No. 31. Ann Arbor, MI: Center for Research on Learning and Teaching, University of Michigan. Retrieved on 30th July 2015 from http://www.crlt.umich.edu/sites /default/files/resource_files/CRLT_no31.pdf

Jahng, N., Chan, E. K. H., & Nielsen, W. S. (2010). Collaborative learning in an online course: A comparison of communication patterns in small and whole group activities. *Journal of Distance Education, 24*(2), 39–58.

Jaques, D., & Salmon, G. (2007). Learning in groups: A handbook for face-to-face and online environments (4th ed.). UK, USA, and Canada: Routledge. doi:10.4324/9780203016459

Kashorda, M., & Waema, T. (2014). E-readiness survey of Kenyan universities (2013) report.Nairobi: Kenya Education Network. Retrieved on 20th August 2014, fromhttp://ereadiness.kenet.or.ke:8080/ereadiness/2013/E-readiness%202013%20Survey%20of%20Kenyan%20Universities_FINAL.pdf

Kim, K. J., Liu, S., & Bonk, C. J. (2005). Online MBA students' perceptions of online learning: Benefits, challenges, and suggestions. *The Internet and Higher Education*, *8*(4), 335–344. doi:10.1016/j.iheduc.2005.09.005

Klosowski, T. (2014, February 20). The Best Add-Ons for Google Drive. Lifehacker (Web blog post), March 12. https://lifehacker.com/the-best-add-ons-for -google-drive-1541643206.

Liu, S., Joy, M., & Griffiths, N. (2010). Students' perceptions of the factors leading to unsuccessful group collaboration. In Advanced Learning Technologies (ICALT), 2010 IEEE 10th International Conference on (pp. 565-569). Sousse, Tunisia, 5-7 July 2010.

Lomas, C., Burke, M., & Page, C. L. (2008). Collaboration Tools. ELI Paper 2. EDUCAUSE Learning Initiative, Retrieved on 30th July 2015 from http://net.educause.edu/ir/library/pdf /eli3020.pdf

Mattar, J. A. (2010). Constructivism and connectivism in education technology: Active, situated, authentic, experiential, and anchored learning. *Technology (Elmsford, N.Y.)*, •••, 1–16.

McDaniel, J., Metcalf, S., & John, M. H. Rossman. Successful Online Teaching Using An Asynchronous Learner Discussion Forum. *JALN*, 3, 2 - November, 1999

Michaelsen, L. K., Knight, A. B., & Fink, L. D. (Eds.). (2004). *Team-based learning: A transformative use of small groups in college teaching*. Stylus.

Moller, L. (1998). Designing communities of learners for asynchronous distance education. *Educational Technology Research and Development*, *46*(4), 115–122. doi:10.1007/BF02299678

North, A. C., Linley, P. A., & Hargreaves, D. J. (2000). Social loafing in a co-operative classroom task. *Educational Psychology*, *20*(4), 389–392. doi:10.1080/01443410020016635

Nyerere, J. A., Gravenir, F. Q., & Mse, G. S. (2012). Delivery of open, distance, and e-learning in Kenya. *International Review of Research in Open and Distance Learning*, *13*(3), 185–205. doi:10.19173/irrodl.v13i3.1120

Palloff, R. M., & Pratt, K. (2005). *Collaborating online: Learning together in community. San Francisco*. Jossey-Bass.

Palloff, R. M., & Pratt, K. (2007). *Building online learning communities: Effective strategies for the virtual classroom*. John Wiley & Sons.

Project-based learning. (2008, November 22). In Wikipedia, the free encyclopedia. Retrieved December 12, 2015, from https://en.wikipedia.org/wiki/Project-based_learning

Roberts, T. S., & McInnerney, J. M. (2007). Seven problems of online group learning (and their solutions). *Journal of Educational Technology & Society*, *10*(4), 257–268.

Salmon, G. (2004). *E-moderating: the key to teaching and learning online* (2nd ed.). Routledge. doi:10.4324/9780203465424

Siemens, G. (2005). Connectivism: A learning theory for the digital age. *International Journal of Instructional Technology and Distance Learning*, *2*(1), 3–10.

Singh, H. K. (2005). Learner satisfaction in a collaborative online learning environment. Retrieved on 26th January 2015, from http://asiapacific-odl.oum.edu.my/C33/F239.pdf

Song, L., Singleton, E. S., Hill, J. R., & Koh, M. H. (2004). Improving online learning: Student perceptions of useful and challenging characteristics. *The Internet and Higher Education*, *7*(1), 59–70. doi:10.1016/j.iheduc.2003.11.003

Sours, T. J., Newbrough, J. R., Shuck, L., & Varma-Nelson, P. (2013). Supporting Student Collaboration in Cyberspace: A cPLTL Study of Web Conferencing Platforms. EDUCAUSE Review Online, Retrieved on 30th July 2015 from. http://www.educause .edu/ero/article/supporting-student. -collaboration-cyberspace-cpltl-study-web -conferencing-platforms.

Zorko, V. (2009). Factors affecting the way students collaborate in a wiki for English language learning. *Australasian Journal of Educational Technology*, *25*(5), 645–665. doi:10.14742/ajet.1113

KEY TERMS AND DEFINITIONS

Academic Library: An academic library is a library which serves an institution of higher learning, such as a college or a university—libraries in secondary and primary schools are called school libraries. These libraries serve two complementary purposes: to support the school's curriculum, and to support the research of the university faculty and students.

Asynchronous Learning Model: In asynchronous mode of elearning generally teacher can post study material, and have announcements and calendar online, assignment posting, submission, and evaluation with feedback etc.

Blended Librarian: The Blended Librarian is the academic professional who offers the best combination of skills and services to help faculty apply technology for enhanced teaching and learning.

Collaborate: To collaborate is "to work jointly with others or together especially in an intellectual endeavor."

Collaboration Suites: An online collaboration suite provides an integrated set of web tools that span a range of collaboration needs.

Collaborative Project Management: Collaborative project management is based on the principle of actively involving all project members in the planning and control process and of networking them using information, communication, and collaboration modules.

Concept Mapping: It is a technique where users externalise their conceptual and propositional knowledge of a domain in a way that can be readily understood by others.

Digital Citizen: A digital citizen is a person utilizing information technology (IT) in order to engage in society, politics, and government.

Digital Instructor: An educator who uses technology in teaching and learning.

E-Learning: It is commonly referred to the intentional use of networked information and communications technology in teaching and learning.

Learning Management System: An LMS is an integrated software environment supporting all these functionalities along with some mechanism for student enrolment into courses, general user management (including administrator, teacher, and student) and some important system functions like backup and restore.

Online Learning/Virtual Learning/Distributed Learning/Web-Based Learning: It refer to educational processes that utilize information and communications technology to mediate asynchronous as well as synchronous mode of learning and teaching activities.

Real-Time Communications: The real time communication is the communication in which sender and receiver exchange their information and data over a channel without any delay.

SlideShare: It is a Web 2.0-based slide hosting service.

Synchronous Learning Model: In the synchronous learning model (Online model), the students can attend 'live' lectures at the scheduled hour from wherever they are irrespective of their location forming a virtual classroom.

Task Management: Task management is the process of managing a task through its life cycle.

Wiki: Wiki is a piece of server software that allows users to freely create and edit web page content using any web browser.

Chapter 3
Open Innovation:
Reaching Out to the Grass Roots Through SMEs - Exploring Concerns of Opportunities and Challenges to Attain Economic Sustainability

Hakikur Rahman
Institute of Computer Management and Science, Bangladesh

ABSTRACT

While talking about successful entrepreneurship and value addition within an enterprise through innovation, one could comprehend that the innovation paradigm has been shifted from simple introduction of new thoughts and products to accumulation of diversified actions, actors, and agents along the process. Furthermore, when the innovation process is not being constrained within the closed nature of it, the process takes many forms during its evolution. Innovations have been seen as closed innovation or open innovation, depending on its nature of action, but contemporary world may have seen many forms of innovation, such as technological innovation, products/service innovation, process/production innovation, operational/management/organizational innovation, business model innovation, or disruptive innovation, though often they are robustly interrelated.

INTRODUCTION

The initial effort of this research was in fact to find a focus area to empower small and medium enterprises (SMEs[1]) through open innovation strategies. Due to the open and collaborative nature of this newly evolved concept, the primary research focus has been kept within crowdsourcing innovation[2], but not limited to other collaborative innovation, though it is not easy to put a restrictive boundary between them. Being new in the research arena, on one hand the concept of open innovation has been flourished very progressively within a short span of time (Chesbrough, 2003a; Chesbrough, Vanhaverbeke & West, 2006; Gassmann, 2006), but at the same time, it has evolved through various growth patterns in

DOI: 10.4018/978-1-7998-5849-2.ch003

diversified directions involving different factors and parameters (Christensen, Olesen & Kjaer, 2005; Chesbrough & Crowther, 2006; Dodgson, Gann & Salter, 2006; Gassmann, 2006; Vanhaverbeke, 2006; West & Gallagher, 2006). Furthermore, as this research is related to SMEs[3], which are the steering factor of economic growth in the European countries, and especially in Portugal where they comprise of almost 99.9% (EC, 2008) of the entrepreneurships, the problem statements were constructed following multiple studies along this aspect, though much work towards the improvement of knowledge factors on SMEs development have not been found.

Admittedly, organizational sustainability increasingly focuses on how to administer new knowledge of ideas and practices that can expand business. Open innovation plays a key role towards effectual strategic sustainable management. Through open innovation, companies can influence knowledge management to an asset that promotes sustainable innovations that influence back organizational sustainability (Lopes, et. al., 2017). Recently, the business model innovation has seen a recent surge in academic research and business practice. Transformations to business models are recognized as a fundamental approach to realize innovations for sustainability. However, little is known about the successful adoption of sustainable business models (SBMs) (Evans, et. al., 2017).

The extent of environmental or social responsibility orientation in the enterprise is assessed on the basis of environmental and social goals and policies, the association of environmental and social management in the enterprise and the interaction of environmental and social issues. The market influence of the enterprise is measured on the basis of market share, sales growth and reactions of competitors (Schaltegger and Wagner, 2011). Research on acquiring innovations includes searching, enabling, filtering, and attaining—each category with its own specific set of mechanisms and conditions. Incorporating innovations has been mostly studied from an absorptive capacity perspective, with less concentration given to the impact of competencies and culture (West and Bogers, 2014).

Henceforth, problem statements or the main purpose of this study lies within the intrinsic definition of innovation itself. Innovation is a way of performing something new. It may refer to incremental and emergent or radical and revolutionary changes in thinking, products, process development, or organizational development. Innovation, as seen by Schumpeter (1934; 1982) incorporates way of producing new products, new methods of production, new sources of supply, opening of new markets, and new ways of organizing businesses. OECD (1992; 1996; 2005), after several adjustment has come into this argument, that innovation is the implementation of a new or significantly improved product (good or service), or process, a new marketing method, or a new organizational method in business practices, workplace organization or external relations.

However, other scholars and researchers in the field of innovation, has put forward definition of innovation in various formats and perspectives. Some definitions or arguments are being included below:

"The creation of new ideas/processes which will lead to change in an enterprise´s economic and social potential" (Drucker, 1998: 149)

This research will look into the economic and social aspect of innovation process, but at the same time look into any technology parameters that are involved within the processes.

Tidd, Bessant & Pavitt (2005) and Bessant & Tidd (2007) argued that there are four types of innovation, i.e., the innovator has four routes of innovation paths, such as product innovation (changes in the products or services [things] which an organization offers), process innovation (changes in the ways in which products or services are created and delivered), positioning innovation (changes in the context in which the products or services are introduced) and paradigm innovation (changes in the underlying mental models which outline what the organization does)[4].

This research would argue that these are the areas of innovation through which innovation takes place in an enterprise.

However, in terms of the types of innovation, Darsø (2001) argued that innovation can be *incremental (improvements of processes, products and methods, often found by technicians or employees during their daily work), radical (novel, surprising and different approach or composition), social (spring from social needs, rather than from technology, and are related to new ways of interaction) and quantum (refers to the emergence of qualitatively new system states brought by small incremental changes).*

Furthermore, talking about types of innovation, Henderson & Clark (1990) identified four; *incremental innovation- improves component knowledge and leaves architectural knowledge unchanged; modular innovation- architectural knowledge unchanged, component knowledge of one or more components reduced in value; architectural innovation- component knowledge unchanged, architectural knowledge reduced in value; and radical innovation*[5]*- both component knowledge and architectural knowledge reduced in value.*

These are the natures of innovation in which innovation belongs.

Thus, innovation can be termed as introduction of new idea into the marketplace in the form of a new product or service, or an improvement in organization or process[6].

This definition enables one to concentrate on demand driven innovation, or innovation may be re-termed as a:

"Process by which an idea or invention is translated into good or service for which people will pay" (Businessdictionary.com[7]*).*

The above definition leads to the very basic parameters of innovation, where there are goods or services between buyer and the seller.

Hence, a functional equation (innovation as the most probable or prominent function of the mentioned parameters in above definitions and arguments) can be deducted:

Figure 1. A functional equation mapping various definitions and arguments

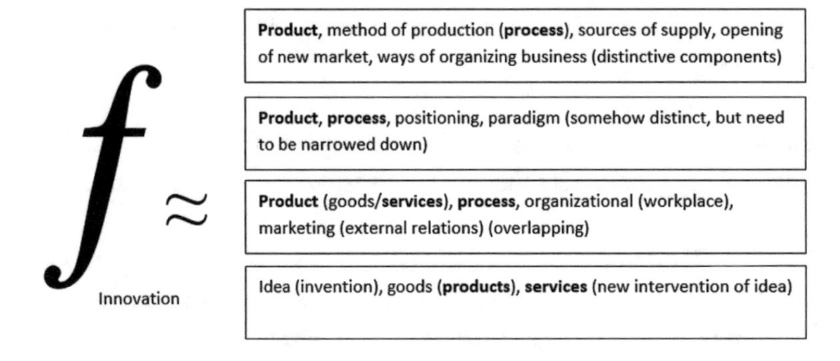

A context diagram from the above equation sets the primary problem statements of this research. If one argues that innovation generates new or improved products, process or services, then the following diagram can be derived:

Figure 2. A context diagram derived from the functional equation

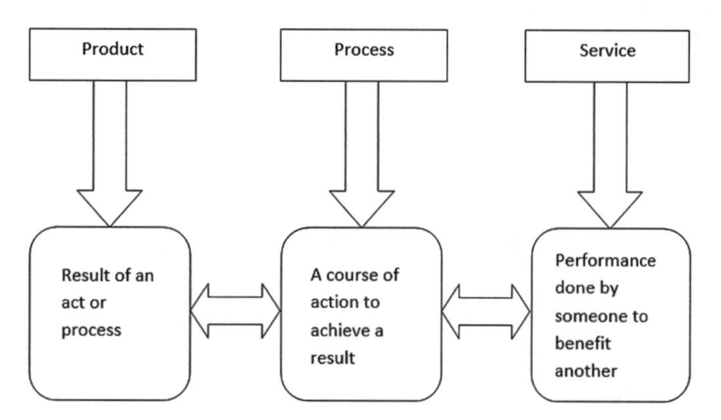

This research would like to carry out investigation into these three aspects of innovation. However, it will set its boundary within the various business processes through organizational transformations and capacity development initiatives incorporating crowdsourcing methods to empower SMEs. Furthermore, it will emphasize on technology diffusion and issues of technical nature to enable the SMEs community to innovate further in reaching out to their grass roots stakeholders.

LITERATURE REVIEW

Open Innovation and SMEs

In search of finding different patterns of innovation and their relationships with SMEs, initial part of the research spent sufficient period of time (over a year or so) in literature review. Review was restricted to definite search strings, but was not confined to specific journals or search engines. However, to keep the credibility of the searched content most of the searches were limited to highly ranked journals and databases. Similarly, case studies were conducted and survey reports were being studied that were being operationalized or established or implemented by well reputed national and international agencies or institutes, such as European Commission, OECD, World Bank, UNDP, Vinnova, Innocentive and others. Moreover, literature review emphasized on empirical studies and cases illustrating activities on SMEs in reaching out to their broad based grass roots clientele.

Studies on open innovation address OI practices in SMEs and how their utilization of OI and the resulting benefits differ from those of large enterprises. The lack of resources in SMEs to engage in looking outward is said to be a obstacle to OI, but at the same time this deficiency is cited as a motive for looking beyond organizational boundaries for technological knowledge (Spithoven, Vanhaverbeke and Roijakkers, 2013). In this context, Brunswicker and Vanhaverbeke (2015) explore how SMEs engage in external knowledge sourcing, a form of inbound open innovation. They draw upon a sample of 1,411

SMEs and empirically conceptualize a typology of strategic categories of external knowledge sourcing, namely minimal, supply-chain, technology-oriented, application-oriented, and full-scope sourcing.

Innovation, being latent within the product, process and service in an entrepreneurship grows naturally if these three could be intermingled further, such as incorporating new idea and changes through product development, process development and service development. In its simplest sense, innovation will grow in an enterprise, if the bonding among them increases. As *f*(P1,P2,S) moves inwards, bonding increases, and innovation grows (see Figure-5).

Figure 3. Bonding of the three parameters of innovation management

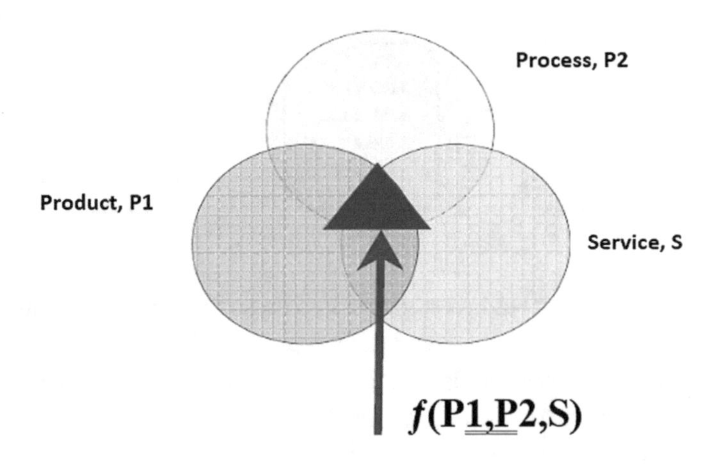

Given the potentiality of open innovation in an enterprise through improvement of the three parameters as shown in Figure-3, especially small and medium scale enterprises (SMEs), literature review gave primary focus to gather knowledge on opportunities and challenges that are being faced by SMEs when they are opening their windows of business processes through open innovation methods. Furthermore, to design an appropriate action plan for empowering them, a thorough literature review has been conducted on existing success cases utilizing open innovation strategies. These eventually assisted in developing the research design and survey methods.

However, before progressing further, this study would like to give an insight into the product, process and service innovation, but not reinventing the wheel. Researchers mention that *product innovation* is linked to the analysis of changes and innovations within the product category. Product innovation allows one to map out changes in one´s enterprise´s product. It compels to determine how a product should evolve to meet needs of the client and be competitive in the future. Essentially, it may include any or some or all of the features of product innovation as mentioned below:

- A product may come out with an entirely new idea;
- A product may offer improved performance against an existing function or desire or need;
- It may provide a new approach for an existing product;
- The product may provide additional functions or features;
- It may be an existing product, but targeting to a new segment of the market;
- The product may have a new price or value mix (promotion or other value addition);

- It may have improved packaging;
- The product may have changes in appearance or forms (Crawford, 1983; Hiebing Jr. & Cooper, 2003).

Process innovation combines the adoption of a new view of the business process with the application of innovation into key processes. The novel and distinctive feature of this combination is its enormous potential to assist an organization in achieving major reductions in process cost or process time, or major improvements in quality, flexibility, and service levels. In this perspective, the business must be viewed not in terms of functions, divisions, or products, but of key processes, they may include redesigning key processes from the beginning to the end, and employing whatever innovative technologies and organizational resources are available. However, a major challenge in process innovation is to make a successful transition to a continuous improvement in environment. Referring to Davenport (1993), this research purview that, if a company that does not institute continuous improvement after implementing process innovation is likely to revert to old ways of doing business. Framework for process innovation may comprise of; identifying processes for innovation, identifying transformation enablers, developing a business vision and process objectives, understanding and measuring existing processes, and designing and creating a prototype of the new process. In essence, the process innovation;

- should respond to the need for better coordination and management of functional interdependencies (relative independence);
- is intended to achieve radical business improvement (notable breakthrough);
- may seem to be a discrete initiative but must be combined with other initiatives for ongoing transformations (integrated actions);
- can seldom be achieved in the absence of a carefully considered combination of both technological and human catalysts (intricately designed), and
- must suit the business culture (localized one) (Zuboff, 1988; Walton, 1989; Davenport, 1993).

Service innovation systems are dynamic configurations of *people, technologies, organizations* and *shared information* that create and deliver value to customers, providers and partners through services. They are forming a growing proportion of the world economy and are becoming central activity for the businesses, governments, families and individuals. Nowadays, firms do not consider themselves to be ´services´ or ´manufacturing´ but providing solutions for customers that involve a combination of products and services. Preferably, service innovation can happen across all service sectors and one should look at all possible *service activities* rather than looking at specific *service sectors* (Van Ark, Broersma & Den Hertog, 2003; Miles, 2005; IfM & IBM, 2008). Service innovation is more closely related to the depth than breadth (vertical than horizontal), and a deep relationship with external sources, such as customer needs of target markets are essential for service innovation (Lee et al., 2010). Service innovation referring SMEs development may:

- establish a common language and shared framework;
- adopt interdisciplinary approach for research and education on service systems;
- promote learning programmes on service science, management and engineering;
- develop modular template-based platform of common interests;
- provide solutions for challenges on service systems research;

- develop appropriate organizational arrangements to enhance industry-academic collaboration; and
- work with partners within a sustainable framework (IfM and IBM, 2008)[8].

Open Innovation Opportunities Among SMEs

In recent years open innovation models have become an integral part of the innovation strategies and business models of enterprises. The most important benefit of open innovation has been seen as it provides a larger base of ideas and technologies. Enterprises look at open innovation as a close collaboration with external partners; customers, consumers, researchers or other intermediaries that may have an input to the future of their company. The main motives for joining forces between companies is to grasp new business opportunities, share risks, join complementary resources and achieve synergies. Thus, enterprises recognize open innovation as a strategic tool to explore new growth opportunities at a lower risk, and open technology sourcing offers them higher flexibility and responsiveness without necessarily incurring huge investments (OECD, 2008).

The open innovation approach assumes that innovating enterprise is no longer the sole locus of innovation, nor it is the only means for reaping the benefits of research, development and innovation (RDI). In comparison to a closed innovation model, where external actors are viewed with suspicion, who could take away useful knowledge to other competitors, an open innovation environment sets a common platform to its users, customers, suppliers, public knowledge institutions, individual inventors and even competitors and regarded as potential contributors of crucial pieces of information. In addition to this, through collaboration, innovating partners can enhance their own technologies or reduce RDI costs, which is an important driver of open innovation (Lemola & Lievonen, 2008). One may term it as economization of RDI.

Another driver of open innovation is knowledge augmentation. As Chesbrough (2006a) mentioned that even larger companies applies a fraction of the total knowledge that is generated in a particular sector of company. In that case, it is better to keep doors open to external knowledge. Increasingly corporate entities are seeking out collaborative relationships with a variety of innovative organizations, universities, research institutes and other intermediaries.

Chesbrough (2006b) further emphasized on obtaining external resources in contrast to internal ones, spin-offs and licensing to gain commercial benefits of innovation in contrast to aim solely at new products by innovating oneself, utilizing available commercial knowledge to eliminate false positives and false negatives in contrast to experiment on an unsuccessful venture, fostering outflows of knowledge and technology to arrange better intellectual property management in contrast to situated under a controlled environment, and promoting innovation intermediaries as cheaper sources of innovation managers in contrast to rely on age old methods and technologies.

Life-Cycle of Technology Exploration and Technology Exploitation

As evolved from the very basic, but comprehensive definition of open innovation (Chesbrough, Vanhaverbeke & West, 2006:1), it is "the use of purposive inflows to accelerate internal innovation" and "the use of purposive outflows to expand the markets for external use of innovation", and thus, it comprises of outside-in and inside-out flow of ideas, knowledge and technologies. The outside-in movements may be termed as technology acquisition or technology exploration, while the inside-out movements may be termed as technology dissemination or technology exploration (Lichtenthaler, 2008; Van de Vrande, De

Jong, Vanhaverbeke & De Rochemont, 2009). Through technology exploration, SMEs can get acquainted with outside information, gain knowledge and utilize them for their empowerment by enhancing existing technology platform. On the other hand, through technology exploitation, SMEs can empower themselves by raising their knowledge platform Chesbrough & Crowther, 2006; Lichtenthaler, 2008). Inter-firm collaborations (strategic alliances and joint ventures) are becoming essential instruments to improve competitiveness of enterprises in complex and critical environments (Hoffman & Schlosser, 2001).

Along these two routes of open innovation, SMEs have the opportunities of increased customer involvement, external participation, networking among other partners, outsourcing of R&D and inward licensing of intellectual property (IP) through technology exploration, and they may have the opportunities of augmented venturing, inclusion of all-out staff involvement within the R&D processes and outward licensing of IP through technology exploitation (Van de Vrande, De Jong, Vanhaverbeke & De Rochemont, 2009).

In the context of technology exploration and technology exploitation, parameters as mentioned above need further investigation. A few important activities under the category of technology exploration are being discussed next.

- Increased customer involvement- Researchers on open innovation recognize increased customer involvement as an essential element to expedite internal innovation process (Gassmann, 2006). Apart from Von Hippel´s (2005) initiating work, this has been supported by many other researchers (Olson & Bakke, 2001; Lilien et al., 2002; Bonner & Walker, 2004). Emphasis has been given to increased involvement of the customers at the beginning of the innovation process (Brockhoff, 2003; Von Hippel, 1998; 2005; Enkel et al., 2005). Firms may benefit from their customers´ ideas, thoughts and innovations by developing better products that are currently offered or by producing products based on the designs of customers (Van de Vrande et al., 2009).
- External networking- Another essential activity to innovate through open and collaborative innovation (Chesbrough, Vanhaverbeke and West, 2006), which includes acquiring and maintaining connections with external sources of social capital, organizations and individuals (Van de Vrande et al., 2009). External networking allows enterprises to acquire specific knowledge without spending time, effort or money but to connect to external partners. Such collaborative network among non-competing entities can be utilized to create R&D alliances and acquire technological capabilities (Gomes-Casseres, 1997; Lee, et al., 2010). Within the limited resources of SMEs, they must find ways to achieve production of economies of scale to market their products effectively, and at the same time provide satisfactory support services to their customers. Lee et al. (2010) observed that SMEs are flexible and more innovative in new areas, but lack in resources and capabilities. On the other hand, larger firms may be less flexible, but have stronger trend to develop inventions into products or processes. These resources often attract SMEs to collaborate with large firms (Barney & Clark, 2007). But stronger ties with larger firms may limit opportunities and alternatives for SMEs, and they prefer to make external networks with other SMEs or institutions, such as universities, private research establishments or other forms of non-competitive intermediaries (Rothwell, 1991; Torkkeli, et al., 2007; Herstad, et al., 2008; Lee, et al., 2010).
- External participation- In fact this may refer to crowdsourcing in terms of open innovation strategies and it enables one to look into the minute details of innovation sequences in an enterprise which may seemingly unimportant or not promising before (Van de Vrande et al., 2009). A com-

pany's competitiveness is increasingly depending on its capabilities beyond the internal boundaries (Prügl & Schreier, 2006).During the start up stages, an enterprise can invest by keeping eyes on potential opportunities (Keil, 2002; Chesbrough, 2006b), and then explore for further increase the knowledge platform through external collaboration (Van de Vrande et al., 2009). Within the innovation process, companies are intensifying relationships and cooperation with resources located outside the firm, ranging from customers, research institutes, intermediaries and business partners to universities (Howells, James and Malek, 2003; Linder, Jarvenpaa and Davenport, 2003).

- Outsourcing R&D- Nowadays, enterprises are outsourcing R&D activities to acquire external knowledge and technical service providers, such as engineering firms or high-tech organizations are taking lead in the open innovation arena (Van de Vrande et al., 2009). The basic assumption is that enterprises may not be able to conduct all R&D activities by themselves, but capitalize on external knowledge through outsourcing, either by collaboration, or taking license or purchasing (Gassmann, 2006). In this context, collaborative R&D is a useful means by which strategic flexibility can be increased and access to new knowledge can be realized (Pisano, 1990; Quinn, 2000; Fritsch and Lukas, 2001). While R&D outsourcing has been targeted to cost savings in most companies, more and more managers are discovering the value of cooperative R&D to achieve higher innovation rates. Collaborative R&D are being utilized to make systematic use of the competences of suppliers, customers, universities and competitors in order to share risks and costs, enlarge the knowledge base, access the complementary tangible and intangible assets, keep up with market developments and meet customer demands (Nalebuff & Brandeburger, 1996; Boiugrain & Haudeville, 2002; Van Gils & Zwart, 2004; Christensen, Olesen and Kjaer, 2005; De Jong, Vanhaverbeke, Van de Vrande and De Rochemont, 2007). Furthermore, the not-invented-here syndrome, a severe barrier to innovation, can also be mitigated if external partners are increasingly involved in the R&D processes (Katz and Allen, 1982) and the most important factor is that, in the open innovation paradigm, it is considered totally acceptable to acquire key development outside the organizational boundary (Prencipe, 2000).
- Inward licensing of Intellectual Property- Intellectual Property (IP) can be termed as creative ideas and expressions of the human mind that have commercial value[9] or it can be seen as an idea, invention, formula, literary work, presentation, or other knowledge asset owned by an organization or individual[10]. The major legal mechanisms for protecting intellectual property rights are copyrights, patents, and trademarks. IP rights enable owners to select who may access and use their property and to protect it from unauthorized use[11]. As Chesbrough (2006a) has mentioned IP as a catalyst of open business model, an enterprise can acquire intellectual property including the licensing of patents, copyrights or trade marks. This has been supported by Van de vrande et al. (2009), as this process may strengthen one's business model and gear up the internal research engines. Inward licensing tend to be less costly than conducting in-house R&D, the licensing payment can be used to control risks by prudent payment scheme, reduces time to bring new products into market and lowers risk when an invention of similar nature has already been commercialized (Box, 2009; Darcy, Kraemer-Eis, Gnellec and Debande, 2009).
- Other issues like user innovation, non-supplier integration or external commercialization of technology fall under this category, but hardly have they formed any definitive trend channeling separate research aspects other than the activities mentioned above. However, as this research progress, efforts will be given to extract contents of similar interest and highlight them.

Before proceeding next, a few activities on technology exploitation are being discussed below:

- Venturing- A venture can be seen as an agreement among people to do things in service of a purpose and according to a set of values, and in this context, an entrepreneur is a venturer that carries primary responsibility for operating a venture[12]. Hence, venturing is the process of establishing and developing a venture, and can be defined as starting up new entrepreneurships drawing on internal knowledge, which implies spin-off and spin-out processes. In most cases, support from the parent organization includes finance, human capital, legal advice, or administrative services (Van de vrande, et al., 2009). Keil (2002) suggests that external corporate venturing unites both physical and intellectual assets of an enterprise by extending and exploiting the internal capabilities for continual regeneration and growth. Though earlier studies on open innovation have primarily focused on venturing activities in large enterprises (Chesbrough, 2003a; Lord, Mandel and Wager, 2002), but the potential of venturing activities is regarded to be as enormous. Chesbrough (2003a) illustrated that the total market value of 11 projects which turned into new ventures exceeding that of their parent company, Xerox, by a factor of two.
- Outward licensing of Intellectual Property- As mentioned earlier, IP plays a crucial role in open innovation as a result of the in-and out flows of knowledge (Arora, 2002; Chesbrough, 2003a, 2006a; Lichtenthaler, 2007). Enterprises have opportunities to out-license their IP through commercialization to obtain more value from it (Gassmann, 2006; Darcy et al., 2009). Darcy et al. (2009) mention that out-licensing of IPs bring opportunity of high profitability, allow multiple licensees to work together at the same time, seem less riskier than Foreign Direct Investment (FDI), simple to operate when less technology components exist, and especially suited for SMEs as this process lowers risk by eliminating need for downstream production facilities. Out-licensing of IPs allow them to profit from their IPs when other firms with different business models find them as profitable, and in this way IPs route to the market through external paths (Van de vrande et al., 2009). By means of out-licensing, many firms have begun to actively commercialize technology. This increase in inward and outward technology transactions reflects the new paradigm of open innovation (Vujovic & Ulhøi, 2008; Lichtenthaler & Ernst, 2009). However, the decision of firms to license out depends on anticipated revenues and profit-dissipation effects (Arora, Fosfuri and Gambardella, 2001), as whether the outward licensing generates revenues in the form of licensing payments, but current profits might decrease when licensees use their technology to compete in the same market (Van de vrande, et al., 2009). Moreover, these new forms of knowledge exploitation and creation of technology markets require the design of appropriate financial instruments to support the circulation and commercialization of knowledge (Darcy et al., 2009).
- Involvement of non-R&D employees- By capitalizing on the initiatives and knowledge of internal employees, active or non-active with the R&D exercises, an enterprise can be benefited (Van de vranda, et al., 2009). Chesbrough et al. (2006) has conducted case studies which illustrate the significance of informal ties of employees of one enterprise with employees of other enterprises, which are crucial in understanding of arrival of new products in the market and also the commercialization processes. Earlier researches (Van de ven, 1986), also support this view that innovation by individual employees within an enterprise who are not involved in R&D, but involved in new idea generation could be effective in fostering the success of an enterprise. Moreover, along the context of currently evolved knowledge society, employees need to be involved in innovation processes in diversified ways, such as by encouraging them to generate new ideas and suggestions,

exempting them to take initiatives beyond organizational boundaries, or introducing suggestion schemes such as idea boxes and internal competitions (Van Dijk and Van den Ende, 2002);

- There are other activities like spin-off (Ndonzuau, Pirnay and Surlemont, 2002), selling of results (OECD, 2008), transfer of technology (OECD, 2008), transfer of know-how (OECD, 2008),

In the perpetual scenario, one can portrait a picture as shown in Figure-4, as SMEs are acting as catalysts in the process of technology exploration and technology exploitation performing the necessary actions. But, in reality, they need to be under a strategically guided and managed environment to achieve their innovativeness and competitiveness to attain a sustainable economic platform (Sautter & Clar, 2008). This research would argue that a third party in between them is essential to take necessary initiatives, strategies and action plans for providing appropriate guidance, support and directives at the field level in tackling issue, difficulties and challenges on behalf of them. Furthermore, this form of support should be carried out from a platform that would be institutional, so that in time of need, SMEs may rely on and feel dependable. In addition to these, if this institutional support comes from a locally generated perspective, they would feel confident and comfortable to be working with.

Figure 4. Life-cycle of technology exploration and technology exploitation in SMEs

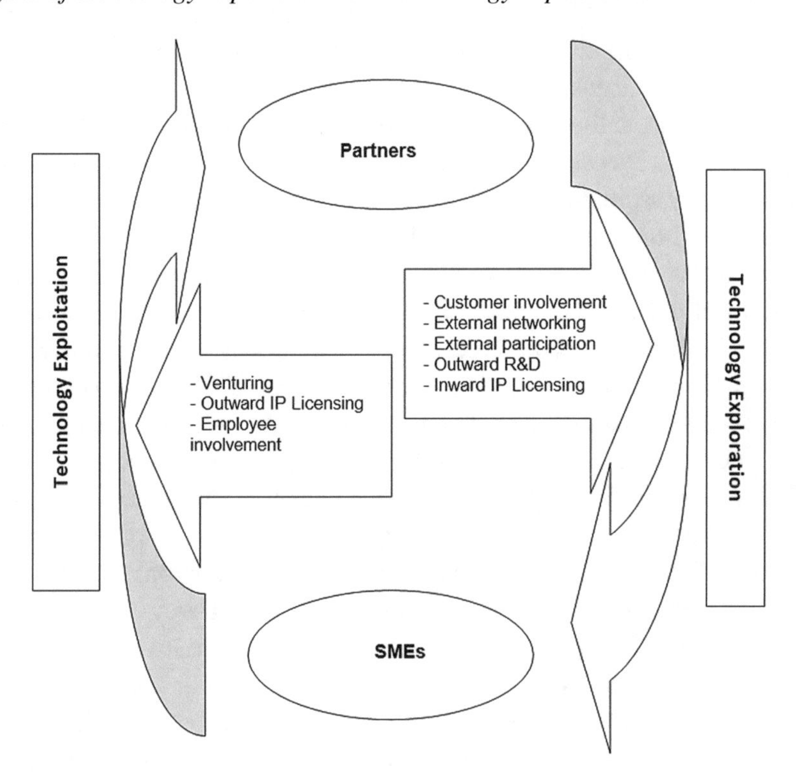

Moreover, the life-cycle of technology exploration and technology exploitation (see Figure-4) is not a perpetual one and demands a catalytic agent to make it roll or put in action. This research would like to put forward action plans in accomplishing the life-cycle management of the mentioned open innovation phenomena. It also affirms that a third party or agent of innovation can play the most important role in ascertaining pre-requisites at each entrepreneurship level, their networked level, or even at the level of their human skills. Among various action plans, establishment of a networked environment is pertinent. Utilizing ICT, an extended platform of communication and knowledge sharing can be established (Walther and D´Adderio, 2001; Antonelli, 2005), which can be physical or virtual.

However, apart from the mentioned benefits or opportunities mentioned above, as Professor Henry Chesbrough (2003a; 2003b; 2006a; 2006b; 2007) claims a fundamental shift in innovation paradigm from closed to open innovation and advocates collaborative and open innovation strategies and open business models to take the full benefit from collaborating with external partners, this research would continue building on the paradigm shift. It would explore to develop a sustainable business model adopting open innovation strategies to be applicable for SMEs operating at the grass roots. Furthermore, as mentioned by Sautter & Clar (2008), this research would support that, it is beneficial to avail the competences of external partners meeting the challenges of increased complexity of research, technological development and innovation (RTDI), and growing global competition. Along this context, a few challenges are being mentioned next.

Open Innovation Challenges for SMEs

Innovation is an essential element in wealth creation and employment generation for enterprises and entrepreneurs in all regions and sub-regions of the world (Tourigny & Le, 2003), and reflects a critical way in which organizations respond to either technological or market challenges (Hage, 1999). However, dispersal of innovation is a classical view of social systems (Rogers, 1962) that has been revisited by many in various aspects of innovation (Moore, 1991; 2006; Evangelista, 2000; Tilley & Tonge, 2003; Toivonen, 2004; Nählinder, 2005; Maes, 2009) as one advances through the era of rapidly changing technology (MacGregor, Bianchi, Harnandez & Mandibil, 2007). But, the real challenge is not to diffuse in increasing amount of products and services, but to ensure that the diffusion process follow a more responsible and sustainable approach throughout the life cycle. It is not easy to achieve (Hoffman et al. 1998). Hoffman et al. (1998) further argues that though innovative effort appears to be widespread, this does not translate directly into improved firm performance and, ultimately, greater profitability.

Moreover, within the context of this research, when applying innovation in a open paradigm, especially dedicated to small and medium enterprises, situation face various challenges often unpredictable and unfamiliar (De Jong, Vanhaverbeke, Kalvet and Chesbrough, 2008; Van de Vrande et al., 2009). SMEs are characterized by stretched resources, which puts them in particular jeopardy from increasing globalization and rapid technological change. One might expect that SMEs would draw extensively on alliances to overcome their resource constrains and increase their viability in hard times. But, studies show that SMEs´ propensity to co-operate is significantly less than that of large companies (Haagedorn & Schakenraad, 1994; Hoffman & Schlosser, 2001). Hoffman & Schlosser (2001), further state that empirical findings show that SMEs do not fully utilize alliances to improve their competitive position. Furthermore, it is assumed that the reported reluctance of SMEs to collaborate is due not only to emotional and cultural barriers, but also to a lack of knowledge about the specific success factors of alliances.

Moreover, the key issue facing many SMEs relates to how they can foster effective innovation using organizational supporting mechanisms (McEvily, Eisenhardt and Prescott, 2004), and their needs including their decision-making processes often differ significantly from those of larger firms (Shrader, Malford and Blackburn, 1989). If one would like to investigate in-depth, there is a dearth of research on what encourages and drives product development, management practices and process technologies deployment from a strategic orientation perspective (Tidd, Bessant and Pavitt, 2001). Also, it is difficult to find empirical research examining the association between strategic orientation and deployment of leading management practices or new process technologies (O´Regan & Ghobadian, 2005).

Tools like ERP (Enterprise Resource Planning) or groupware, which are considered crucial to increase process transparency and gain control on distributed networks at global level, are less diffused in SMEs (Di Maria & Micelli, 2008). Often the grass roots entrepreneurs are not aware of complexities of the global market (Tapscot & Wimmiams, 2007), which lead innovation researchers unaware of the local economic and social systems (Di Maria & Micelli, 2008). In terms of collaborative innovation, modularity and codification widens circulation and utilization of knowledge across contexts, but at the same time complicates the codification for promoting sophisticated sharing strategies based on pragmatic collaboration (Helper, MacDuffie & Sabel, 2000; Di Maria & Micelli, 2008).

Apart from these, Chesbrough & Crowther (2006) identified the not-invented-here (NIH) syndrome and lack of internal commitment as main hampering factors. The NIH syndrome has been previously found to be a prominent barrier for external knowledge acquisition (Katz & Allen, 1982). Although open innovation focuses on the external acquisition of knowledge, its underlying antecedents are also applicable to technology exploitation, leading to the only-used-here´ (OUH) syndrome (Lichtenthaler & Ernst, 2006). Boschma (2005) identified various forms of ´proximity´ which are essential for effective collaboration. These include cognitive, organizational, cultural and institutional differences between collaboration partners, implying that potential problems may arise due to insufficient knowledge, cultures or modes of organization, or bureaucratic elements. Other potential challenges include lacking of financial resources, scant opportunities to recruit specialized workers, lack of appropriate information inflow, market entry barriers, increased competition, dynamic shifting of market, free-riding behavior, and problems with contracts (Boer, Hill & Krabbendam, 1990; Mohr & Spekman, 1994; Coughlan, Harbison, Dromgoole & Duff, 2001; Hoffman & Schlosser, 2001; Van de Vrande, De Jong, Vanhaverbeke & De Rochemont, 2009; Lee et al., 2010).

Role of Intermediaries

Systems of innovation are conceptualized as a complete set of institutions and organizations along with sub-sets of relationships among them. These relationships could be among firms, universities, industry associations, scientific societies, professional bodies, regulatory agencies, higher education institutes, research organizations, support entities and dedicated intermediaries (Lopez-vega, 2009) on sub-sets of common habits and practices, routine or established practices, or established rules or laws regulating the relationships and interactions (Edquist & Johnson, 1997). Intermediaries can take the role of adapting specialized solutions on the market as per demand of individual user firms by linking players within a technological system and on transforming relations (Stankiewicz, 1995), or encourage creation of organizations through collaboration and create technology centers or other forms of inter-fir clearing houses for exchange of innovations (Lundvall, Johnson, Andersen and Dalum, 2002), or play a role in policy terms within the innovation system by increasing the connectedness and designing new possibilities (Howells, 2006).

Within the context of open innovation, emphasizing SMEs development highlight the role of a intermediary (Watkins & Horley, 1986) or third party (Mantel & Rosegger, 1987) or broker (Aldrich & von Glinow, 1992) or consultant (Bessant & Rush, 1995) or knowledge mediator (Millar & Choi, 2003), though the role of intermediary for innovation and technology development was already existed in other sectors (Howells, 2006). The introducer of open innovation (Chesbrough, 2003a) has supported the idea of collaborating with an intermediary and mentioned that the time for developing a new technology can be drastically reduced through this process. Diener & Piller (2009), put forward an extensive literature review on the role, functionalities and characteristics of intermediaries, and mentioned that intermediaries connect an enterprise with different sources of knowledge. However, Diener & Piller (2009) also mentioned that, the brokering job as an intermediary is complex, and requires translation of information, communication, and manipulation of diversified perspectives, including legitimacy. But, Datta (2007) added another feature for them to act as multiple value-added service providers, which this research supports.

Intermediaries can act as an interface between the demand side (end-users or firms or other intermediaries) and supply side (knowledge intensive business services, KIBS or other R&D providers) (Klerkx, 2008). Innovation intermediaries are an integral component of the innovation chain (Mehra, 2009), and as mentioned before, in the paradigm of open innovation they could be private entities, NGOs, professional bodies, research organizations, academic institutes, trade unions and individuals (Clarke, 2007). Howells (2006) termed them as an organization or body that acts as agent or broker in any aspect of the innovation process between two or more parties, and Klerkx (2008) further elaborated their functionaries, which include providing information about potential collaborators; brokering a transaction between two or more parties acting as a mediator, or go-between, bodies or organizations that are already collaborating; and helping find advice, funding and support for the innovation outcomes of such collaborations.

Other literatures on intermediaries indicates their role through fulfilling a variety of functions from connecting companies that accelerate technological outputs to others engaged in transfer of technologies (Bessant & Rush, 1995; Mehra, 2009); knowledge co-creation (Turpin, Garrett-Jones and Rankin, 1996); mediation between science, policy and governance (Cash, 2001; Gereffi, Humphrey and Sturgeon, 2005); promotion of neutral space for development of new technologies (Winch & Courtney, 2007); product or functionality upgradation through acquiring new functions in the value chain (Humphrey & Schmitz, 2000; Pietrobelli & Rabellotti, 2006; Rahman and Ramos, 2010); communities of practice (Brown & Duguid, 2000; 2001); R&D collaboration (Narula, 2004); networking (Kirkels & Duysters, 2010); and creation of new possibilities and dynamism with the system (Howells, 2006) by maximizing chances of innovation and increase the likelihood of success in developing new products and services (Lee, et al., 2010).

Development of A Sustainable Business Model

A business model may be seen as the totality of how a company selects its clients, defines and differentiates its responses; classifies those tasks it will perform itself and those it will outsource; configures its resources, goes to market, creates utility for clients; and get hold of profits. It is the entire system for delivering utility to clients and gaining a profit from that activity (Pourdehnad, 2007). Figure-7 shows a relationship diagram with the various actors or stakeholders involved in a business model. This evidently envisages the clear bonding among visible groups of stakeholders among the business communities.

Triple Helix Model (see Figure 6) is another highly discussing model in this arena. According to this model, a spiral of innovation involves government, university, and industry in multiple reciprocal relationships, to create a flexible overlapping innovation system (CSR Europe, 2008b).

This research would like to point out to another business model that may be utilized in SMEs OI process, which has been developed incorporating mixed approach (Shorthouse, 2008). Figure-9 shows

Figure 5. Relationship with the stakeholders in a business model
(Adopted from Pourdehnad, 2007)

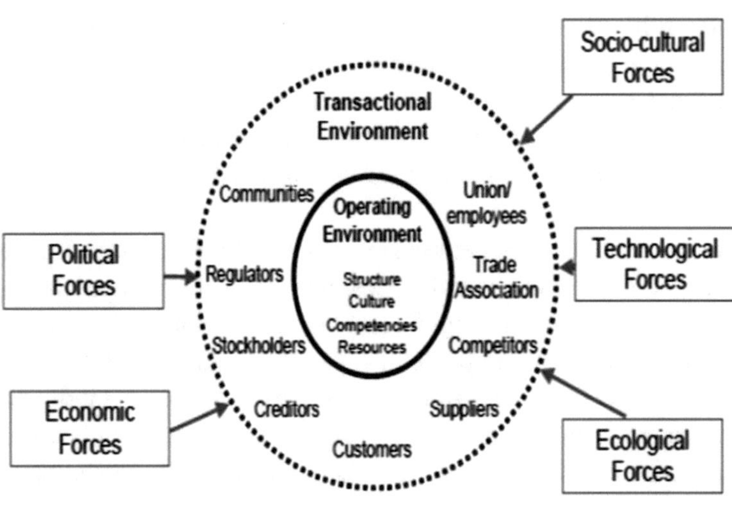

Figure 6. The triple helix model
(Adopted from CSR Europe, 2008b)

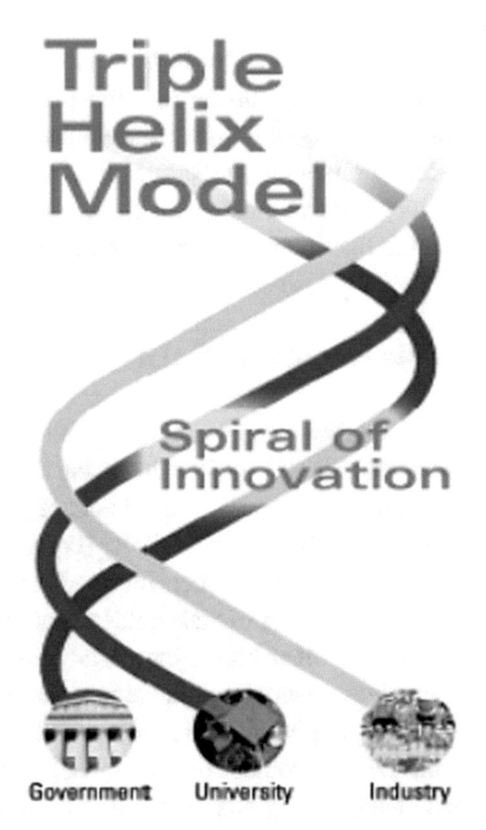

a mixed approach (closed and open innovation) business model. Shorthouse (2008) has adopted a joint effort to reach the niche market through using both closed and open innovation business model.

Figure 7. A mixed approach business model
(Adopted from Shorthouse, 2008)

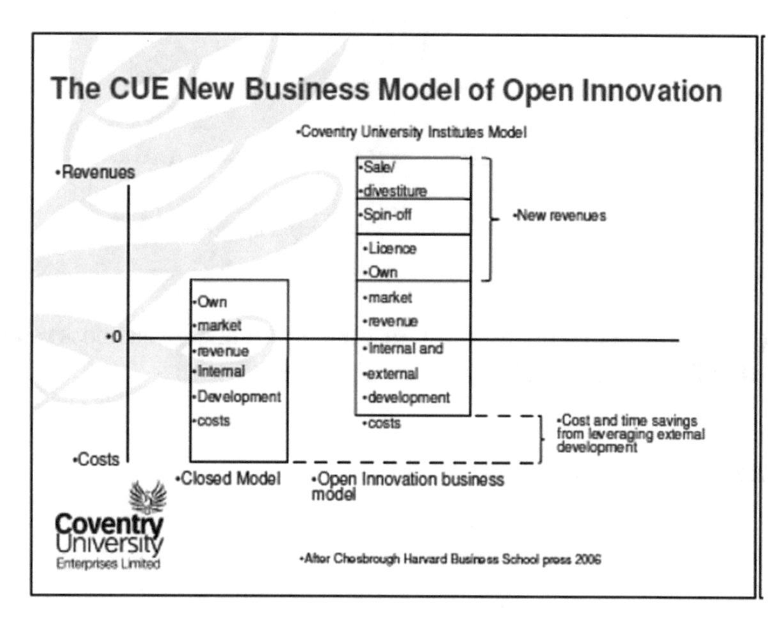

For a business model, using partnership approach as mentioned earlier, Fredberg, Elmquist and Ollila (2008) underline the following flow chart, illustrating the partnership approach in an enterprise (see Flow chart-1):

Figure 8. Illustrating the partnership approach in an entrepreneurship.

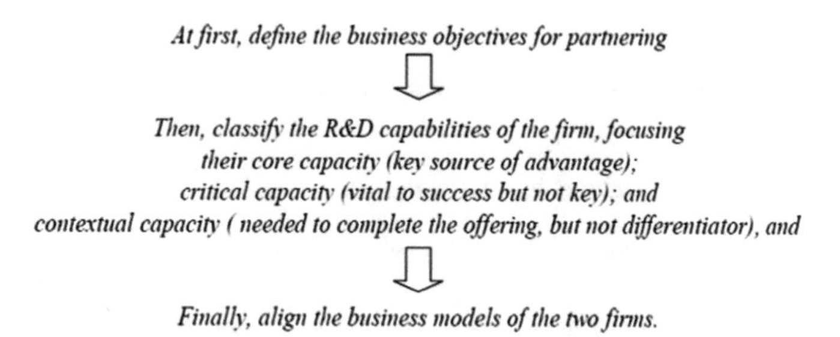

Based on the above arguments and arguments made by Chesbrough and Rosenbloom (2002), authors like to introduce another flow chart (see Flow chart-2) illustrating the formulation of innovation strategy in an open innovation business model:

Figure 9. Shows the formulation of innovation strategies

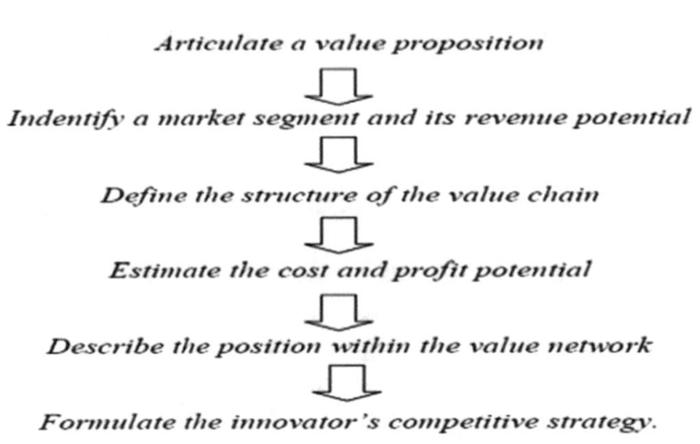

Provided the appropriate articulation, however, to attain a sustainable business model in SMEs open innovation, one need to follow the flowchart indicated in Figure 10.

Figure 10. Empowerment of the stakeholders

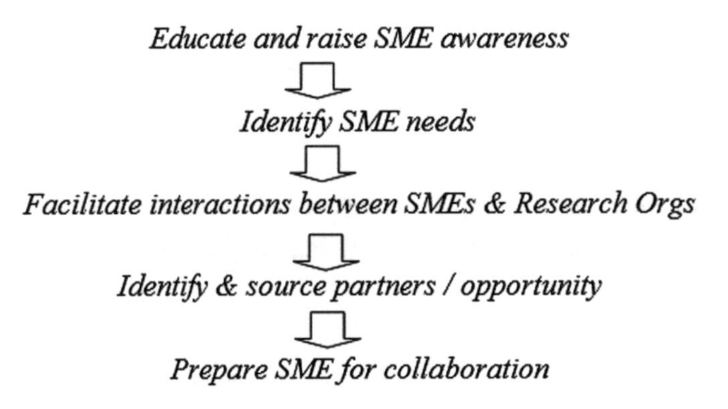

Finally, a sustainable business model should also follow as shown in Figure 11.

Figure 11. Implementation of a sustained business model

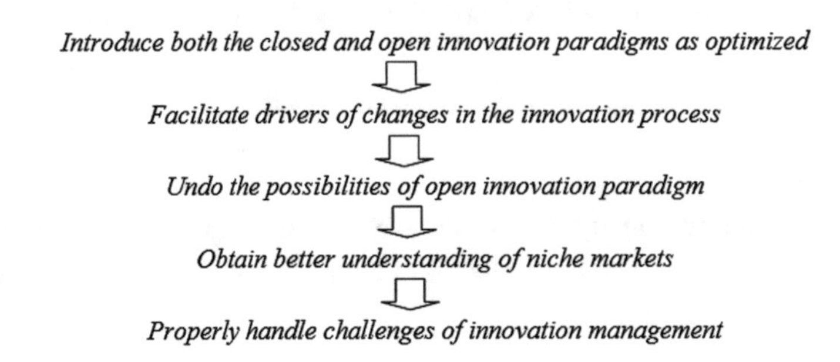

Hence, the business model should incorporate inclusion of SMEs inclusive in the dimension shown in Figure 12.

Figure 12. Building block of the proposed business model (Rahman and Ramos model)

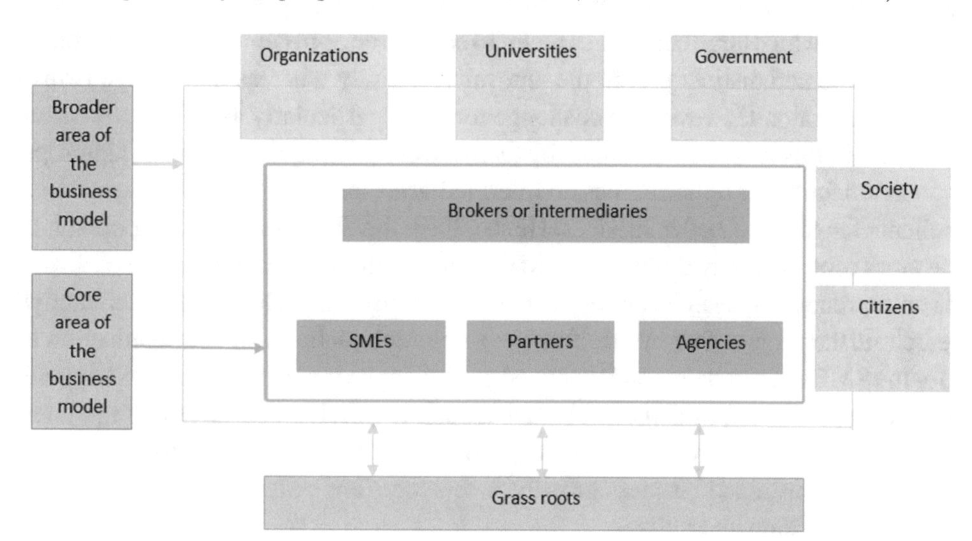

The proposed business model (to be adopted in a few selected SMEs) emphasizes on two other tiers of relationship; among the core partners in the network and among the peripheral partners in the network. However, both the segments need not to be isolated from each other. Rather they may remain as active member of the entire community. Eventually, for sustained entrepreneurship the entire group must interact to the grass roots for effective dissemination of open innovation strategies promoting economic and value gain.

Development of Tools for SMEs Empowerment

Most of the researches in open innovation paradigm focus targeting common stakeholders through major global entrepreneurs or their alliances. In addition, a few of those global business houses are controlling the entire market or system of open innovation development through process modification or diversification of resources or other activities. Despite immense potentiality to reach out the stakeholders at the grass roots through open ended demand, diversity of product variation and scale of economic capacity, major contemporary researches are confined towards generic pattern-oriented clients (Rahman & Ramos, 2010).

In these contexts, the SMEs who always deal with the clients at the grass roots and they have to satisfy the client base the most, though seldom they produce the product. Moreover, despite the globalization that offers unprecedented opportunities and challenges for SMEs, but seemingly they are thinking of mere survival in the context of global economy, marketing, value promotion, job creation and expansion (CSR Europe, 2008a; b). SMEs in Europe comprises of about 23M€ investment market that account for 99% of all businesses and represent 2/3rd of the total employment (Renaud, 2008). However, in spite of being key contributor to the global economy accounting for approximately 50% of local and national

GDP, 30% of export and 10% of FDI[13] most of the SMEs communities are lagging behind promoting their products at the national level, and at the global level (OECD, 2006).

This research argues that, time has come to transform the process of innovation system (through organized deregulation, wider knowledge distribution, focused training, and capacity development) to put SMEs at the heart of the technology transfers as they represent a privileged source for the innovations in a competitive context in which large companies prefer to concentrate on their core competencies. So, SMEs play an important role and distinct part in the innovative activity whereas many large firms are acting as just a systems integrator. However, the challenge remains, particularly as delicate as before, because SMEs do not exists in this global environment by themselves, they require stronghold interoperability to larger entrepreneurs for better opportunities, to intermediaries for improving their capacities, and to the grass roots clients for offering better services. Hence, there are scope of promoting open innovation to reach out the grass roots, not only through the SMEs, but also through a combination of SMEs, agencies of corporate entrepreneurs, research centres, team of university researchers, and other catalytic factors.

The research further argues that by establishing a common platform of communication as a tool of interaction for the SMEs, can enhance their knowledge about open innovation practices happening around their communities and abroad; establish dialogues among themselves, partners and the intermediaries for initiating improved open innovation strategies; and engage themselves by introducing adoptive open innovation models for sustained business growth. A few tools are being discussed next, which this research would like to develop and utilize.

The Survey

Although it is rationally clear what innovation is, but it is obviously not a simple thing to measure precisely, especially the effect of it on SMEs. Most people would naturally grasp after a few minutes of reflection that innovative ideas usually emerge in the minds of people and then go through some degree of the following: further elaboration, communication, refinement, implementation and replication. Attempts to measure innovation can be broadly classified into measures of inputs to innovation (finance, human skills, R&D, etc.), measures of outputs of innovation (finance, human skills, number of patents, copyright, etc.), and measures of by-products or symptoms of innovation (knowledge, value addition, etc.) (Warner, 2001). Porter and Bond (1999) presented an innovation measure that combined several of these input and output measures.

With these contextual approaches, to measure the impact of open innovation and learn about the challenges the SMEs are facing at the grass roots, a survey questionnaire has been developed. It is the result of an ongoing research work on Open Innovation and Small and Medium Enterprises (SMEs) development. The survey has been conducted in a pilot form to a few selected SMEs in Portugal and intention is there to conduct the survey in several phases. During the preliminary phase, it basically enquires about the general characteristics of firms that are active in the sector of SMEs, insights on business constraints, competition, human resources problem and data on innovation. The survey may be divided into two broad categories; the questionnaire and the interview. The survey questionnaire has been tested through the pilot one and work is going on to improve the questionnaire as per the feedback from the respondents and the survey outcome.

Broad objectives of the survey is to acquire knowledge about the current status of SMEs active in open innovation practices in Portugal. The focus is to know the general and financial characteristics of the enterprises belonging to this category, and learn about their inclusion in innovation in terms of financial, technological, managerial, policy issues and other relevant contexts. This survey could lead to;

- know about the actual state of affairs in the business sector focusing SMEs,
- foster understanding of developments taking place in various sectors of SMEs in business development within the country,
- analyze basic characteristics of those enterprises,
- acquire knowledge on best practices and innovation,
- identify and recommend the best practices for policy makers and other beneficiaries,
- write a report justifying those analysis to bring up a coherent environment conducive to national SME innovation, and
- put forward recommendations suggesting efforts and activities to resolve impediments of SME innovation.

Similar survey has been carried out in Turkey and a comparative study will soon be published. Efforts are there to conduct such surveys in India and Israel.

FUTURE RESEARCH DIRECTIONS

Despite lack of empirical evidence of its relevance (Batterlink, 2009), benefit (Chesbrough and Crowther, 2006), extent (Batterlink, 2009) and geographical spread (Chesbrough, 2006a) to small and medium enterprises, open innovation has launched a new paradigm (Gassmann, Enkel and Chesbrough, 2010) in business development. Batterlink (2009) mentions about its relevance and extent related mainly to large and high tech industries, and Chesbrough and Crowther (2006) supported them. Moreover, Chesbrough (2006a) pointed about its relevance and geographical spread more or less within the US-based firms.

At this moment, if one investigates over the question, as how far will open innovation go reaching the grass roots clientele of SMEs and what would be the long term impact in terms of economic sustainability, this research observes that future research works need to be carried out to find out above mentioned aspects and issues in relation to SMEs in Europe and for this case, in Portugal.

Gassmann et al. (2010) suggested a few parameters of investigation that would set a new baseline for future research, such as to observe the extent of

- industry penetration focusing pioneers to mainstream
- R&D intensity focusing high to low tech industries
- structure and composition of partnerships
- content and other value added incentives.

It has been observed that the strength of the European economy during the past century has mostly been dependent on large and established firms. Furthermore, most of these firms operate multi-nationally, moving their activities where knowledge, technology and market conditions are most favorable. This implies a new competitive landscape for Europe. In this aspect, SMEs are becoming increasingly important as transformational agents in the economy, as important channels for commercialization of research and as sources of new growth companies. However, the solutions addressing these challenges are predominantly based on a European Added Value perspective, especially when it comes to public support to SMEs, since assessment of growth potential and competencies of SMEs applying for support often requires local knowledge and proximity (Brogren, 2009). Further researches are desired in this area to learn about the behavior of exposed SMEs and implement similar ideas in other regions or countries.

Future research would also need to examine the conditions for change or transformations with the effect that proactive agents have on the reminder of the population. This would constitute a significant research effort, and while future research should aim to uncover clues in this area, and the focus would be on disseminating and improving innovation tool for SMEs at the grass roots (MacGregor, Bianchi, Hernandez. & Mendibil, 2007).

CONCLUSION

It is apparent that there are a range of potential contribution that SMEs can make to economic development at the national and regional levels, that include employment generation (new job creation), acting as supplier to larger companies (in terms of skills, knowledge or physical products or services) and contributing to a more diversified economic strength through the development of new activities, especially through the formation of newly established firms in service sector (Smallbone et. al., 2003). However, as it has been observed, SMEs are in lack of supports, such as finance, organizational or knowledge, in addition to policy issues.

In terms of policy function, it has been observed that there are several un-chartered areas (predominantly visible), where policy involvements are necessary, such as user innovation at the grass roots, technology marketing at the enterprise level, corporate entrepreneurship in incumbent enterprises, balanced (career, work, knowledge, skill) incentives for scientific researches and setting of standard for various innovation processes. There are a cluster of SMEs, who are in lack of conceptual basis of open innovation strategies and for them this sort of sophisticated interventions would not be the ideal approach. Rather to be start with, they could be involved in basic open innovation and interaction skills, and as they grows under a managed innovation umbrella, other strategies could be applied to them. In addition to these, there are also remote policy areas (often seems invisible), such as labour market and skill development, where policy initiation could lead to rapid adoptability to open innovation strategies (De Jong, et. al, 2008).

Furthermore, for majority of the SMEs, especially those operating in an increased dynamic and digitized environment, mere survival becomes the prime factor, rather than being added with knowledge parameters, which are indispensable and important resource for innovation. These include establishment of trusted relations to the community or society (including all partners), and formation of collaborative network among all the stakeholders (Hafkesbrink and Scholl, (2010). Undoubtedly the ability to innovate and to bring innovation successfully to market is the crucial determinant of the global competitiveness of nations over the coming decade. Fortunately, there is a growing awareness among policymakers that innovative activity is the main driver of economic advancement and organizational well-being, as well as a latent factor in meeting the dynamic global challenges (OECD, 2007).

REFERENCES

Aldrich, H. E., & von Glinow, M. A. (1992). Business start-ups: the HRM imperative. In S. Birley & I. C. MacMillan (Eds.), *International Perspectives on Entrepreneurial Research* (pp. 233–253). North-Holland.

Antonelli, C. (2005). Models of knowledge and systems of governance. *Journal of Institutional Economics, 1*(1), 51–73. doi:10.1017/S1744137405000044

Arora, A. (2002). Licensing tacit knowledge: Intellectual property rights and the market for know-how. *Economics of Innovation and New Technology, 4*(1), 41–59. doi:10.1080/10438599500000013

Arora, A., Fosfuri, A., & Gambardella, A. (2001). Markets for technology and their implications for corporate strategy. *Industrial and Corporate Change, 10*(2), 419–450. doi:10.1093/icc/10.2.419

Artemis. (2009). *Multi-Annual Strategic Plan and Research Agenda 2009*. ARTEMIS-I ARTEMIS-IRC-02-V061008-draft (MASP). ARTEMISIA.

Barney, J., & Clark, D. (2007). *Resource-based Theory: Creating and Sustaining Competitive Advantage*. Oxford University Prass.

Batterlink, M. (2009). *Profiting from external knowledge: How companies use different knowledge acquisition strategies to improve their innovation performance* (PhD thesis). Wageningen University.

Bessant, J., & Rush, H. (1995). Building bridges for innovation: The role of consultants in technology transfer. *Research Policy, 24*(1), 97–114. doi:10.1016/0048-7333(93)00751-E

Bessant, J., & Tidd, J. (2007). *Innovation and Entrepreneurship*. Wiley.

Boer, H., Hill, M., & Krabbendam, K. (1990). FMS implementation management: Promise and performance. *International Journal of Operations & Production Management, 10*(1), 5–20. doi:10.1108/01443579010004994

Boiugrain, F., & Haudeville, B. (2002). Innovation, collaboration and SMEs internal research capacities. *Research Policy, 31*(5), 735–747. doi:10.1016/S0048-7333(01)00144-5

Bonner, J., & Walker, O. (2004). Selecting influential business-to-business customers in new product development: Relational embeddedness and knowledge heterogeneity considerations. *Journal of Product Innovation Management, 21*(3), 155–169. doi:10.1111/j.0737-6782.2004.00067.x

Boschma, R. A. (2005). Proximity and innovation: A critical assessment. *Regional Studies, 39*(1), 61–74. doi:10.1080/0034340052000320887

Box, S. (2009). *OECD Work on Innovation- A Stocktaking of Existing Work*. OECD Science, Technology & Industry Working Papers, 2009/2. OECD Publishing.

Brockhoff, K. (2003). Customers' perspectives of involvement in new product development. *International Journal of Technology Management, 26*(5/6), 464–481. doi:10.1504/IJTM.2003.003418

Brogren, C. (2009). *Comments from VINNOVA: Public consultation on Community Innovation Policy, VINNOVA*. Reg. No.

Brown, J. S., & Duguid, P. (2000). *The Social Life of Information*. Harvard Business School Press.

Brown, J. S., & Duguid, P. (2001). Knowledge and Organization: A Social-Practice Perspective. *Organization Science*, *12*(2), 198–213. doi:10.1287/orsc.12.2.198.10116

Brunswicker, S., & Vanhaverbeke, W. (2015). Open innovation in small and medium-sized enterprises (SMEs): External knowledge sourcing strategies and internal organizational facilitators. *Journal of Small Business Management*, *53*(4), 1241–1263. doi:10.1111/jsbm.12120

Cash, D. W. (2001). In order to aid in diffusion useful and practical information: Agricultural extension and boundary organizations. *Science, Technology & Human Values*, *26*(4), 431–453. doi:10.1177/016224390102600403

Chesbrough, H. (2003a). *Open Innovation: The New Imperative for Creating and Profiting from Technology*. Harvard Business School Press.

Chesbrough, H. (2003b). The era of open innovation. *MIT Sloan Management Review*, *44*(3), 35–41.

Chesbrough, H. (2006a). Open Innovation: A New Paradigm for Understanding Industrial Innovation. In Open Innovation: Researching a New Paradigm. Oxford University Press.

Chesbrough, H. (2006b). *Open Business Models: How to Thrive in the New Innovation Landscape*. Harvard Business School Press.

Chesbrough, H. (2007). The market for innovation: Implications for corporate strategy. *California Management Review*, *49*(3), 45–66. doi:10.1177/000812560704900301

Chesbrough, H., & Crowther, A. K. (2006). Beyond high tech: Early adopters of open innovation in other industries. *R & D Management*, *36*(3), 229–236. doi:10.1111/j.1467-9310.2006.00428.x

Chesbrough, H., Vanhaverbeke, W., & West, J. (Eds.). (2006). Open innovation: Researching a new paradigm. Oxford University Press.

Christensen, J. F., Olesen, M. H., & Kjær, J. S. (2005). The Industrial Dynamics of Open Innovation: Evidence from the transformation of consumer electronics. *Research Policy*, *34*(10), 1533–1549. doi:10.1016/j.respol.2005.07.002

Clarke, I. (2007). *The role of intermediaries in promoting knowledge flows within global value chains*. Paper no. 12, PhD Doctoral Symposium, Brunel University, UK.

Coughlan, P., Harbison, A., Dromgoole, T., & Duff, D. (2001). Continuous improvement through collaborative action learning. *International Journal of Technology Management*, *22*(4), 285–301. doi:10.1504/IJTM.2001.002965

Crawford, C. M. (1983). New Products Management, Richard D. Irwin, Inc.

CSR Europe. (2008a). The European Alliance for CSR Progress Review 2007: Making Europe a Pole of Excellence on CSR. European Commission.

Darcy, J., Kraemer-Eis, H., Guellec, D., & Debande, O. (2009). *Financing Technology Transfer*. Working paper 2009/002, EIF Research & Market Analysis, European Investment Fund, Luxemburg.

Darsø, L. (2001). *Innovation in the Making*. Samfundslitteratur.

Datta, P. (2007). An agent-mediated knowledge-in-motion model. *Journal of the Association for Information Systems*, *8*(5), 288–311. doi:10.17705/1jais.00130

Davenport, T. H. (1993). *Process innovation: reengineering work through information technology*. Ernst & Young.

De Jong, J. P. J., Vanhaverbeke, W., Kalvet, T., & Chesbrough, H. (2008). *Policies for Open Innovation: Theory, Framework and Cases*. VISION Era-Net.

De Jong, J. P. J., Vanhaverbeke, W., Van de Vrande, V., & De Rochemont, M. (2007). Open innovation in SMEs: trends, motives and management challenges. *Proceeding of EURAM Conference*.

Di Maria, E., & Micelli, S. (2008). *SMEs and Competitive Advantage: A Mix of Innovation, Marketing and ICT. The Case of "Made in Italy"*. Marco Fanno Working Paper No. 70, Università degli Studi di Padova, Padova, Italy.

Diener, K. & Piller, F. (2009). *The Market for Open Innovation: Increasing the efficiency and effectiveness of the innovation process*. Open Innovation Accelerator Survey 2009, RWTH Aachen University, TIM Group.

Dodgson, M., Gann, D., & Salter, A. (2006). The role of technology in the shift towards open innovation: The case of Procter & Gamble. *R & D Management*, *36*(3), 333–346. doi:10.1111/j.1467-9310.2006.00429.x

Drucker, P. (1998, November-December). The Discipline of Innovation. *Harvard Business Review*, 149. PMID:10187245

EC. (2008). *SBA Fact Sheet Portugal, European Commission: Enterprise and Industry*. EC.

Edquist, C., & Johnson, B. (1997). Institutions and organizations in systems of innovation. In C. Edquist (Ed.), *Systems of Innovation: Technologies, Institutions and Organizations*. Routledge.

Enkel, E., Kausch, C., & Gassmann, O. (2005). Managing the risk of customer integration. *European Management Journal*, *23*(2), 203–213. doi:10.1016/j.emj.2005.02.005

CSR Europe. (2008b). *R&D Open Innovation: Networks with SME*. Open Innovation Network.

Evangelista, R. (2000). Sectoral Patterns of Technological Change in Services. *Economics of Innovation and New Technology*, *9*(3), 183–221. doi:10.1080/10438590000000008

Evans, S., Vladimirova, D., Holgado, M., Van Fossen, K., Yang, M., Silva, E. A., & Barlow, C. Y. (2017). Business model innovation for sustainability: Towards a unified perspective for creation of sustainable business models. *Business Strategy and the Environment*, *26*(5), 597–608. doi:10.1002/bse.1939

Fritsch, M., & Lukas, R. (2001). Who cooperates on R&D? *Research Policy*, *30*(2), 297–312. doi:10.1016/S0048-7333(99)00115-8

Gassmann, O. (2006). Opening up the innovation process: Towards an agenda. *R & D Management*, *36*(3), 223–228. doi:10.1111/j.1467-9310.2006.00437.x

Gassmann, O., Enkel, E., & Chesbrough, H. (2010). The future of open innovation. *R & D Management*, *40*(3), 213–221. doi:10.1111/j.1467-9310.2010.00605.x

Gereffi, G., Humphrey, J., & Sturgeon, T. (2005). The governance of global value chains. *Review of International Political Economy*, *12*(1), 78–104. doi:10.1080/09692290500049805

Gomes-Casseres, B. (1997). Alliance strategies of small firms. *Small Business Economics*, *9*(1), 33–44. doi:10.1023/A:1007947629435

Haagedorn, J., & Schakenraad, J. (1994). The effect of strategic technology alliances on company performance. *Strategic Management Journal*, *15*(4), 291–309. doi:10.1002mj.4250150404

Hafkesbrink, J., & Scholl, H. (2010). Web 2.0 Learning- A Case Study on Organizational Competences in Open Content Innovation. In Competence Management for Open Innovation- Tools and IT-support to unlock the potential of Open Innovation. Eul Verlag.

Hage, J. T. (1999). Organizational Innovation and Organizational Change. *Annual Review of Sociology*, *25*(1), 597–622. doi:10.1146/annurev.soc.25.1.597

Helper, S., MacDuffie, J. P., & Sabel, C. M. (2000). Pragmatic collaboration: Advancing knowledge while controlling opportunism. *Industrial and Corporate Change*, *9*(3), 443–488. doi:10.1093/icc/9.3.443

Henderson, R. M., & Clark, K. B. (1990). Architectural Innovation: The Reconfiguration of Existing Product Technologies and the Failure of Established Firms. *Administrative Science Quarterly*, *35*(1), 9–30. doi:10.2307/2393549

Herstad, S.J., Bloch, C., Ebersberger, B. & van de Velde, E. (2008). *Open innovation and globalisation: Theory, evidence and implications*. VISION Era.net.

Hiebing, R. G. Jr, & Cooper, S. W. (2003). *The Successful Marketing Plan: A Disciplined and Comprehensive Approach*. The McGraw-Hill Companies, Inc.

Hoffman, H., Parejo, M., Bessant, J., & Perren, L. (1998). Small firms, R&D, technology and innovation in the UK: A literature review. *Technovation*, *18*(1), 39–55. doi:10.1016/S0166-4972(97)00102-8

Hoffman, W. H., & Schlosser, R. (2001). Success factors of strategic alliances in small and medium-sized enterprises: An empirical study. *Long Range Planning*, *34*(3), 357–381. doi:10.1016/S0024-6301(01)00041-3

Howells, J. (2006). Intermediation and the role of intermediaries in innovation. *Research Policy*, *35*(5), 715–728. doi:10.1016/j.respol.2006.03.005

Howells, J., James, A., & Malek, K. (2003). The sourcing of technological knowledge: Distributed innovation processes and dynamic change. *R & D Management*, *33*(4), 395–409. doi:10.1111/1467-9310.00306

Humphrey, J., & Schmitz, H. (2000). *Governance and Upgrading: Linking industrial cluster and global value chain research*. IDS Working Paper 120.

IfM & IBM. (2008). *Succeeding through service innovation: A service perspective for education, research, business and government*. Cambridge, UK: University of Cambridge Institute for Manufacturing.

Katz, R., & Allen, T. J. (1982). Investigating the not-invented-here (NIH)- syndrome: A look at performance, tenure and communication patterns of 50 R&D project groups. *R & D Management*, *12*, 7–19.

Keil, T. (2002). *External Corporate Venturing: Strategic Renewal in Rapidly Changing Industries*. Quorum.

Kirkels, Y., & Duysters, G. (2010). Brokerage in SME networks. *Research Policy*, *39*(3), 375–385. doi:10.1016/j.respol.2010.01.005

Klerkx, L. W. A. (2008). Establishment and embedding of innovation brokers at different innovation system levels: insights from the Dutch agricultural sector. In *Proceedings of the Conference Transitions towards sustainable agriculture food chains and peri-urban areas*. Wageningen University.

Lee, S., Park, G., Yoon, B., & Park, J. (2010). Open innovation in SMEs- An intermediated network model. *Research Policy*, *39*(2), 290–300. doi:10.1016/j.respol.2009.12.009

Lemola, T., & Lievonen, J. (2008). The role of Innovation Policy in Fostering Open Innovation Activities Among Companies. *Proceedings from "European Perspectives on Innovation and Policy" Programme for the VISION Era-Net Workshop*.

Lichtenthaler, U. (2007). The drivers of technology licensing: An industry comparison. *California Management Review*, *49*(4), 67–89. doi:10.2307/41166406

Lichtenthaler, U. (2008). Open innovation in practice: An analysis of strategic approaches to technology transactions. *IEEE Transactions on Engineering Management*, *55*(1), 148–157. doi:10.1109/TEM.2007.912932

Lichtenthaler, U., & Ernst, H. (2006). Attitudes to externally organizing knowledge management tasks: A review, reconsideration and extension of the NIH syndrome. *R & D Management*, *36*(4), 367–386. doi:10.1111/j.1467-9310.2006.00443.x

Lichtenthaler, U., & Ernst, H. (2009). Opening up the Innovation Process: The Role of Technology Aggressiveness. *R & D Management*, *39*(1), 38–54. doi:10.1111/j.1467-9310.2008.00522.x

Lilien, G. L., Morrison, P. D., Searls, K., Sonnack, M., & von Hippel, E. (2002). Performance assessment of the lead user idea-generation process for new product development. *Management Science*, *48*(8), 1042–1059. doi:10.1287/mnsc.48.8.1042.171

Linder, J. C., Jarvenpaa, S., & Davenport, T. H. (2003). Toward an Innovation Sourcing Strategy. *MIT Sloan Management Review. Reprint*, *4447*, 43–49.

Lopes, C. M., Scavarda, A., Hofmeister, L. F., Thomé, A. M. T., & Vaccaro, G. L. R. (2017). An analysis of the interplay between organizational sustainability, knowledge management, and open innovation. *Journal of Cleaner Production*, *142*, 476–488. doi:10.1016/j.jclepro.2016.10.083

Lopez-vega, H. (2009, January 22-24). *How demand-driven technological systems of innovation work?: The role of intermediary organizations*. A paper from the DRUD-DIME Academy Winter 2009 PhD Conference on Economics and Management of Innovation Technology and Organizational Change, Aalborg, Denmark.

Lord, M. D., Mandel, S. W., & Wager, J. D. (2002). Spinning out a star. *Harvard Business Review, 80*, 115–121. PMID:12048993

Lundvall, B. A., Johnson, B., Andersen, E. S., & Dalum, B. (2002). National Systems of production and competence building. *Research Policy, 31*(2), 213–231. doi:10.1016/S0048-7333(01)00137-8

MacGregor, S., Bianchi, M., Hernandez, J. L., & Mendibil, K. (2007). Towards the tipping point for social innovation. *Proceedings of the 12th International Conference on Towards Sustainable Product Design (Sustainable Innovation 07),* 145-152.

Maes, J. (2009). *SMEs´ Radical Product Innovation: the Role of the Internal and External Absorptive Capacity Spheres. Job Market Paper July 2009, Katholieke Universiteit Leuven.* Faculty of Business and Economics.

Mantel, S. J., & Rosegger, G. (1987). The role of third-parties in the diffusion of innovations: a survey. In R. Rothwell & J. Bessant (Eds.), *Innovation: Adaptation and Growth* (pp. 123–134). Elsevier.

McEvily, S. K., Eisenhardt, K. M. M., & Prescott, J. E. (2004). The global acquisition, leverage, and protection of technological competencies. *Strategic Management Journal, 25*(8/9), 713–722. doi:10.1002mj.425

Mehra, K. (2009). Role of Intermediary Organisations in Innovation Systems- A Case from India. *Proceedings from the 6th Asialics International Conference.*

Miles, I. (2005). Innovation in Services. In The Oxford Handbook of Innovation. Oxford University Press.

Millar, C. C. J. M., & Choi, C. J. (2003). Advertising and knowledge intermediaries: Managing the ethical challenges of intangibles. *Journal of Business Ethics, 48*(3), 267–277. doi:10.1023/B:BUSI.0000005788.90079.5d

Mohr, J., & Spekman, R. (1994). Characteristics of partnership success: Partnership attributes, communication behavior and conflict resolution techniques. *Strategic Management Journal, 15*(2), 135–152. doi:10.1002mj.4250150205

Moore, G. A. (1991). *Crossing the Chasm.* HarperBusiness.

Moore, G. A. (2006). *Dealing with Darwin: How Great Companies Innovate at Every Phase of Their Evolution.* UK Capstone Publishing.

Nählinder, J. (2005). *Innovation and Employment in Services: The case of Knowledge Intensive Business Services in Sweden* (PhD Thesis). Department of Technology and Social Change, Linköping University.

Nalebuff, B. J., & Brandeburger, A. M. (1996). *Co-opetition.* Harper Collins.

Narula, R. (2004). R&D collaboration by SMEs: New opportunities and limitations in the face of globalization. *Technovation, 24*(2), 153–161. doi:10.1016/S0166-4972(02)00045-7

Ndonzuau, F. N., Pirnay, F., & Surlemont, B. (2002). A stage model of academic spin-off creation. *Technovation, 22*(5), 281–289. doi:10.1016/S0166-4972(01)00019-0

O'Regan, N., & Ghobadian, A. (2005). Innovation in SMEs: The impact of strategic orientation and environmental perceptions. *International Journal of Productivity and Performance Management*, *54*(2), 81–97. doi:10.1108/17410400510576595

OECD. (1992). Oslo Manual. DSTI, Organization for Economic Co-operation and Development (OECD).

OECD. (1996). Oslo Manual (2nd ed.). DSTI, OECD.

OECD. (2005). Oslo Manual (3rd ed.). DSTI, OECD.

OECD. (2006). *The Athens Action Plans for Removing Barriers to SME Access to International Markets*, Adopted at the OECD-APEC Global Conference in Athens.

OECD. (2007). *Innovation and Growth: Rationale for an Innovation Strategy, OECD Report*. OECD.

OECD. (2008). *Open Innovation in Global Networks, Policy Brief, OECD Observer*. OECD.

Olson, E., & Bakke, G. (2001). Implementing the lead user method in a high technology firm: A longitudinal study of intentions versus actions. *Journal of Product Innovation Management*, *18*(2), 388–395. doi:10.1111/1540-5885.1860388

Pietrobelli, C., & Rabellotti, R. (Eds.). (2006). *Upgrading to Compete:Global Value Chains, Clusters, and SMEs in Latin America*. Inter-American Development Bank.

Pisano, G. P. (1990). The R&D boundaries of the firm: An empirical analysis. *Administrative Science Quarterly*, *35*(1), 153–176. doi:10.2307/2393554

Porter, M. E., & Bond, G. C. (1999). *The Global Competitiveness Report 1999*. World Economic Forum.

Pourdehnad, J. (2007). *Idealized Design - An "Open Innovation" Process*. Presentation from the annual W. *Edwards Deming Annual Conference*, Purdue University, West Lafayette, IN.

Prencipe, A. (2000). Breadth and depth of technological capabilities in CoPS: The case of the aircraft engine control system. *Research Policy*, *29*(7-8), 895–911. doi:10.1016/S0048-7333(00)00111-6

Prügl, R., & Schreier, M. (2006). Learning from leading-edge customers at *The Sims*: Opening up the innovation process using toolkits. *R & D Management*, *36*(3), 237–250. doi:10.1111/j.1467-9310.2006.00433.x

Quinn, J. B. (2000). Outsourcing innovation: The new engine of growth. *Sloan Management Review*, *41*, 4, 13–28.

Rahman, H., & Ramos, I. (2010). Open Innovation in SMEs: From Closed Boundaries to Networked Paradigm. *Issues in Informing Science and Information Technology*, *7*, 471–487. doi:10.28945/1221

Renaud, P. (2008). *Open Innovation at Oseo Innovation: Example of the Passerelle Programme (A tool to support RDI collaboration between innovative SME's and large enterprises)*. Presentation at the OECD Business Symposium on Open Innovation in Global Networks, Copenhagen.

Rogers, E. M. (1962). *Diffusion of Innovations*. Free Pres.

Rothwell, R. (1991). External networking and innovation in small and medium-sized manufacturing firms in Europe. *Technovation*, *11*(2), 93–112. doi:10.1016/0166-4972(91)90040-B

Sautter, B., & Clar, G. (2008). *Strategic Capacity Building in Clusters to Enhance Future-oriented Open Innovation Processes.* Foresight Brief No. 150, The European Foresight Monitoring Network. www.efmn.info

Schaltegger, S., & Wagner, M. (2011). Sustainable entrepreneurship and sustainability innovation: Categories and interactions. *Business Strategy and the Environment, 20*(4), 222–237. doi:10.1002/bse.682

Schumpeter, J. A. (1934). *The Theory of Economic Development.* Harvard University Press.

Schumpeter, J. A. (1982). *The Theory of Economic Development: An Inquiry into Profits, Capital, Credit, Interest, and the Business Cycle (1912/1934).* Transaction Publishers.

Shorthouse, S. (2008). *Innovation and Technology Transfer.* Presentation from the International Conference DISTRICT 2008. International Conference Centre, Dresden.

Shrader, C., Mulford, C., & Blackburn, V. (1989). Strategic and operational planning, uncertainty and performance in small firms. *Journal of Small Business Management, 27*(4), 45–60.

Smallbone, D., North, D., & Vickers, I. (2003). The role and characteristics of SMEs. In *Regional Innovation Policy for Small-Medium Enterprises.* Edward Elgar. doi:10.4337/9781781009659.00010

Spithoven, A., Vanhaverbeke, W., & Roijakkers, N. (2013). Open innovation practices in SMEs and large enterprises. *Small Business Economics, 41*(3), 537–562. doi:10.100711187-012-9453-9

Stankiewics, R. (1995). The role of the science and technology infrastructure in the development and diffusion of industrial automation in Sweden. In B. Carlsson (Ed.), *Technological systems and economic performance: The case of the factory automation* (pp. 165–210). Kluwer Academic Publishers. doi:10.1007/978-94-011-0145-5_6

Tapscott, D., & Williams, A. D. (2007). *Wikinomics: How mass collaboration changes everything.* Penguin Book.

Tidd, J., Bessant, J., & Pavitt, K. (2001). *Managing Innovation.* John Wiley.

Tidd, J., Bessant, J., & Pavitt, K. (2005). *Managing Innovation: Integrating Technological, Market and Organizational Change* (3rd ed.). John Wiley & Sons Ltd.

Tilley, F., & Tonge, J. (2003). Introduction. In O. Jones & F. Tilley (Eds.), *Competitive Advantage in SME's: Organising for Innovation and Change.* John Wiley & Sons.

Toivonen, M. (2004). *Expertise as Business: Long-term development and future prospects of knowledge-intensive business services (KIBS).* Laboratory of Industrial Management, Helsinki University of Technology.

Torkkeli, M., Tiina Kotonen, T., & Pasi Ahonen, P. (2007). Regional open innovation system as a platform for SMEs: A survey. *International Journal of Foresight and Innovation Policy, 3*(4), 2007. doi:10.1504/IJFIP.2007.016456

Tourigny, D., & Le, C. D. (2004). Impediments to Innovation Faced by Canadian Manufacturing Firms. *Economics of Innovation and New Technology, 13*(3), 217–250. doi:10.1080/10438590410001628387

Turpin, T., Garrett-Jones, S., & Rankin, N. (1996). Bricoleurs and boundary riders: Managing basic research and innovation knowledge networks. *R & D Management*, *26*(3), 267–282. doi:10.1111/j.1467-9310.1996.tb00961.x

Van Ark, B., Broersma, L. & Den Hertog, P. (2003). *Service Innovation, Performance and Policy: A Review*. Research Series no. 6, Synthesis report in the framework of the project on Structural Information Provision on Innovation in Services SIID for the Ministry of Economic Affairs of the Netherlands.

Van de Ven, A. H. (1986). Central problems in the management of innovation. *Management Science*, *32*(5), 590–607. doi:10.1287/mnsc.32.5.590

Van de Vrande, V., De Jong, J. P. J., Vanhaverbeke, W., & De Rochemont, M. (2009). Open innovation in SMEs: Trends, motives and management challenges. *Technovation*, *29*(6-7), 423–437. doi:10.1016/j.technovation.2008.10.001

Van Dijk, C., & Van den Ende, J. (2002). Suggestion systems: Transferring employee creativity into practicable ideas. *R & D Management*, *32*(5), 387–395. doi:10.1111/1467-9310.00270

Van Gils, A., & Zwart, P. (2004). Knowledge Acquisition and Learning in Dutch and Belgian SMEs: The Role of Strategic Alliances. *European Management Journal*, *22*(6), 685–692. doi:10.1016/j.emj.2004.09.031

Vanhaverbeke, W. (2006). The Inter-organizational Context of Open Innovation. In H. Chesbrough, W. Vanhaverbeke, & J. West (Eds.), *Open Innovation: Researching a New Paradigm* (pp. 205–219). Oxford University Press.

Von Hippel, E. (1998). Economics of product development by users: The impact of 'Sticky' local information. *Management Science*, *44*(5), 629–644. doi:10.1287/mnsc.44.5.629

Von Hippel, E. (2005). *Democratizing Innovation*. MIT Press. doi:10.7551/mitpress/2333.001.0001

Vujovic, S., & Ulhøi, J. P. (2008). Opening up the Innovation Process: Different Organizational Strategies. Information and Organization Design Series, Vol. 7, Springer US. doi:10.1007/978-0-387-77776-4_8

Walther, J., & D'Addario, K. (2001). The Impacts of Emoticons on Message Interpretation in Computer-Mediated Communication. *Social Science Computer Review, Volume*, *19*(3), 324–347. doi:10.1177/089443930101900307

Walton, R. E. (1989). *Up and Running: Integrating Information Technology and the Organization*. Harvard Business School Press.

Warner, A. (2001). *Small and Medium Sized Enterprises and Economic Creativity*. Paper presented at UNCTAD's intergovernmental Expert Meeting on "Improving the Competitiveness of SMEs in Developing Countries: the Role of Finance, Including E-finance, to Enhance Enterprise Development", Geneva.

Watkins, D., & Horley, G. (1986). Transferring technology from large to small firms: the role of intermediaries. In T. Webb, T. Quince, & D. Watkins (Eds.), *Small Business Research* (pp. 215–251). Gower.

West, J., & Bogers, M. (2014). Leveraging external sources of innovation: A review of research on open innovation. *Journal of Product Innovation Management*, *31*(4), 814–831. doi:10.1111/jpim.12125

West, J., & Gallagher, S. (2006). Challenges of open innovation: The paradox of firm investment in open-source software. *R & D Management, 36*(3), 319–331. doi:10.1111/j.1467-9310.2006.00436.x

Winch, G. M., & Courtney, R. (2007). The organization of innovation brokers: An international review. *Technology Analysis and Strategic Management, 19*(6), 747–763. doi:10.1080/09537320701711223

Zhang, J., & Zhang, Y. (2009). Research on the Process Model of Open Innovation Based on Enterprise Sustainable Growth. *Proceeding of the International Conference on Electronic Commerce and Business Intelligence*, 318-322.

Zuboff, S. (1988). *In the Age of the Smart Machine: The Future of Work and Power.* Basic Books.

ADDITIONAL READING

Ackerman, M. (1994). Definitional and contextual issues in organizational and group memories. In *Proceedings of the 27th annual Hawaii international conference on system sciences* (pp. 191–200). IEEE Computer Society Press. 10.1109/HICSS.1994.323444

Boschma, R. A., & ter Wal, A. L. J. (2007, May). in an industrial district: The case of a footwear district in the South of Italy. *Industry and Innovation, 14*(2), 177–199. doi:10.1080/13662710701253441

Chan, I., & Chao, C. (2008). Knowledge management in small and medium-sized enterprises. *Communications of the ACM, 51*(4), 83–88. doi:10.1145/1330311.1330328

Chesbrough, H. (2002). Making sense of corporate venture capital. *Harvard Business Review*, 4–11. PMID:11894386

Coenen, L., Moodysson, J., & Asheim, B. T. (2004). Nodes, networks and proximities: On the knowledge dynamics of the Medicon Valley biotech cluster. *European Planning Studies, 12*(7), 1003–1018. doi:10.1080/0965431042000267876

Desouza, K. C., & Awazu, Y. (2006). Knowledge management at SMEs: Five peculiarities. *Journal of Knowledge Management, 10*(1), 32–43. doi:10.1108/13673270610650085

Gertler, M. S., & Levitte, Y. M. (2005). Local nodes in global networks: The geography of knowledge flows in biotechnology innovation. *Industry and Innovation, 12*(4), 487–507. doi:10.1080/13662710500361981

Henkel, J. (2006). Selective revealing in open innovation processes: The case of embedded Linux. *Research Policy, 35*(7), 953–969. doi:10.1016/j.respol.2006.04.010

Jennex, M., & Olfman, L. (2002). Organizational memory/knowledge effects on productivity, a longitudinal study. In *Proceedings of the 35th Hawaii international conference on system sciences, HICSS35.* IEEE Computer Society. 10.1109/HICSS.2002.994053

Koruna, S. (2004). External technology commercialization policy guideline. *International Journal of Technology Management, 27*(2/3), 241–254. doi:10.1504/IJTM.2004.003954

Laursen, K., & Salter, A. (2006). Open for innovation: The role of openness in explaining innovation performance among UK manufacturing firms. *Strategic Management Journal, 27*(2), 131–150. doi:10.1002mj.507

Lecocq, X., & Demil, B. (2006). Strategizing industry structure: The case of open systems in low-tech industry. *Strategic Management Journal, 27*(9), 891–898. doi:10.1002mj.544

Niosi, J. (2003). Alliances are not enough explaining rapid growth in biotechnology firms. *Research Policy, 32*(5), 737–750. doi:10.1016/S0048-7333(02)00083-5

Nonaka, I., Toyama, R., & Nagata, A. (2000). A firm as a knowledge creating entity: A new perspective on the theory of the firm. *Industrial and Corporate Change, 9*(1), 1–20. doi:10.1093/icc/9.1.1

Owen-Smith, J., & Powell, W. W. (2004). Knowledge networks as channels and conduits: The effects of spillovers in the Boston biotechnology community. *Organization Science, 15*(1), 5–21. doi:10.1287/orsc.1030.0054

Ruggles, R. (1998). The state of the notion: Knowledge management in practice. *California Management Review, 40*(3), 80–89. doi:10.2307/41165944

Swedish presidency of the EU (June 12, 2009). Industry views on Research, Innovation and Education, Conference summary, Government of Sweden.

Traveter, K., & Wagner, G. (2005). Towards Radical Agent-Oriented Software Engineering Processes Based on AOR Modeling. In B. Henderson-Sellers & P. Giorgini (Eds.), *Agent-Oriented Methodologies* (pp. 277–316). IGI Publishing. doi:10.4018/978-1-59140-581-8.ch010

Valentin, F., & Jensen, R. L. (2003). Discontinuities and distributed innovation: The case of biotechnology in food processing. *Industry and Innovation, 10*(3), 275–310. doi:10.1080/1366271032000141652

Visvizi, A., Lytras, M. D., Damiani, E., & Mathkour, H. (2018). Policy making for smart cities: Innovation and social inclusive economic growth for sustainability. *Journal of Science and Technology Policy Management, 9*(2), 126–133. doi:10.1108/JSTPM-07-2018-079

Vrgovic, P., Vidicki, P., Glassman, B., & Walton, A. (2012). Open innovation for SMEs in developing countries–An intermediated communication network model for collaboration beyond obstacles. *Innovation, 14*(3), 290–302. doi:10.5172/impp.2012.14.3.290

Wensley, A. (2000). Tools for knowledge management. In BPRC conference on knowledge management: Concepts and controversies. doi:10.1016/B978-0-7506-7247-4.50008-2

KEY TERMS AND DEFINITIONS

Collaborative Innovation: This sort of innovation takes place in collaboration with clients, suppliers, universities, research houses, intermediaries and other partners in a collaborative ways with shared ideas, but envisioned goals.

Collaborative Innovation Network: A collaborative innovation network is a social construct that is used to describe innovative teams. It can be seen as a cyber-team of self-motivated people with a collective vision, enabled by the Web technologies to collaborate in achieving a common goal by sharing ideas, information, knowledge and work.

Crowdsourcing Innovation: Crowdsourcing innovation is the practice of obtaining needed services, ideas, or content by imploring contributions from a large group of people or individual, and especially from an online community, rather than from traditional employees or suppliers. This form of innovation is often used to subdivide tedious work or to fund-raise startup companies and charities, and this process may take place both online and offline.

Open Innovation: Open innovation is the use of useful inflows and outflows of knowledge to accelerate internal innovation, and at the same time, expand the markets for external use of innovation.

Organizational Well-Being: Well-being can be seen as a state of complete physical, mental, and social health and not merely the absence of disease or infirmity of an individual. Organizational well-being, along this way can be seen as the organization's ability to promote and maintain the physical, psychological and social workers wellbeing at all levels and for every job.

SMEs Empowerment: It is meant by the real connotation to enhance the knowledge dimension of the enterprises belonging to this sector in terms of knowledge, capacity, and human skills and thereby empowering themselves to face various obstacles or making them capable to tackle challenges to reach out to the grass roots communities.

ENDNOTES

1. There is no single agreed definition of a SME. A variety of definitions are applied among OECD and APEC economies, and employee number is not the sole defined criterion. SMEs are considered to be non-subsidiary, independent firms which employ less than a given number of employees.
2. Crowdsourcing is a form of open innovation (and in some cases, user innovation) that attempts to involve a large pool of outsiders to solve a problem. Product recommendations at Amazon etc. are probably the most often seen example (Chesbrough, Vanhaverbeke & West, 2006)
3. Small and médium enterprises (also SMEs, small and medium businesses, SMBs, and variations thereof) are companies whose headcount or turnover falls below certain limits (https://en.wikipedia.org/wiki/SMEs)
4. About Types of Innovation: Tidd et al (2005) argue that there are four types of innovation; consequently the innovator has four pathways to investigate when searching for good ideas, such as: a) Product Innovation - new products or improvements on products (example- the new Mini or the updated VX Beetle, new models of mobile phones and so on.); b) Process Innovation - where some part of the process is improved to bring benefit (example- Just in Time); c) Positioning Innovation – for example- Lucozade used to be a medicinal drink but the was repositioned as a sports drink; and d) Paradigm Innovation - where major shifts in thinking cause change (example- during the time of the expensive mainframe, Bill Gates and others aimed to provide a home computer for everyone) (http://ezinearticles.com/?Types-of-Innovation&id=38384).
5. Radical innovation concerns technology breakthrough, user embracement as well as new business models (Artemis, 2009).

[6] https://www.mjward.co.uk/Businesses-phrases-terms-jargon/Business-Phrases-Terms-1.html
[7] http://www.businessdictionary.com/definition/innovation.html#
[8] Examples of service innovation may include: On-line tax returns, e-commerce, helpdesk outsourcing, music download, loyalty programs, home medical test kits, mobile phones, money market funds, ATMs and ticket kiosks, bar code, credit cards, binding arbitration, franchise chains, instalment payment plans, leasing, patent system, public education and compound interest saving accounts (IfM and IBM, 2008: 17).
[9] https://www.c7.ca/glossary
[10] www.uen.org/core/edtech/glossary.shtml
[11] www.c7.ca/glossary
[12] http://igniter.com/post56
[13] This refers to firms in the formal sector only.

Chapter 4
Dimensions of Researches for Open Innovation in SMEs

Hakikur Rahman
Institute of Computer Management and Science, Bangladesh

ABSTRACT

Innovation has been treated as a recognized driver of economic prosperity of a country through the sustained growth of its entrepreneurships. Moreover, the recently coined term open innovation is increasingly taking the lead in enterprise management in terms of value addition and knowledge building. Foci of academics, researchers, and practitioners nowadays are revolving around various innovation models, comprising innovation techniques, processes, and strategies. This chapter seeks to find out open innovation research and practices that are being carried out circumscribing development of entrepreneurships, particularly the sector belonging to the small and medium enterprises (SMEs) through a longitudinal study.

INTRODUCTION

Innovation is not any more just a research topics, but it has become a significant driver for prosperity, growth and sustained profitability to global entrepreneurships. Innovation along its route to the current period exhaled new methods or tools in terms of products, processes or organizational management. As far as this literature review and research go, from its early inception inscribing issues of economic development (Schumpeter, 1934; 1942; 1950), patents and licensing (Von Hippel, 1988), organizational networking (Powell, 1990), process innovation (Davenport, 1993), co-opetition (Brandenburger and Nalebuff, 1996), management of intellectual capital (Grindley and Teece, 1997) till the coining up of its features in more familiar ways framing on the utilization of information technologies, such as open innovation (Chesbrough 2003a; 2003b), innovation never stayed stalled. Furthermore, due to opening up the innovation processes and combining internally and externally developed technologies and strategies to create economic value the innovation has crossed the boundary of closed innovation to open innovation (Rahman and Ramos, 2010; 2012).

The perception of open innovation has attracted considerable interest since Henry Chesbrough first coined it to capture the increasing reliance of enterprises on external sources of innovation. Although

DOI: 10.4018/978-1-7998-5849-2.ch004

open innovation has flourished as a topic within innovation management research, at the same time, it has also triggered debates about the coherence of the research endeavors pursued under this umbrella, including its theoretical foundations (Kovacs, Van Looy and Cassiman, 2015). However, we need to endow open collaboration researchers with new datasets that extent different contexts, as well as novel computational models and analytical techniques (Brunswicker, Bertino and Matei, 2015).

Enterprises may open up their innovation processes on two dimensions. While inbound open innovation refers to the acquisition of external technology through open exploration processes, outbound open innovation describes the outward transfer of technology through the open exploitation processes. However, prior open innovation researches have focused on the inbound dimension, whereas the outbound dimensions have been relatively neglected (Lichtenthaler, 2009c). The concept of 'open innovation' has received a substantial amount of coverage within the academic literature and beyond. But, much of this seems to have been without sufficient critical analysis of the evidence (Trott and Hartmann, 2009).

Enterprises have to innovate to stay competitive, and they have to collaborate with other entities to innovate effectively. Although the benefits of open innovation have been described in many researches, however, those underlying mechanisms, as how companies can be successful open innovators have not be understood well. Progressively, a growing community of innovation management researchers started to develop different frameworks to understand open innovation in a more systematic method. Vanhaverbeke, Chesbrough & West (2014) provide a thorough assessment of research conducted on open innovation, as well as a comprehensive overview of what will be the most important, most promising and most relevant research topics in this arena. Chesbrough, Vanhaverbeke and West (2006), was the initial initiative to bring open innovation closer to the academic community. Since then, open innovation research has been growing in an exponential way and research has evolved in different and unexpected directions. As the research field has grown, it becomes increasingly difficult for young (and even experienced scholars) to keep an overview of the most important trends in open innovation research, of the research topics that are most potential for the coming years, and of the most interesting management challenges that are emerging in enterprises practicing open innovation.

Traditionally, firms used to prefer the so-called, closed innovation strategies in developing their own products internally, and with limited interactions with the external world (Lichtenthaler, 2011). In recent years, researchers and practitioners are showing interests in open innovation research and practices that are visible during the literature review in various publications, and conference proceedings. This has also been observed in contemporary literatures that innovation researches are shifting from the closed and controlled environment of the corporate entrepreneurs towards more open and flexible model, based on cooperation and coordination among various parties. Knowledge and new technologies are no longer remaining sole properties of major monopoly corporations (Caetano and Amaral, 2011; Westergren and Holmstrom, 2012).

In this aspect, the business sector belonging to the small and small enterprises (SMEs[1]) play important role in networking and making innovation clusters in association with universities and research houses, being recognized as major driving forces in the open innovation paradigm.

SMEs also play a crucial role in raising investments in spin offs, start ups, or research and development (R&D) and making countries more competitive, which is true for not only the European Union but also in other countries (European Union, 2005). Moreover, the majority of the developing and transitional economies have acknowledged that SMEs are the potential engine of economic growth and source of sustainable development, which are essential for industrial reformation, new job creation, and revenue generation of the population at large (Koyuncugil & Ozgulbas, 2009).

However, this research observes that utilization of open innovation strategies for the development of SMEs remains low in terms of researches and practices (Chesbrough 2003a; 2003b; West, Vanhaverbeke & Chesbrough, 2006; Lichtenthaler & Ernst, 2009; Lindermann, Valcareel, Schaarschmidt & Von Kortzfleisch, 2009; Van de Vrande, de Jong, Vanhaverbeke & de Rochemont, 2008; 2009), especially finding interpretative results justifying through empirical studies. Only a limited number of literatures are there to support the introduction of OI strategies in SMEs.

This study while conducting an empirical study in Portugal among some selected SMEs, has tried to synthesize various research dimensions by carrying out a longitudinal study. In doing so, a thorough literature review has been conducted emphasizing researches conducted by leading researchers and practitioners through two most comprehensive search engines (Sciencedirect, and Scopus), though hardly these could be recognized as cent percent contribution towards SMEs growth. While investigating into the open innovation aspects of SMEs, the study covered characteristics of individual firms, and group of firms (by human aspect, financial aspect, and issues of challenges in adopting OI strategies) taken at national or regional contexts. The intention is to prepare a report by mapping the issues of challenges, and on adoption of OI strategies in Portugal. However, this chapter restrains only within the theoretical contexts on various research dimensions that the study has encountered during the initial stage of the research, including the conduction of the pilot survey and extension of the survey in a few other countries of similar socio-economic status.

The chapter has been divided into five sections. Introducing about generic perspectives of open innovation in SMEs, the next section discusses specific research aspects of open innovation. The third section is the main thrust of the chapter, which discusses about the literature search using two search engines with relevant observation from the research group. Thereafter, before concluding this particular aspect of the study with the findings, it put forward a few research hints for future research and dialogue in this particular area of research.

BACKGROUND

Open Innovation and SMEs

Innovative entrepreneurship need to be treated as a utility with purpose, meaning, and accountability; not a glitch. The task can be fulfilled by an individual alone or by teaming up with one or more partners, or with the support from other firms or similar ventures. If the task is performed in a collaborative platform, even a large group of firms can function as partners in the entrepreneurship. The entrepreneur is the one who brings together the necessary resources (financial, logistic, managerial and personnel) that the innovation calls for. The entrepreneur is the one who finds the place of application and directs the execution of the alteration. Sometimes a long time may pass before a promising invention is taken up by a true entrepreneur. Probably it may happen that an invention or discovery and an entrepreneur do not match immediately. Fortunately in the realm of technology advancement, it is quite frequent that the match is made without much difficulty. However, in most cases the Schumpeterian entrepreneur drives the innovation process during the first realization of any revolutionary innovation. Furthermore, the process following the pioneering innovation (also known as diffusion), is also mostly driven by entrepreneurs and majority of the initiative appears at the beginning of the entrepreneurship sequence (Kornai, 2010). The entire collaborative functions can be familiarized as a part and parcel of innovation, or rather open innovation. Also, in recent days, not only innovative entrepreneurs are there who are initiating innovative ventures, or looking for a partner of similar attitude, but also there are successful intermediaries with knowledge and expertise in this field, and assisting each individual firms or group of firms utilizing the innovative information technology platforms.

The most important benefit of open innovation to entrepreneurship is that it provides a flexible and extended base of knowledge; with new ideas and technologies based on sources internal or external partners, such as clients, suppliers, researchers, practitioners, staffs and all others who are involved in producing productive input in the value addition process. Main motive of joining forces is not to compete, but to seize new business opportunities by sharing risks and resources (OECD, 2008b; Sousa, 2008). The open innovation approach assumes that the firm is no longer the sole place of innovation, nor the lone entity in reaping the benefits of research, development and innovation. In contrast to the closed innovation where each entity used to treat others with either suspicion, who could take away useful knowledge; or at least thought to be another competitor, who could reduce the profit margin; in the open innovation paradigm, each and every entities are being taken as partners, and regarded as potential contributors of crucial piece of information needed to the value addition process. This process is open ended and collaborative, thus assists in reducing RDI costs, risks, and technology costs, and essentially becoming important driver of economic growth (Lemola and Lievonen, 2008; Rahman and Ramos, 2012-research model).

The supporters of open innovation are universally positive, but researchers suggest that the specific mechanisms and outcomes of open innovation models are very sensitive to the context and the contingency. This is not surprising because the open or closed nature of innovation is historically contingent and does not involve a simple shift from closed to open as often suggested in the literature. Research has shown that patterns of innovation differ fundamentally by sector, firm and strategy (Tidd, 2014).

Researches on open innovation suggest that enterprises benefit differentially from adopting open innovation strategies. However, it is indistinguishable why this is so. One possible explanation could be that enterprises' business models are not attuned to open strategies. Saebi and Foss (2015) propose a contingency model of open business models by systematically linking open innovation strategies to core business model dimensions, especially the content, structure, and governance of transactions.

Various school of thoughts support the idea of benefitting from open innovation for the small and medium scale business firms, which may lead to product innovation or process innovation including employment generation (Roper and Hewitt-Dundas, 2004). But, in reality this newly opened paradigm is relatively challenging, as the society is yet to confirm the impact of open innovation in SMEs, in spite of claims on enhanced collaboration and employment generation (Nahlinder, 2005). Thus, not only improving the competitive advantage of SMEs is essential to the individual firm or the group of firms, but also the impact at the national economy is need to be visible (Tilley and Tonge, 2003). However, except surveys conducted at the institutional levels, like OECD Science Technology and Industry Scoreboard, OECD SME and Entrepreneurship Outlook, only a few literatures exist illustrating empirical studies on SMEs and the impact of open innovation in collaboration of more than one researchers, such as de Jong, Orietta Marsili, 2006; van de Vrande, de Jong, Vanhaverbeke and de Rochemont, 2009; Dahlander and Gann, 2010; Gassmann, Enkel and Chesbrough, 2010; Lee, Park, Yoon& Park, 2010; and others. Apart from them several researches are being conducted at individual level, such as Lichtenthaler, 2006; 2007a; 2007b; 2008a; 2008b; 2008c; 2008d; 2008e; 2008f; 2009a; 2009b; 2009c; 2010a; 2010b; 2010c; 2011a; 2011b; von Hippel, 1975; 1978; 1986; 1987; 1990; 1994; 1998; 2001a; 2001b; 2005; 2007; 2010. Hence, this study suggests that further researches are required to map the impact of open innovation at the group level, individual level, country level, or cross-country level.

Open Innovation Researches on SMEs

Innovation is an essential element among other functionalities for enterprises and entrepreneurships, especially surviving within the current economic situation, and at the same time planning for sustainable growth relative to their competitors, locally and globally. It is not a luxury, but essentiality. Though not plenty, but tools exist to assist the entrepreneurs to measure their propensity to innovate and increase their capability for innovation or their innovation performance. However, this research finds that the situation is scanty for SMEs, due to the fact that majority of those tools are being applicable at corporate level, not for small enterprises who could not afford such tools due to various challenges, such as a lack of awareness, funding and capacity which causes apprehension about innovation, open innovation, intellectual property and other strategies (Gassmann, 2006; Van de Vrande, de Jong, Vanhaverbeke & de Rochemont, 2009). Furthermore, amongst SMEs who have been subjected to relevant researches there is persuasive evidence that innovation tends to be a domestic affair with more developments coming from existing resources rather than coming from outside sources (Bevis and Cole, 2010).

It has been observed that most of the researches on open innovation remain restricted towards targeting common stakeholders through major corporate houses or their alliances. In addition, it is a fact that a few of those corporate are controlling the entire global market or system of open innovation chain through process modification, product differentiation, or diversification of resources. Moreover, despite immense potentiality to reach out the stakeholders at the grass roots through open ended demand, diversity of product variation, and scale of economic capacity majority of the contemporary researches are confined towards generic pattern-oriented clients (Rahman & Ramos, 2010). It is good sign that the scenario is rapidly transforming in the recent decade. Innovation is no longer remains within a vertically integrated company with everything remaining in-house. With the advent of open innovation concept and new information technologies, open and flexible cooperation among business houses, research centers and universities is being treated as the most beneficial approach for business development, especially start-offs, spin-offs, or kick-starting joint ventures. In this new business model different actors are applying their principles in addition to other partners through interactive participations to bring out an acceptable outcome for value addition (Chesbrough, 2003a; Maijers, Vokurka, van Uffelen, & Ravensbergen, 2005; Wijffels, 2009; Rahman and Ramos, 2012).

The other fact is that SMEs are being accepted as the global drivers of technological innovation and economic development and represent the deep, broad and fertile platform that nourishes, sustains, and regenerates the global economic ecosystem (Kowalski, 2009). At the same time, to engage the open innovation strategies, research and development (R&D, taken as a key innovation indicator) is increasingly being outsourced to lower the cost of production (Dehoff & Sehgal, 2006; 2008). Apart from the cooperation, coordination and collaboration, open innovation strategies involve vigorous networking with partner companies, interaction with start-up ventures, public research institutes, universities and external suppliers; thus sharing and accessing of outside information and technology, IP management, knowledge management, creative entrepreneurship thinking, making strategic alliances and above all leading to be global visionary (Kowalski, 2009; Lichtenthaler, 2011a). All these factors resonate for integrated and enhanced researches on the aspect of SME development through open innovation strategies.

The current section serves as a broad background on the concept of open innovation that has been observed by this research along while carrying out the literature review. However, the research hypothesis is to find out contemporary researches following a respectable search from a dependable repository. The study has taken various approaches in this aspect, but mainly depended on contents from the

ScienceDirect, a concern of Elsevier B.V, Scopus, Google Scholar, and the university's integrated online library search. Due to the subscription status of the researchers own institute, it was easy to obtain cross reference materials easily from other subscribed sources. The main literature review has been conducted across various research dimensions, and they have been presented in the next section. The next section discusses about various research dimensions that have been carried out by contemporary researchers in the field of open innovation for the development of SMEs.

RESEARCHES DIMENSIONS: THE REVIEW

By far the open innovation (authors prefer it to be seen as a form of collaborative innovation) is becoming the central topics surrounding business strategy and innovation (Huizingh, 2009) and open innovation is being claimed to be the new breed of innovation requiring enterprises to look beyond the boundaries of their organizations, thus using external and internal knowledge for successful value creation (Thoben, 2008). In the eyes of an open innovator on entrepreneurship development, economic prosperity is expected to result from exploiting the innovation capacity of an enterprise by improving the competitiveness and thus enhancing the productivity (BVCA, 2005). However, as observed during this longitudinal study, open innovation has so far been adopted mainly in high-tech and multinational companies. Though open innovation has received increasing attention in the scientific research arena, but so far it has mainly been investigated in corporate and high-tech multinational enterprises (MNEs) based on instruments, such as in-depth interviews and case studies (Chesbrough, 2003b; Kirschbaum, 2005; Lichtenthaler, 2010a). Moreover, when searching for cases or examples, most of them are found to be focusing on very specific industries, for example open source software (Henkel, 2006) or tabletop role-playing games (Lecocq & Demil, 2006) or crafts industries (Santisteban, 2006) or tourism (Novelli, Schmitz & Spencer, 2006; Hjalager, 2010). Even if a large sample of enterprises is being explored by various researchers, their focuses remain on specific issues rather than the full open innovation model (Van de Vrande, de Jong, Vanhaverbeke & de Rochemont, 2009). Perhaps, the reason could be that the open innovation strategies depend on the very specific cases, applications, sectors or environments, and cannot be generalized, as such. Or, the open innovation maturity model has not been reached at such a stage that can be generalized.

The review has, however observed that open innovation researches also exist in smaller organizations (Van de Vrande, de Jong, Vanhaverbeke & de Rochemont, 2009) and this trend is increasing (Gassmann, Enkel & Chesbrough, 2010; Saarikoski, 2006; Dahlander & Gann, 2010; Fredberg, Elmquist & Ollila, 2008). To find out further detail about the research trends on open innovation for SMEs development, several searches were conducted among the contemporary researchers, their researches, research and research practices carried out by reputed organizations like OECD, European Union, European Commission and others in the field of open innovation, especially targeted for SMEs development. Following the search methodology of Saarikoski (2006:24) and supported by similar methodology on structured literature review of Fredberg, Elmquist and Ollila (2008:10), a search into the Internet (Google) with the search string 'open innovation research AND SMEs development' (empirical setting of the research) was conducted and it yield 3,580,000 hits (though the string ["open innovation research" AND "SMEs development"] resulted only 2 hits). With the later string, Google Scholar returns only 1 hit. These hits included contents on this aspect incorporating all those mentioned entities (researchers, researches, research organizations, and academia, national and international organizations). However, to keep the search less generalized and focused to specific search settings and foremost to have an overview on the contemporary research works including practices on open innovation, the search for this study was conducted on a content provider namely, ScienceDirect, and Scopus. This search has been carried out on ScienceDirect by using the search formulae set_1.

Search in ScienceDirect

The Search Formulae set_1:

1. (open AND innovation for all fields and research AND SMEs for all fields) [(All Sources) (All Sciences) (All Years*)]
2. (open AND innovation for all fields and research AND SMEs for all fields) [(All Sources) (Business, Management and Accounting) (All Years*)]
3. (open AND innovation AND research for all fields and SMEs for all fields) [All Sources (Business, Management and Accounting) (All Years*)]
4. (open AND innovation for titles and research AND SMEs for all fields) [All Sources (Business, Management and Accounting) (All Years*)]

* Since 2002.

The search string (i) brings 2,394 counts; search string (ii) brings 1,496 counts and search string (iii) brings 1,496 counts; while search string (iv) brings only 17 counts.

Search string (iii) (with 1,496 counts) has been taken as the entry point of this longitudinal study and among them 1,353 were journal articles, 143 books and 4 were tagged as reference works. Table-1 shows their publication pattern considering major number of publication (here the minimum count is 40) and table-2 shows their years of publication (here the data has been given from 2000 till the date of the search, which is October 21, 2010). Noteworthy to mention that the search was conducted applying to all fields. But, when the search was modified with 'open innovation' in the title and 'research+smes' within the fields, the result returned only 10 journal articles and books.

Table 1. Number of entries in different Journals (with minimum count of 40) (counts were taken since 2002 till the date of the search, which is October 08, 2012)

Search String # iii With Minimum Count of 40	
Name of Journal	**Number of Entries**
Technovation	205
Research Policy	191
Technological Forecasting and Social Changes	70
Industrial Marketing Management	66
International Business Review	64
European Management Journal	61
Journal of Business Venturing	53
World Patent Information	46
Journal of Business Research	40

Table 2. Number of publications in various years on OI researches for SMEs (counts were taken since 2003 till the date of the search, which is October 08, 2012)

Search String # iii	
Year of Publication	**Number of Publications**
2012	248
2011	164
2010	131
2009	122
2008	124
2007	107
2006	97
2005	93
2004	56
2003	55
2002 and earlier since 1992	303

Table-2 shows that the trend of researches using open innovation strategies for SMEs development is growing after the term 'open innovation' has been coined by Prof. Henry Chesbrough. This table also reveals that the trend of using OI strategies was there as number of articles before 2002 with the available data from the ScienceDirect is significant. However, the main purpose of this study is to find out thematic patterns or themes of researches obtained from most relevant contents of these searches. To be more specific, the most relevant contents were separated from these search using search string (iii). Also, the main notion of using ScienceDirect is to provide the first impression of freely available content without being subscribed, notwithstanding other arguments.

Search in Scopus

On Scopus, the following search string has been used (termed as the Search Formulae set_2):

1. (open AND innovation AND research for all Titles, Abstracts and Keywords) [All documents) (All Years*)]

* Since 2003.

The Search Formulae set_2 resulted in 42 counts with ALL documents type and Subject Areas belonging to Life Sciences (with more than 4,300 titles), Physical Sciences (>7,200 titles), health Sciences (>6,800 titles), and Social Sciences and Humanities (>5,300 titles).

Another set of search was carried out among the publication of the leading researchers and practitioners in this field. Among them, the book, "Open Innovation: The New Imperative for Creating and Profiting from Technology" (Chesbrough, 2003a, being the most cited author on 'open innovation' (Fredberg, Elmquist & Ollila, 2008)); "Open Innovation: Practice, Trends, Motives and bottlenecks in the SMEs" (De Jong, 2006); "Open Innovation: Researching a new Paradigm" (Chesbrough, Vanhaverbeke & West,

2006); Journal articles written by Van de Vrande, de Jong, Vanhaverbeke & de Rochemont, 2008; 2009; De Jong, Vanhaverbeke, Kalvet & Chesbrough, 2008; Gassman, 2006; and Gasmann, Enkel & Chesbrough, 2010 were included. The first book was selected as the most cited book in this sector, the second article was selected as the most relevant search return each time made on search engines, and the rest were selected by the authors after careful observations from various search strings and citation index.

As a fourth check, contribution of forerunners on the concept of open innovation in the form of books were also included in the categorization, such as Schumpeter (1934; 1942; 1950), Von Hippel (1986; 1987; 1988) and Davenport (1993a; 1993b; 1994). Noteworthy to mention that there were several others books (available in the references), but they are not being included here, as separate entities. And, unless they provide fundamental concepts on open innovation research, literatures earlier than 2003 were less emphasized.

Finally, to avail the information about open innovation researches in SMEs, this study looked at various publications from international organizations like, OECD, European Union, and European Commission; individual organizations like, Vinnova, Vision Era-Net; portals like, OpenInnovation dot net, Innocentive dot com, Ideaconnection dot com; and articles from special issues from journals like, MIT Sloan Management Review, Technovation, Research Policy, and Harvard Business Review.

After several round of iterations, the following research themes were taken into consideration for exploration, pending further research impact on these themes and extended debate on their substances in relation to the development of smaller firms serving at the grass roots. Table-3 is showing the selected research themes, which are being discussed in the next sub-section in terms of their relevancies.

Table 3. Synthesized research themes on OI for the SMEs

Research Themes	Literatures From the Search
Conceptualization of open innovation (and policy initiation)	Savioz & Blum, 2002; Amara, Landry, Becheikh & Ouimet, 2008; Chesbrough, 2003a; Lee, Park, Yoon & Park, 2010; Huizingh, 2011
Establishment of research model (and action research)	Major & Cordey-Hayes, 2000; Chesbrough, 2003b; 2006; Edwards, Delbridge & Munday, 2005; Lawson, Longhurst & Ivey, 2006; Thorgren, Wincent & Örtqvist, 2009; Raymond & St-Pierre, 2010; Rhee, Park & Lee, 2010
Development of business model (and framework development)	Cooke, 2005; Chesbrough, 2006; De Jong, 2006; Partanen, Möller, Westerlund, Rajala & Rajala, 2008; Freel & De Jong, 2009; Lee, et al., 2010; Belussi, Sammarra & Sedita, 2010; Carlos de Oliveira and Kaminski, 2012
Opportunities and challenges of open innovation (and action plan to combat the situation)	Hoffman, Parejo, Bessant & Perren, 1998; Levy, Powell & Galliers, 1999; Del Brío & Junquera, 2003; Tödtling & Trippl, 2005; Van de Vrande, de Jong, Vanhaverbeke, & Rochemont, 2009; Groen & Linton, 2010; Knudsen & Mortensen, 2010; Rahman & Ramos, 2010
Adoption of OI strategies and technologies (and the impact at the ground reality)	Bougrain & Haudeville, 2002; Izushi, 2003; De Jong & Marsili, 2006; Dickson, Weaver & Hoy, 2006; Ferneley & Bell, 2006; O´Regan, Ghobadian & Sims, 2006; Laforet, 2008; Van de Vrande et al., 2009; Leiponen & Byma, 2009; Zeng, Xie & Tam, 2010; Lichtenthaler, 2010; Lee & Lan, 2011; Mention, 2010; Hjalager, 2010; O'Regan, De Jong & Hippel, 2009; Spithoven, Clarysse and Knockaert, 2010
Measuring the impact of OI strategies (and taking actions to improve the situation)	Huang, Soutar & Brown 2004; Massa & Testa, 2008; Woodhams & Lupton, 2009; Grupp & Schubert, 2010; Liao and Rice, 2010; Raymond and St-Pierre, 2010; Caetano and Amaral, 2011; Gredel, Kramer and Bend, 2012; Fu, 2012; Nunes, Serrasqueiro and Leitão, 2012
Development of tools or instruments based on OI strategies (and piggybacking on any existing tools or instruments that are available in the market)	Kaufmann & Tödtling, 2002; Kohn & Hüsig, 2006; Descotes, Walliser, Holzmüller and Guo, 2011; Ramos, Acedo and Gonzalez, 2011; Hervas-Oliver, Garrigos and Gil-Pechuan, 2011; Kang and Park, 2012; Love and Ganotakis, 2012;

Researches and Researchers: Research Themes

Ranging from the conceptualization, policy initiation, and establishment of research model to development of business model, adaptation of strategies, measurement of the impact and development of tools for use and dissemination of the strategies, this study emphasizes on seven distinct research themes. One may argue about this setting of the research theme, but this study observed that without an appropriate conceptualization of the innovation process (and certainly without the involvement of the policy initiators), it cannot be established, and similarly next stages are invariably dependent (even inter-dependent within the internal processes) on the previous stages, such as without learning about the opportunities and challenges behind the strategies, the methodologies cannot be adopted, or without learning about the impact of the open innovation strategies, the tools cannot be developed, and so forth. Another school of thoughts could be how much these themes are relevant or appropriate to the SMEs. These researchers argue that adoption of open innovation strategies to small scale enterprises are yet to reach the level of maturity in even developed nations or nations who are leaders in doing so, hence these themes are also need to be investigated extensively at the grass roots level and these call for comprehensive research, especially in marginal or critically affected socio-economic environments.

Conceptualization of Open Innovation Strategies in Entrepreneurship Development

Though the concept of open innovation was in the market for many years before the term was newly introduced and popularized by Prof. Henry Chesbrough in 2003, this study finds many researchers are still researching on modernizing the concepts of open innovation strategies or advancing the degree of novelty of innovation. With the adoption of OI strategies in the entrepreneurship development, many researchers have initiated extended studies to reach out to the market by breaking the boundary of the firm. Terms like, innovation merchants, innovation architects, innovation missionaries, innovation intermediaries, business angels, or outsourcing R&Ds, use of venture capital, technology intelligence, licensing management, intellectual property management started gaining their acceptance and popularity to fill an ever-existing gap between the producer and the user (Chesbrough, 2003a; Lee, Park, Yoon & Park, 2010; Savioz & Blum, 2002; Amara, Landry, Becheikh & Ouimet, 2008). Apart from the mentioned terms, such as 'Users-as-innovators', 'customers-as-innovators', 'suppliers-as-innovators' (von Hippel 1986; 1988), 'networked coordination' (Powell, 1990), 'co-opetition' (Brandeburger & Nalebuff 1996), 'communities of practice' (Wenger 1998), and the 'private-collective innovation model' (Von Hippel & Von Krogh 2003), there are many other visible concepts that have been observed and they need to be scrutinized according to their importance to contribute for the entrepreneurship development. At the same time the research dimension has also been shifted from 'closed boundaries to networked paradigm' (Livieratos & Papoulias, 2009; Rahman & Ramos, 2010), which need extended investigation, especially at the local level of the value chain.

It has also been observed that due to lack of appropriate interventions from the national level, open innovation strategies are not being flourished at the local level, in spite of several distinct regional and global agencies are acting in this arena. Furthermore, due to lack of ability to access external resources with minimal technology assets (Narula, 2004), more inclined towards external intervention than engage themselves within to look for external sources (Edwards, Delbridge and Munday, 2005), lack of knowledge in open innovation acquisition process (Vanhaverbeke and Cloodt, 2006) keep them away from open innovation incentives at the early stage of their development.

Establishment of Research Model

With the term 'open innovation' is being popularized, majority of the researchers are trying to launch various research models, especially incorporating the role of external partners (SMEs, academia, research house, universities and intermediaries) to achieve improved performance and efficiency on product, process, service and organizational innovation. Along the path to the paradigm shift, as indicated by majority of the researchers, various research models have tried to validate the role of internal R&D, effect of the firm's size, the linkage between R&D activities and innovation exposition, and looked into various channels of open innovation thus mainstreaming open innovation research in the entrepreneurships through illustrating the conceptual arguments and conceptual frameworks incorporating various drivers of innovativeness (Chesbrough, 2003b; 2006; Raymond & St-Pierre, 2010; Edwards, Delbridge & Munday, 2005; Thorgren, Wincent & Örtqvist, 2009; Lawson, Longhurst & Ivey, 2006; Major & Cordey-Hayes, 2000; Rhee, Park & Lee, 2010).

Furthermore, SME networking and alliances have attracted considerable research attention based on collaboration models such as, bi-firm networks and include alliances with and outsourcing to other firms (Lee et al., 2010). Researchers also claim that SMEs are more flexible to adopt to open innovation due to their ability to utilize external networks more efficiently (Rothwell and Dodgson, 1994). More importantly, inter-firm collaboration is particularly important for SMEs with their limited technology assets (Lichtenthaler, 2005). In this aspect, Lee et al. (2010) proposes a collaborative model based on three distinct patterns, such as outsourcing depending on customer and supplier to explore funding and licensing; partnership depending on strategic alliance to explore joint-venture and R&D partnership; and networking depending on inter-firm alliance to explore networking and collaboration.

Development of Business Model

Since the mid-1980s a novel systemic model of innovation has emerged by incorporating a number of factors, such as externality, transferability, modularity, and network structure, which are not included in the previously dominant linear model (Livieratos & Papoulias, 2009). In this respect, innovation is considered as a systemic, path dependent and knowledge-centric social process influenced by the institutional environment (Chesbrough, 2003a). However, Livieratos and Papoulias (2009) argue that, within the open innovation model the innovation process becomes more complex and fragmented, actors are increasingly heterogeneous and more interdependent, and the period from conceptualization to commercialization is shorter (shorter life time cycle). Livieratos & Papoulias (2009) further argue that, this model has created porous boundaries between the an innovative firm and its surrounding environment by changing the inter- and intra-organizational modes of coordination and triggering new answers to Coase's (1937) question as to, 'what determines the boundaries of the organization'.

Similar to finding an integrated open innovation research model, there is a lack in finding an integrated open innovation business model. Chesbrough (2003a) while introducing this open innovation model, however, argued on the fact that larger firms have better capability to develop and commercialize technologies internally, though due to labor mobility, abundant venture capital, and widely dispersed knowledge across industries lead them to go beyond their peripheries. Hence, open innovation business model adopts both internal and external pathways to exploit technologies and at the same time acquire knowledge from outside (van de Vrande et al., 2009; Lee et al., 2010).

Lee et al. (2010) mentioned about various business models according to their nature, such as product innovation, process innovation, radical innovation, incremental innovation, systemic innovation, component innovation, technology-push and market-pull, including closed innovation and open innovation. Lee et al. (2010) also referred to other business models according to their innovation processes (such as, linear models, or chain-linked models), or according to the fitness for developed or developing countries. Hence, open innovation business models no longer remain restricted by the simple boundary of its own firm, or inter-intra-connected firms.

While investigating the existence and the performance of an Open Regional Innovation System (ORIS model) characterized by the firms' adoption of an open innovation strategy Belussi, Sammarra & Sedita (2010) argue that in terms of adopting open innovation strategies, it overcomes not only the boundaries of the firms, but also the boundaries of the region. Furthermore, Damaskopoulos & Evgeniou (2003) find that open innovation strategy based business models adopt frameworks comprising three interrelated levels of analysis, such as the level of the firm, the level of the market and industrial structures and the regulatory environment.

Literatures portrayed adoption of framework or business model incorporating OI strategies, such as Triple Helix or ORIS, Open Business Model articulating value creation or value addition by emphasizing the role of social capital (Wang, Jaring & Wallin, 2009; West, 2006; Chesbrough, 2006; De Jong, 2006; Partanen, Möller, Westerlund, Rajala & Rajala, 2008; Freel & De Jong, 2009; Cooke, 2005; Belussi, Sammarra & Sedita, 2010; Mortara and Minshall, 2011).

Opportunities and Challenges on Adopting Open Innovation Strategies

Among many opportunities the availability of public research seems to be the best opportunity factor for the SMEs, as this not only provides chances for innovation, but also produces a pool of experts (researchers, academics, staffs, and consulting firms) with strongly localized knowledge appropriation (Autant-Bernard, Fadairo and Massard, 2012). However, creation and administration of an effective innovative network often remain as critical challenges for SMEs (Lazzarotti, Manzini & Pizzurno, 2008). It has been observed that SMEs are more open to open innovation due to their limited size and resources. At the same time, intense competition and more demanding customers are also tend to be the major motivation for open innovation. But, the most important bottleneck for open innovation is differences in organization and culture between the individual partners (De Jong, 2006).

While researchers are carrying out researches to develop innovation opportunity framework (Levy, Powell & Galliers, 1999; Wang, Jaring & Wallin, 2009; Rahman & Ramos, 2010), but at the same time, the review observes that some other researchers are finding open innovation as challenges for the SMEs development. Groen & Linto (2010) raised a challenge, as whether the term open innovation hindering any growth in research and understanding about the entrepreneurship, and if so should the term be used as it is currently, or be it renamed? Knudsen & Mortensen (2010) chart an unnoticed theme in the current debate on open innovation, as a foundational question, as whether increasing openness is beneficial at all. They further investigate that, with increasing degrees of openness the product development projects are slower than the average in the industry, and these projects are slower than what is usual for the firm's traditional projects and had higher cost than the average in the industry and the firm's usual projects, as well.

In terms of capacity development, Hoffman, Parejo, Bessant & Perren (1998) mention that, despite the strong commitment to support innovation within SMEs at both regional and local level, the actual processes whereby small firms undertake innovative activity remain unclear at the grass roots. Further in this context, Tödtling & Trippl (2005) argue that, there is no "ideal model" or "generic model" so far to be adopted within the innovation policy as innovation activities differ strongly between central, peripheral and old industrial areas. Foremost, the context dependency of open innovation has found to be one of the least understood topics in this arena and further research is needed on the internal and external environment characteristics affecting the overall performance of an organization (Huizingh, 2011).

Strategies and Technologies

During the recent economic crisis, many firms are attempting to capture additional value from their technologies through utilization of open innovation strategies (Lichtenthaler, 2010c). Majority of the research documents are found to be discussing on adoption of open innovation strategies in the form of practices or applications incorporating innovation technologies. These strategies include inter-firm cooperation, cooperation with intermediary institutions, and cooperation with research organization (Zeng, Xie & Tam, 2010; Izushi, 2003; Leiponen & Byman, 2009; Mention, 2010); changes in management attitude, planning and external orientation (De Jong & Marsili, 2006); technology licensing (Teece, 2006; Huston and sakkab, 2006); technology-product integration (Caetano and Amaral, 2011); linking technology intelligence to open innovation (Veugelers, Bury and Viaene, 2010); creation of absorptive capacity (Wang, Vanhaverbeke and Roijakkers, 2012); venture capital investment (Ferrary, 2010); R&D outsourcing through acquisition strategy (Ferrary, 2011) and enhanced R&D alliances (Dickson, Weaver & Hoy, 2006). It has been observed that these strategies are providing significant impact on the innovation performances for SMEs. There are researches on the introduction of tools, such as, bricolage (Ferneley & Bell, 2006); or taxonomies, like, fruit flies approach (De Jong & Marsili, 2006); or terminologies, like, technology exploitation (van de Vrande, de Jong, Vanhaverbeke, & Rochemont, 2009; Lichtenthaler, 2010c) or technology exploration (van de Vrande et al., 2009).

In these contexts, O'Regan, Ghobadian and Sims (2006) emphasize that close association between strategy, organizational culture, leadership and novelty plays important role in achieving successful innovation. However, Lee & Lan (2011) argue that adoption of knowledge management is becoming an emerging agenda in developing company strategies, and thus leading towards the knowledge based economy. They further argue that, implementation depends on a harmonious amalgamation of infrastructure and process capabilities, including technology, culture and organizational formation. Furthermore, based on a random sample of 500 South Yorkshire non-hi-tech manufacturing SMEs, Laforet (2008) finds that the size, strategy and market orientation are also associated with innovation. Hence, strategies and technologies behind the open innovation are getting complex, dynamic and time dependent.

Measuring the Impact of OI Strategies

Strategies to optimize investments in the recognized technologies becomes of vital importance if an organization would like to match knowledge and ideas that are originating from outside of the organization but with internal core competences (Veugelers, Bury and Viaene, 2010). However, as the OI strategies are being increasingly adopted at all levels of the entrepreneurships, especially for the SMEs, researchers engaged themselves in finding the ultimate benefits of their utilization by measuring their impact. Massa and Testa (2008) points out to various indicators of the innovation measurements. Rejeb, Morel-Guimarães, Boly and Assiélou (2009) argue that innovation, being a competitive economic factor, is a process that compels a continuous, evolving and matured management. Therefore, innovative companies need to measure their innovation capacity as they grow further. In this aspect, innovation indicators have been used largely by "innovation scholars", a community that comprised of researchers from a variety of disciplines (ranging from engineering, and information science to sociology, and political science), who have a common research focus on technological innovation (Grupp & Schubert, 2010). However, Huang, Soutar and Brown (2004) indicated that four major factors underline the commonly used success measurement in an organization, such as financial performance, objective market acceptance, subjective market acceptance and product-level measures. They also mention that these four factors are interrelated and can be used as well to predict the overall measurement.

Tools and Instruments

Regarding use of software in the innovation process for SMEs, Kohn & Hüsig (2006) during their investigation have found that a large variety of software products are available in the market. Their research, while trying to address the issue of how far these products are specifically being used in practice, they find that these software products are rarely used to support the innovation process in German SMEs. Kaufmann & Tödtling (2002) mention that, the problem that most SMEs hardly interact with knowledge providers from outside the business sector (for example, universities, research houses or intermediaries) and the interaction is not reduced by the support instruments. Further, SMEs perform insufficiently the function of interfaces to innovation-related resources and information from outside the environment. Kaufmann & Tödtling (2002) also argue that, there is a lack of proactive consultancy concerning strategic, organizational and technological weaknesses which is necessary because most of the time, the firms are not aware of such deficiencies within themselves.

Caetano and Amaral (2011) mention about technology road mapping (TRM) as a method that assists organizations plan their technologies by describing the path to be followed in order to integrate a given technology into products and services. However, they state that other road mapping methods that are found in the literature were created to suit the large corporations, which combine R&D and product development structures, such as organizations that mainly adopt the market pull strategy and closed innovation to define technologies to be developed based on very specific market needs. A generic search on Google with string "open innovation tool" yields over 88,000 counts. However, a few are being cited here after going through their home pages. A list of those tools is being given below:

- innogetCloud[2]:- Mentioned as an open innovation tool that is marketed branding as "Software as a Service" (SaaS). It has been stated that using innogetCloud, organizations are able to build an open innovation marketplace where the members can interact and collaborate by posting technology offers and requests;
- Qmarkets[3]:- Described as a idea management software that allows their users to effectively manage their innovation process end to end - from idea generation and collection to idea selection and implementation;
- IdeaNet[4]:- Indicated to provide a combination of software tools for innovation and consultants by facilitating idea challenges, problem solving challenges and knowledge challenges to generate vibrant innovation and knowledge communities; and

- Innovator[5]:- Mentioned as an Enterprise Management System with enterprise innovation and intellectual property management solution to enhance idea management, decision management, invention disclosure/IP management and innovation and IP security;

Furthermore, there exist companies, like NineSigma, Ideaconnection, and Innocentive who are providing various supports to firms and individuals through open innovation tools and technologies.

However, in spite of the emergence of outsourcing R&D and it's emphasize by early introducers of open innovation in the entrepreneurship enhancement, this study does not find significant contributions from the researchers or practitioners illustrating the application of R&D outsourcing for SMEs development. Outsourcing R&D is mainly adopted by corporate houses. Hence, there is a visible gap in applying OI strategies in practice, especially for SMEs. Van de Vrande et al. (2009) conducted a study, which they claim as the first, explorative one to address this gap by focusing on SMEs. Their study tried to measure the extent of application of OI practices by SMEs and find out whether there is any positive trend on adoption of OI model over the time. Along this route, this study has tried to the above mentioned seven themes that have been found to be distinct and visible, and it is expected that future research on these perspectives will promote open innovation practices among SMEs. Next a few future research hints have been discussed for further dialogues and debates.

FUTURE RESEARCH

Within the research areas of open innovation in SMEs, awareness on the effective utilization of OI strategies, the timely intervention of opportunities for technologies outside the core business process, and skilled management of the OI strategies are essential for the successful entrepreneurship. However, these are particularly challenging for SMEs due to their lack of specialized knowledge base and also limited financial resources that can be devoted to innovation activities (Bianchi, Campodall'Orto, Frattini & Vercesi, 2010).

Furthermore, in spite of the enormous growth in research on open innovation, there are several openings of further research that this study has observed; such as linking open innovation research with other management areas, like marketing, human resources, change process, product diversification, and especially intellectual property management (Van de Vrande, Vanhaverbeke & Gassman, 2010; Rahman and Ramos, 2012). In addition to those factors, to match the global demand, increased competition and increasing supply of innovation, businesses need to internationalize their innovation activities through collaboration with external partners, such as customers, suppliers, universities, research houses and intermediaries (De Backer & Cervantes, 2008). Foremost, the understanding of open innovation at the outermost periphery of the business chain, driving factors of global innovation networking across different SMEs sub-sectors, accessibility and relationship of open innovation strategies with the implementing firms deserves further attention and in-depth research.

As mentioned in the problem statement of this study, and reiterated within the texts in the literature review, it is worthy to mention for the sake of future research that the concentration on open innovation researches are primarily focused to corporate businesses or large entrepreneurships and it is yet unclear whether these findings can be generalized to SMEs (Pedersen, Sondergaard & Esbjerg, 2009). Following a few success cases, small sized pilot projects may be initiated in a few backlogging countries with similar social and economic patterns to learn about the incubation of OI strategies before making a further leap to standardize a common platform for a larger number of countries within a region, or prescribe a generalized model at the global level.

Based on the study findings, this research sets out a future research framework involving products, processes, services and organizational transformations. Table 4 summarizes the framework.

Table 4.

Future Research Framework
- need to improve the current understanding of open innovation in SMEs, especially while adopting the different strategies
- need to focus on the nature of innovation and the extent to which open innovation is embedded in SMEs
- should investigate how entrepreneurs engage in open innovation during their growth phases, and particularly what managerial implication can be derived
- need to find out the characteristics of SMEs that are more likely to get benefit from collaboration, particularly via an intermediary, whether it could be a non-profit, or not-for-profit entity
- should be carried out by minimizing the screening questions which may opt out, especially start-ups and micro-enterprises, as these enterprises have been identified as the sources of breakthrough innovations and challengers of contemporary innovation actors
- should attempt to survey open innovation in broader samples of enterprises in more detailed and exploratory way
- should focus on the requirement of OI on differences in culture, structure and decision making among partners of different sizes and sectors
- may incorporates findings of OI strategies in improving innovation cooperation for SMEs in emerging economies and developing countries and extend the generalizations of the findings
- need to focus in identifying different segments within the population of every stakeholder (the entrepreneurs could be segmented by industry or geographic location and the academics by discipline)
- need to pay more attention to the outflows of knowledge, which is intellectual property management
- should emphasize on National Innovation System (NIS) inter-relating the national, economic, institutional and social environment
- should restructure to establish the communication channels in facilitating knowledge sharing with the external entities such as business partners and government agencies

Ref: Massa & Testa, 2008; Van de Vrande, de Jong, Vanhaberbeke & Rochemont, 2009; Lee, Park, Yoon & Park, 2010; Zeng, Xie & Tam, 2010; Rahman and Ramos, 2010; Lee and Lan, 2011; Samara, Georgiadis and Bakouros, 2012

CONCLUSION

In the recent economic crisis, many firms attempt to capture additional value by adopting open innovation strategies (Lichtenthaler, 2010c). It has also been observed that the small firms that do innovate are among those who could successfully increase their chances of survival (Cofis and Marsili, 2003) and growth (De Jong, Vermeulen & O'Shaughnessy, 2004). However, the behavior of small firms can vary substantially. Some small firms survive by competing in a market niche, while others pursue more radical innovations and eventually, become the market leaders (De Jong & Marsili, 2006).

This study finds the evidence of clustered researches in various segments of SMEs sectors (especially, manufacturing), but likes to deduce that a continued research covering major categories of SMEs sectors (such as service, process, or product) is a demanding field of open innovation. Along the study, the significant research themes have been explored, including their nature, number, exposition and effectiveness. The study concludes that a huge research gap exists at the periphery of the entrepreneurship where most of the businesses nourish among the developed economies, which are the SMEs. The study also concludes that there is a broadening gap in terms of applying results of empirical researches in the form of practices aiming at SMEs in the arena of open innovation. This study has already conducted a test survey among a few selected Portuguese SMEs (due to the nature of the title of the chapter, it has not been include here, but it forms a separate chapter in this book) and finds that majority of the surveyed SMEs are adopting OI strategies just by replicating others, but not solely dependent on any concrete empirical researches. This validates the conclusion about further enhanced researches among the peripheral SMEs on the adoption, utilization, and impact of open innovation.

REFERENCES

Amara, N., Landry, R., Becheikh, N., & Ouimet, M. (2008). Learning and novelty of innovation in established manufacturing SMEs. *Technovation*, *28*(7), 450–463. doi:10.1016/j.technovation.2008.02.001

Autant-Bernard, C., Fadairo, M. & Massard, N. (2012). *Knowledge diffusion and innovation policies within the European regions: Challenges based on recent empirical evidence*. Original Research Article, Research Policy, In Press, Corrected Proof, Available online 10 August 2012. doi:10.1016/j.respol.2012.07.009

Belussi, F., Sammarra, A., & Sedita, S. R. (2010). Learning at the boundaries in an "Open Regional Innovation System": A focus on firms' innovation strategies in the Emilia Romagna life science industry. *Research Policy*, *39*(6), 710–721. doi:10.1016/j.respol.2010.01.014

Bianchi, M., Campodall'Orto, S., Frattini, F., & Vercesi, P. (2010). Enabling open innovation in small- and medium-sized enterprises: How to find alternative applications for your technologies. *R & D Management*, *40*(4), 414–431. doi:10.1111/j.1467-9310.2010.00613.x

Bougrain, F., & Haudeville, B. (2002). Innovation, collaboration and SMEs internal research capacities. *Research Policy*, *31*(5), 735–747. doi:10.1016/S0048-7333(01)00144-5

Brandenburger, A., & Nalebuff, B. (1996). *Co-Opetition*. Doubleday.

Brunswicker, S., Bertino, E., & Matei, S. (2015). Big data for open digital innovation–a research roadmap. *Big Data Research, 2*(2), 53-58.

Caetano, M., & Amaral, D. C. (2011, July). Original Research Article. *Technovation*, *31*(7), 320–335. doi:10.1016/j.technovation.2011.01.005

Chesbrough, H. W. (2003a). The era of open innovation. *MIT Sloan Management Review*, *44*(3), 35–41.

Chesbrough, H. W. (2003b). *Open Innovation: The New Imperative for Creating and Profiting from Technology*. Harvard Business School Press.

Chesbrough, H. W. (2006). *Open Business Models: How to Thrive in the New Innovation Landscape*. Harvard Busines School Press Books.

Chesbrough, H. W., Vanhaverbeke, W., & West, J. (2006). *Open Innovation: Researching a new paradigm*. Oxford University Press.

Chesbrough, H. W., & Bogers, M. (2014). Chapter 1: Explicating Open Innovation: Clarifying an Emerging Paradigm for Understanding Innovation. In New Frontiers in Open Innovation. Oxford University Press.

Coase, R. (1937). The Nature of the Firm. *Economica*, *4*(16), 386–405. doi:10.1111/j.1468-0335.1937.tb00002.x

Cofis, E., & Marsili, O. (2003). *Survivor: The role of innovation in firm's survival, No. 03-18. WPT. Koopmans Institute, USE*. Utrecht University.

Cooke, P. (2005). Regionally asymmetric knowledge capabilities and open innovation: Exploring 'Globalisation 2'—A new model of industry organisation. *Research Policy*, *34*(8), 1128–1149. doi:10.1016/j.respol.2004.12.005

Dahlander, L., & Gann, D. M. (2010). How open is innovation? *Research Policy*.

Damaskopoulos, P., & Evgeniou, T. (2003). Adoption of New Economy Practices by SMEs in Eastern Europe. *European Management Journal*, *21*(2), 133–145. doi:10.1016/S0263-2373(03)00009-4

Davenport, T. H. (1993a). *Process Innovation*. Harvard Business School Press.

Davenport, T. H. (1993b). Process Innovation: reengineering work through information technology, Ernst & Young, Harvard Business School Press.

Davenport, T. H. (1994). Managing in the New World of Process. *Public Productivity & Management Review*, *18*(2), 133–147. doi:10.2307/3380643

De Backer, K., & Cervantes, M. (2008). *Open innovation in global networks*. OECD.

Dehoff, K. & Sehgal, V. (2006) Innovators without Borders. *Strategy+Business*.

Dehoff, K. & Sehgal, V. (2008). Beyond Borders: The Global Innovation 1000. *Strategy+Business*.

Del Brío, J. Á., & Junquera, B. (2003, December). A review of the literature on environmental innovation management in SMEs: Implications for public policies. *Technovation*, *23*(12), 939–948. doi:10.1016/S0166-4972(02)00036-6

De Jong, J. P. J. (2006). *Open Innovation: Practice, Trends, Motives and bottlenecks in the SMEs [Meer Open Innovatie: Praktijk, Ontwikkelingen, Motieven en Knelpunten in het MKB]*. EIM.

De Jong, J.P.J., Vermeulen, P.A.M. & O'Shaughnessy, K.C. (2004) Effects of Innovation in Small Firms. *M & O*, *58*(1), 21-38.

De Jong, J. P. J., & Marsili, O. (2006). The fruit flies of innovations: A taxonomy of innovative small firms. *Research Policy*, *35*(2), 213–229. doi:10.1016/j.respol.2005.09.007

De Jong, J. P. J., Vanhaverbeke, W., Kalvet, T., & Chesbrough, H. (2008). Policies for Open Innovation: Theory, Framework and Cases. Research project funded by VISION Era-Net.

De Jong, J. P. J., & von Hippel, E. (2009, September). Transfers of user process innovations to process equipment producers: A study of Dutch high-tech firms. *Research Policy*, *38*(7), 1181–1191. doi:10.1016/j.respol.2009.04.005

Descotes, R. M., Walliser, B., Holzmüller, H., & Guo, X. (2011, December). Original Research Article. *Journal of Business Research*, *64*(12), 1303–1310.

Dickson, P. H., Weaver, K. M., & Hoy, F. (2006, July). Opportunism in the R&D alliances of SMES: The roles of the institutional environment and SME size. *Journal of Business Venturing*, *21*(4), 487–513. doi:10.1016/j.jbusvent.2005.02.003

Edwards, T., Delbridge, R., & Munday, M. (2005). Understanding innovation in small and medium-sized enterprises: A process manifest. *Technovation*, *25*(10), 1119–1127. doi:10.1016/j.technovation.2004.04.005

European Union. (2005). *EU-Summary*. The joint Japan-EU Seminar on R&D and Innovation in Small and Medium size Enterprises (SMEs), Tokyo.

Ferrary, M. (2011). Specialized organizations and ambidextrous clusters in the open innovation paradigm. *European Management Journal*, *2011*(29), 181–192. doi:10.1016/j.emj.2010.10.007

Ferneley, E., & Bell, F. (2006). Using bricolage to integrate business and information technology innovation in SMEs. *Technovation*, *26*(2), 232–241. doi:10.1016/j.technovation.2005.03.005

Fredberg, T., Elmquist, M., & Ollila, S. (2008). Managing Open Innovation: Present Findings and Future Directions. Report VR 2008:02, VINNOVA - Verket för Innovationssystem/Swedish Governmental Agency for Innovation Systems.

Freel, M., & De Jong, J. P. J. (2009). Market novelty, competence-seeking and innovation networking. *Technovation*, *29*(12), 873–884.

Fu, X. (2012, April). How does openness affect the importance of incentives for innovation? Original Research Article. *Research Policy*, *41*(3), 512–523. doi:10.1016/j.respol.2011.12.011

Gassmann, O. (2006). Opening up the innovation process: Towards an agenda. *R & D Management*, *36*(3), 223–228. doi:10.1111/j.1467-9310.2006.00437.x

Gassmann, O., Enkel, E., & Chesbrough, H. (2010). The future of open innovation. *R & D Management*, *40*(3), 213–221. doi:10.1111/j.1467-9310.2010.00605.x

Gredel, D., Kramer, M., & Bend, B. (2012, September–October). Original Research Article. *Technovation*, *32*(9–10), 536–549. doi:10.1016/j.technovation.2011.09.008

Grindley, P. C., & Teece, D. J. (1997). Managing intellectual capital: Licensing and cross-licensing in semiconductors and electronics. *California Management Review*, *39*(2), 1–34. doi:10.2307/41165885

Groen, A. J., & Linton, J. D. (2010). Is open innovation a field of study or a communication barrier to theory development? *Technovation*, *30*(11-12), 554. doi:10.1016/j.technovation.2010.09.002

Grupp, H., & Schubert, T. (2010, February). Review and new evidence on composite innovation indicators for evaluating national performance. *Research Policy*, *39*(1), 67–78. doi:10.1016/j.respol.2009.10.002

Huang, X., Soutar, G. N., & Brown, A. (2004). Measuring new product success: An empirical investigation of Australian SMEs. *Industrial Marketing Management*, *33*(2), 117–123. doi:10.1016/S0019-8501(03)00034-8

Izushi, H. (2003). Impact of the length of relationships upon the use of research institutes by SMEs. *Research Policy*, *32*(5), 771–788. doi:10.1016/S0048-7333(02)00085-9

Henkel, J. (2006). Selective revealing in open innovation processes: The case of embedded Linux. *Research Policy*, *35*(7), 953–969. doi:10.1016/j.respol.2006.04.010

Hervas-Oliver, J. L., Garrigos, J. A., & Gil-Pechuan, I. (2011, September). Original Research Article. *Technovation*, *31*(9), 427–446. doi:10.1016/j.technovation.2011.06.006

Hjalager, A.-M. (2010, February). A review of innovation research in tourism. *Tourism Management*, *31*(1), 1–12. doi:10.1016/j.tourman.2009.08.012

Huang, X., Soutar, G.N. & Brown, A. (2004) Review and new evidence on composite innovation indicators for evaluating national performance. *Research Policy, 39*(1), 67-78.

Hoffman, K., Parejo, M., Bessant, J., & Perren, L. (1998). Small firms, R&D, technology and innovation in the UK: A literature review. *Technovation*, *18*(1), 39–55. doi:10.1016/S0166-4972(97)00102-8

Huizingh, E. K. R. E. (2009). The future of innovation. *The XX International Society for Professional Innovation Management - ISPIM Conference 2009. Proceedings*, 21.

Huston, L., & Sakkab, N. (2006). Connect and develop: Inside Procter & Gamble's new model for innovation. *Harvard Business Review*, *84*, 58–66.

Huizingh, E. K. R. E. (2011, January). Original Research Article. *Technovation*, *31*(1), 2–9. doi:10.1016/j.technovation.2010.10.002

Kang, K-N., & Park, H. (2012). Original Research Article. *Technovation*, *32*(1), 68–78.

Kaufmann, A., & Tödtling, F. (2002, March). How effective is innovation support for SMEs? An analysis of the region of Upper Austria. *Technovation*, *22*(3), 147–159. doi:10.1016/S0166-4972(00)00081-X

Kirschbaum, R. (2005). Open innovation in practice. *Research Technology Management*, *48*(4), 24–28. doi:10.1080/08956308.2005.11657321

Knudsen, M.P. & Mortensen, T.B. (2010). Some immediate – but negative – effects of openness on product development performance. *Technovation.*

Kohn, S., & Hüsig, S. (2006, August). Potential benefits, current supply, utilization and barriers to adoption: An exploratory study on German SMEs and innovation software. *Technovation*, *26*(8), 988–998. doi:10.1016/j.technovation.2005.08.003

Kovacs, A., Van Looy, B., & Cassiman, B. (2015). Exploring the scope of open innovation: A bibliometric review of a decade of research. *Scientometrics*, *104*(3), 951–983. doi:10.100711192-015-1628-0

Kowalski, S. P. (2009). *SMES, Open Innovation and IP Management: Advancing Global Development.* A presentation at the WIPO-Italy International Convention on Intellectual Property and Competitiveness of Micro, Small and Medium-Sized Enterprises (MSMEs), Rome, Italy.

Koyuncugil, A. S., & Ozgulbas, N. (2009). Risk modeling by CHAID decision tree algorithm. *ICCES*, *11*(2), 39–46.

Laforet, S. (2008). Size, strategic, and market orientation affects on innovation. *Technovation, 25*(10), 1119-1127.

Laursen, K., & Salter, A. (2006). Open for innovation: The role of openness in explaining innovation performance among UK manufacturing firms. *Strategic Management Journal, 27*(2), 131–150. doi:10.1002mj.507

Lawson, C. P., Longhurst, P. J., & Ivey, P. C. (2006). The application of a new research and development project selection model in SMEs. *Technovation, 26*(2), 242–250. doi:10.1016/j.technovation.2004.07.017

Lazzarotti, V., Manzini, R., & Pizzurno, E. (2008). Managing innovation networks of SMEs: a case study. *Proceeding of the International Engineering Management Conference: managing engineering, technology and innovation for growth (IEMC Europe 2008)*, 521-525. 10.1109/IEMCE.2008.4618024

Lecocq, X., & Demil, B. (2006). Strategizing industry structure: The case of open systems in low-tech industry. *Strategic Management Journal, 27*(9), 891–898. doi:10.1002mj.544

Lee, S., Park, G., Yoon, B., & Park, J. (2010). Open innovation in SMEs—An intermediated network model. *Research Policy, 39*(2), 290–300. doi:10.1016/j.respol.2009.12.009

Lee, M. R., & Lan, Y.-C. (2011, January). Toward a unified knowledge management model for SMEs. *Expert Systems with Applications, 38*(1), 729–735. doi:10.1016/j.eswa.2010.07.025

Leiponen, A., & Byma, J. (2009, November). If you cannot block, you better run: Small firms, cooperative innovation, and appropriation strategies. *Research Policy, 38*(9), 1478–1488. doi:10.1016/j.respol.2009.06.003

Lemola, T. & Lievonen, J. (2008). *The role of innovation policy in fostering open innovation activities among companies*. Vision ERAnet.

Levy, M., Powell, P., & Galliers, R. (1998). Assessing information systems strategy development frameworks in SMEs. *Information & Management, 36*(5), 247–261. doi:10.1016/S0378-7206(99)00020-8

(2010, February). Liao, Tung-Shan & Rice, J. (2010). *Original Research Article Research Policy, 39*(1), 117–125.

Lichtenthaler, U. (2005). External commercialization of knowledge: Review and research agenda. *International Journal of Management Reviews, 7*(4), 231–255. doi:10.1111/j.1468-2370.2005.00115.x

Lichtenthaler, U. (2006). Technology exploitation strategies in the context of open innovation, International Journal of Technology Intelligence and Planning, Volume 2. *Number, 1/2006*, 1–21.

Lichtenthaler, U. (2007a). Hierarchical strategies and strategic fit in the keep-or-sell decision. *Management Decision, 45*(3), 340–359. doi:10.1108/00251740710744990

Lichtenthaler, U. (2007b). The drivers of technology licensing: An industry comparison. *California Management Review, 49*(4), 67–89. doi:10.2307/41166406

Lichtenthaler, U. (2008a). Open Innovation in Practice: An Analysis of Strategic Approaches to Technology Transactions. *Engineering Management, IEEE Transactions on, 55*(1), 148 – 157.

Lichtenthaler, U. (2008b, May/June). Integrated roadmaps for open innovation. *Research Technology Management, 51*(3), 45–49. doi:10.1080/08956308.2008.11657504

Lichtenthaler, U. (2008c, September). Relative capacity: Retaining knowledge outside a firm's boundaries. *Journal of Engineering and Technology Management, 25*(3), 200–212. doi:10.1016/j.jengtecman.2008.07.001

Lichtenthaler, U. (2008d, April). Leveraging technology assets in the presence of markets for knowledge. *European Management Journal, 26*(2), 122–134. doi:10.1016/j.emj.2007.09.002

Lichtenthaler, U. (2008e, July). Externally commercializing technology assets: An examination of different process stages. *Journal of Business Venturing, 23*(4), 445–464. doi:10.1016/j.jbusvent.2007.06.002

Lichtenthaler, U. (2008f, July). Externally commercializing technology assets: An examination of different process stages Original Research Article. *Journal of Business Venturing, 23*(4), 445–464. doi:10.1016/j.jbusvent.2007.06.002

Ulrich Lichtenthaler Lichtenthaler, U. (2009a, April). Retracted: The role of corporate technology strategy and patent portfolios in low-, medium- and high-technology firms . *Research Policy, 38*(3), 559–569. doi:10.1016/j.respol.2008.10.009

Lichtenthaler, U. (2009b, August). Absorptive capacity, environmental turbulence, and the complementarity of organizational learning processes. *Academy of Management Journal, 52*(4), 822–846. doi:10.5465/amj.2009.43670902

Lichtenthaler, U. (2009c, September). Outbound open innovation and its effect on firm performance: Examining environmental influences. *R & D Management, 39*(4), 317–330. doi:10.1111/j.1467-9310.2009.00561.x

Lichtenthaler, U. (2010a, July–August). Original Research Article. *Technovation, 30*(7–8), 429–435. doi:10.1016/j.technovation.2010.04.001

Lichtenthaler, U. (2010b, November). Organizing for external technology exploitation in diversified firms. *Journal of Business Research, 63*(11), 1245–1253. doi:10.1016/j.jbusres.2009.11.005

Lichtenthaler, U. (2010c, July-August). Technology exploitation in the context of open innovation: Finding the right 'job' for your technology. *Technovation, 30*(7-8), 429–435. doi:10.1016/j.technovation.2010.04.001

Lichtenthaler, U. (2011a, February–March). 'Is open innovation a field of study or a communication barrier to theory development?' A contribution to the current debate. *Technovation, 31*(2–3), 138–139. doi:10.1016/j.technovation.2010.12.001

Lichtenthaler, U. (2011b, February). Open Innovation: Past Research, Current Debates, and Future Directions. *The Academy of Management Perspectives, 25*(1), 75–93.

Lichtenthaler, U., & Ernst, H. (2009). Opening Up the Innovation Process: The Role of Technology Aggressiveness. *R & D Management, 39*(1), 38–54. doi:10.1111/j.1467-9310.2008.00522.x

Lindermann, N., Valcareel, S., Schaarschmidt, M., & Von Kortzfleisch, H. (2009). SME 2.0: Roadmap towards Web 2.0- Based Open Innovation in SME-Network- A Case Study Based Research Framework. CreativeSME2009, IFIP International Federation for Information Processing, 28-41.

Livieratos, A. D., & Papoulias, D. B. (2009). *Towards an Open Innovation Growth Strategy for New, Technology-Based Firms*. National Technical University of Athens. Retrieved October 30, 2010 from http://www.ltp.ntua.gr/uploads/GJ/eK/GJeKw8Jf5RqjCO9CWapm4w/Growth.pdf

Love, J.H. & Ganotakis, P. (2012). Original Research Article. *International Business Review*.

Lundvall, B. (1995). *National systems of innovation: Towards a theory of innovation and interactive learning*. Biddles Ltd.

Maijers, W., Vokurka, L., van Uffelen, R. & Ravensbergen, P. (2005). Open innovation: symbiotic network, Knowledge circulation and competencies for the benefit of innovation in the Horticulture delta. *Presentation IAMA Chicago 2005.*

Major, E. J., & Cordey-Hayes, M. (2000). Engaging the business support network to give SMEs the benefit of foresight. *Technovation*, *20*(11), 589–602. doi:10.1016/S0166-4972(00)00006-7

Massa, S., & Testa, S. (2008, July). Innovation and SMEs: Misaligned perspectives and goals among entrepreneurs, academics, and policy makers. *Technovation*, *28*(7), 393–407. doi:10.1016/j.technovation.2008.01.002

Mention, A-L. (2010). Co-operation and co-opetition as open innovation practices in the service sector: Which influence on innovation novelty? *Technovation*.

Mortara, L., & Minshall, T. (2011). How do large multinational companies implement open innovation? *Technovation*, *31*(10-11), 586–597. doi:10.1016/j.technovation.2011.05.002

Nahlinder, J. (2005). *Innovation and Employment in Services: The Case of Knowledge Intensive Business Services in Sweden* (Doctoral Thesis). Department of technology and Social Change, Linköping University, Sweden.

Nagaoka, S., & Kwon, H. U. (2006). The incidence of cross-licensing: A theory and new evidence on the firm and contract level determinants. *Research Policy*, *35*(9), 1347–1361. doi:10.1016/j.respol.2006.05.007

Narula, R. (2004). R&D collaboration by SMEs: New opportunities and limitations in the face of globalisation. *Technovation*, *25*(2), 153–161. doi:10.1016/S0166-4972(02)00045-7

Novelli, M., Schmitz, B., & Spencer, T. (2006, December). Networks, clusters and innovation in tourism: A UK experience. *Tourism Management*, *27*(6), 1141–1152. doi:10.1016/j.tourman.2005.11.011

Nunes,P.M., Serrasqueiro, Z. & Leitão, J. (2010). Is there a linear relationship between R&D intensity and growth? Empirical evidence of non-high-tech vs. high-tech SMEs. *Research Policy, 41*(1), 36-53.

OECD. (2008). *Open Innovation in Global Networks, Policy Brief, OECD Observer*. OECD.

O'Regan, N., Ghobadian, A., & Sims, M. (2006, February). Fast tracking innovation in manufacturing SMEs. *Technovation*, *26*(2), 251–261. doi:10.1016/j.technovation.2005.01.003

Partanen, J., Möller, K., Westerlund, M., Rajala, R., & Rajala, A. (2008, July). Social capital in the growth of science-and-technology-based SMEs. *Industrial Marketing Management*, *37*(5), 513–522. doi:10.1016/j.indmarman.2007.09.012

Pedersen, M., Sondergaard, H.A., & Esbjerg, L. (2009). Network characteristics and open innovation in SMEs. *Proceedings of The XX ISPIM Conference*.

Powell, W. (1990). Neither market nor hierarchy: network forms of organization. In B. Stow & L. L. Cummings (Eds.), *Research in Organizational Behavior*. JAI Press.

Rahman, H., & Ramos, I. (2010). Open Innovation in SMEs: From Closed Boundaries to Networked Paradigm. *Issues in Informing Science and Information Technology*, *7*, 471–487. doi:10.28945/1221

Rahman, H., & Ramos, I. (2012). Open Innovation in Entrepreneurships: Agents of Transformation towards the Knowledge-Based Economy. *Proceedings of the Issues in Informing Science and Information Technology Education Conference*.

Ramos, E., Acedo, F. J., & Gonzalez, M. A. (2011, October–November). Internationalisation speed and technological patterns: A panel data study on Spanish SMEs. *Technovation*, *31*(10–11), 560–572. doi:10.1016/j.technovation.2011.06.008

Rejeb, H. B., Morel-Guimarães, L., Boly, V., & Assiélou, N. G. (2009). Measuring innovation best practices: Improvement of an innovation index integrating threshold and synergy effects. *Technovation*, *28*(12), 838–854. doi:10.1016/j.technovation.2008.08.005

Raymond, L., & St-Pierre, J. (2010). R&D as a determinant of innovation in manufacturing SMEs: An attempt at empirical clarification. *Technovation*, *30*(1), 48–56. doi:10.1016/j.technovation.2009.05.005

Rhee, J., Park, T., & Lee, D. H. (2010). Drivers of innovativeness and performance for innovative SMEs in South Korea: Mediation of learning orientation. *Technovation*, *30*(1), 65–75. doi:10.1016/j.technovation.2009.04.008

Rothwell, R., & Dodgson, M. M. (1994). The Handbook of Industrial Innovation. Edward Elgar.

Roper, S., & Hewitt-Dundas, N. (2004). *Innovation persistence: survey and case - study evidence*. Working Paper, Aston Business School, Birmingham, UK.

Saarikoski, V. (2006). *The Odyssey of the Mobile Internet- the emergence of a networking attribute in a multidisciplinary study* (Academic dissertation). TIEKE, Helsinki.

Saebi, T., & Foss, N. J. (2015). Business models for open innovation: Matching heterogeneous open innovation strategies with business model dimensions. *European Management Journal*, *33*(3), 201–213. doi:10.1016/j.emj.2014.11.002

Samara, E., Georgiadis, P., & Bakouros, I. (2012). The impact of innovation policies on the performance of national innovation systems: A system dynamics analysis. *Technovation*, *32*(11), 624–638. doi:10.1016/j.technovation.2012.06.002

Santisteban, M. A. (2006). Business Systems and Cluster Policies in the Basque Country and Catalonia (1990-2004). *European Urban and Regional Studies*, *13*(1), 25–39. doi:10.1177/0969776406059227

Savioz, P., & Blum, M. (2002). Strategic forecast tool for SMEs: How the opportunity landscape interacts with business strategy to anticipate technological trends. *Technovation*, *22*(2), 91–100. doi:10.1016/S0166-4972(01)00082-7

Schumpeter, J. A. (1934). *The Theory of Economic Development*. Harvard University Press.

Schumpeter, J. A. (1942). *Capitalism, Socialism, and Democracy*. Harper & Row.

Schumpeter, J. A. (1950). *Capitalism, Socialism, and Democracy* (3rd ed.). Harper.

Sousa, M. C. (2008). Open innovation models and the role of knowledge brokers. *Inside Knowledge Magazine*, *11*(6), 1–5.

Spithoven, A., Clarysse, B., & Knockaert, M. (2010). Building absorptive capacity to organize inbound open innovation in traditional industries. *Technovation*, *30*(2), 130–141. doi:10.1016/j.technovation.2009.08.004

Teece, D. J. (2006). Reflections on “Profiting from Innovation”. *Research Policy*, *35*(8), 1131–1146. doi:10.1016/j.respol.2006.09.009

Tidd, J. (2014) Open innovation research, management and practice. Imperial College Press.

Thoben, K. D. (2008) A new wave of innovation in collaborative networks. *Proceedings of the 14th international conference on concurrent enterprising: ICE 2008*, 1091-1100.

Thorgren, S., Wincent, J., & Örtqvist, D. (2009). Designing interorganizational networks for innovation: An empirical examination of network configuration, formation and governance. *Journal of Engineering and Technology Management*, *26*(3), 148–166. doi:10.1016/j.jengtecman.2009.06.006

Tilley, F., & Tonge, J. (2003). Introduction. In O. Jones & F. Tilley (Eds.), *Competitive Advantage in SME's: Organising for Innovation and Change*. John Wiley & Sons.

Tödtling, F., & Trippl, M. (2005). One size fits all? Towards a differentiated regional innovation policy approach. *Research Policy*, *34*(8), 1203–1219. doi:10.1016/j.respol.2005.01.018

Trott, P., & Hartmann, D. A. P. (2009). Why ‘open innovation’ is old wine in new bottles. *International Journal of Innovation Management*, *13*(04), 715–736. doi:10.1142/S1363919609002509

Van de Vrande, V., de Jong, J. P. J., Vanhaverbeke, W., & de Rochemont, M. (2008). Open innovation in SMEs: Trends, motives and management challenges. A report published under the SCALES-initiative (SCientific AnaLysis of Entrepreneurship and SMEs), as part of the ‘SMEs and Entrepreneurship programme’ financed by the Netherlands Ministry of Economic Affairs, Zoetermeer.

Van de Vrande, V., de Jong, J. P. J., Vanhaverbeke, W., & de Rochemont, M. (2009, June-July). Open innovation in SMEs: Trends, motives and management challenges. *Technovation*, *29*(6-7), 423–437. doi:10.1016/j.technovation.2008.10.001

Van de Vrande, V., Vanhaverbeke, W., & Gassman, O. (2010). Broadening the scope of open innovation: Past research, current state and future directions. *International Journal of Technology Management*, *52*(3-4), 221–235. doi:10.1504/IJTM.2010.035974

Van Hemel, C., & Cramer, J. (2002, October). Barriers and stimuli for ecodesign in SMEs. *Journal of Cleaner Production, 10*(5), 439–453. doi:10.1016/S0959-6526(02)00013-6

Vanhaverbeke, W., & Cloodt, M. (2006). Open innovation in value networks. In H. Chesbrough, W. Vanhaverbeke, & J. West (Eds.), *Open Innovation: Researching a New Paradigm.* Oxford University Press.

Vanhaverbeke, W., Chesbrough, H., & West, J. (2014). Surfing the new wave of open innovation research. *New Frontiers in Open Innovation, 281*, 287-288.

Veugelers, M., Bury, J., & Viaene, S. (2010). Linking technology intelligence to open innovation. *Technological Forecasting and Social Change, 77*(2), 335–343. doi:10.1016/j.techfore.2009.09.003

Hippel, V. (1975). *The Dominant Role of Users in the Scientific Instrument Innovation Process, WP 764-75*. NSF.

Hippel, V. (1978, January). Successful Industrial Products from Customer Ideas. *Eric von Hippel Journal of Marketing, 42*(1), 39–49.

Von Hippel, E. (1986). Lead users: A source of novel product concepts. *Management Science, 32*(7), 791–805. doi:10.1287/mnsc.32.7.791

Von Hippel, . (1987). Cooperation between rivals: Informal Know-how trading. *Research Policy, 16*(6), 291–302. doi:10.1016/0048-7333(87)90015-1

Von Hippel, E. (1988). The Sources of Innovation. Oxford University Press.

Von Hippel. (1994). "Sticky Information" and the Locus of Problem Solving: Implications for Innovation. Management Science, 40(4), 429-439.

Von Hippel. (1998). Economics of product development by users: The impact of `sticky' local information. *Management Science, 44*(5), 629-644.

Hippel, V. (2001a, Summer). Innovation by User Communities: Learning from Open-Source Software. *MIT Sloan Management Review, 42*(4), 82–86.

Hippel, V. (2001b). Perspective: User toolkits for innovation. *Journal of Product Innovation Management, 18*(4), 247–257. doi:10.1111/1540-5885.1840247

von Hippel, E. (2005). Democratizing innovation: The evolving phenomenon of user innovation. *Journal für Betriebswirtschaft, 55*(1), 63–78. doi:10.100711301-004-0002-8

Hippel, V. (2007, April). 2007 Horizontal Innovation Networks—By and for Users. *Industrial and Corporate Change, 16*(2), 293–315. doi:10.1093/icc/dtm005

Hippel, V. (2010). Open User Innovation. Handbook of the Economics of Innovation, 1, 411-427.

Von Hippel, E., & Von Krogh, G. (2003). Open source software and the "private – collective" innovation model: Issues for organization science. *Organization Science, 14*(2), 209–223. doi:10.1287/orsc.14.2.209.14992

Wang, L., Jaring, P., & Wallin, A. (2009) Developing a Conceptual Framework for Business Model Innovation in the Context of Open Innovation. *Proceedings of the Third IEEE International Conference on Digital Ecosystems and Technologies (IEEE DEST 2009)*, 460-465.

Wang, Y., Vanhaverbeke, W., & Roijakkers, N. (2012). Exploring the impact of open innovation on national systems of innovation - A theoretical analysis. *Technological Forecasting and Social Change*, *79*(3), 419–428. doi:10.1016/j.techfore.2011.08.009

Wenger, E. (1998). *Communities of Practice: Learning, Meaning and Identity*. Cambridge University Press. doi:10.1017/CBO9780511803932

West, J. (2006). Does Appropriability Enable or Retard Open Innovation? In Open Innovation: Researching a New Paradigm. Oxford University Press.

West, J., Vanhaverbeke, W., & Chesbrough, H. (2006). Open Innovation: A Research Agenda. In Open Innovation: Researching a New Paradigm. Oxford University Press.

Westergren, U. H., & Holmström, J. (2012). Exploring preconditions for open innovation: Value networks in industrial firms. *Information and Organization*, *22*(4), 209–226. doi:10.1016/j.infoandorg.2012.05.001

Wijffels, H. (2009). Leadership, sustainability and levels of consciousness. A speech at the conference on the Leadership for a Sustainable World, Den Haag, The Netherlands.

Woodhams, C., & Lupton, B. (2009, June). Analysing gender-based diversity in SMEs. *Scandinavian Journal of Management*, *25*(2), 203–213. doi:10.1016/j.scaman.2009.02.006

Zeng, S. X., Xie, X. M., & Tam, C. M. (2010). Relationship between cooperation networks and innovation performance of SMEs. *Technovation*, *30*(3), 181–194. doi:10.1016/j.technovation.2009.08.003

ADDITIONAL READING

Alshawi, S., Missi, F. & Irani, Z. (2010) Organisational, technical and data quality factors in CRM adoption - SMEs perspective, *Industrial Marketing Management, In Press, Corrected Proof, Available online 15 September 2010.*

Bayraktar, E., Demirbag, M., Koh, S. C. L., Tatoglu, E., & Zaim, H. (2009, November). A causal analysis of the impact of information systems and supply chain management practices on operational performance: Evidence from manufacturing SMEs in Turkey. *International Journal of Production Economics*, *122*(1), 133–149. doi:10.1016/j.ijpe.2009.05.011

Bevis, K. I., & Cole, A. (2010) Open Innovation Readiness: a Tool, *ISPIM XXIth International Conference*, Bilbao, June 2010, International Society for Professional Innovation Management.

Bianchi, M., Cavaliere, A., Chiaroni, D., Frattini, F. & Chiesa, V. (2010) Organisational modes for Open Innovation in the bio-pharmaceutical industry: An exploratory analysis, *Technovation, In Press, Corrected Proof, Available online 25 March 2010*

Bidault, F. (2004). Global licensing strategies and technology pricing. *International Journal of Technology Management, 27*(2/3), 295–305. doi:10.1504/IJTM.2004.003959

Bougrain, F., & Haudeville, B. (2002, July). Original Research Article. *Research Policy, 31*(5), 735–747. doi:10.1016/S0048-7333(01)00144-5

BVCA (2005) *Creating success from University Spin-outs*, A Review conducted by Library House on behalf of the BVCA, BVCA (British Venture Capital Association)/ Library House

Chesbrough, H.W. (2003c) The Era of Open Innovation. *Sloan Management Review*, 44, 3 (Spring): 35-41.

Chesbrough, H.W. (2003d) "Open Innovation: How Companies Actually Do It," *Harvard Business Review*, 81, 7 (July): 12-14.

Chesbrough, H.W. (2003e) "Open Platform Innovation: Creating Value from Internal and External Innovation," *Intel Technology Journal*, 7, 3 (August): 5-9.

Chesbrough, H.W. (2004), "Managing Open Innovation: Chess and Poker," Research-Technology Management, 47, 1 (January): 23-26.

Chesbrough, H. W. (2006). Open Innovation: A New Paradigm for Understanding Industrial Innovation. In H. Chesbrough, W. Vanhaverbeke, & J. West (Eds.), *Open Innovation: Researching a New Paradigm* (pp. 1–12). Oxford University Press.

Chesbrough, H. W. (2006). New Puzzles and New Findings. In H. Chesbrough, W. Vanhaverbeke, & J. West (Eds.), *Open Innovation: Researching a New Paradigm* (pp. 15–34). Oxford University Press.

Chesbrough, H. W. (2006). *Open Business Models: How to Thrive in the New Innovation Landscape.* Harvard Business School Press.

Chesbrough, H. W., Vanhaverbeke, W., & West, J. (2006) (Eds.). Open innovation: Researching a new paradigm. Oxford University Press: London.

Chesbrough, H.W. (2007) "The market for innovation: implications for corporate strategy." California Management Review, 49, 3 (Spring): 45–66.

Chesbrough, H.W. and A. K. Crowther (2006) "Beyond high tech: early adopters of open innovation in other industries," R&D Management, 36, 3 (June): 229-236

Chesbrough, H. W. (2007). Business model innovation: It's not just about technology anymore. *Strategy and Leadership, 35*(6), 12–17. doi:10.1108/10878570710833714

Chiaroni, D., Chiesa, V., & Frattini, F. (2011). The Open Innovation Journey: How firms dynamically implement the emerging innovation management paradigm. *Technovation, 31*(1), 34–43. doi:10.1016/j.technovation.2009.08.007

Davenport, T. H., & Short, J. E. (1990, Summer). The New Industrial Engineering: Information Technology and Business Process Redesign. *Sloan Management Review*, •••, 11–27.

Davenport, T. H., & Stoddard, D. B. (1994, July). Reengineering: Business Change of Mythic Proportions? *Management Information Systems Quarterly, 18*(2), 121–127. doi:10.2307/249760

Davenport, T. H., & Beers, M. C. (1995). Managing Information About Processes. *Journal of Management Information Systems*, *12*(1), 57–80. doi:10.1080/07421222.1995.11518070

Davenport, T. H., & Harris, J. G. (2005). Automated decision making comes of age. *MIT Sloan Management Review*, *46*(4), 83–89.

Davenport, T.H. (1998) Putting the enterprise into the enterprise system, *Harvard Business Review*, July–August, 76(4): 121–131

EIRMA (2005) *Responsible Partnering: A Guide to Better Practices for Collaborative Research and Knowledge Transfer between Science and Industry*, EIRMA, EUA, EARTO, ProTon Europe, version 1.0 January 2005

EIRMA (2009) *Joining Forces in a World of Open Innovation: Guidelines for Collaborative Research and Knowledge Transfer between Science and Industry*, EIRMA, EUA, EARTO, ProTon Europe, version 1.1 October 2009

Ferrary, M. (2010). Syndication of venture capital investment: The art of resource pooling. *Entrepreneurship Theory and Practice*, *34*(5), 885–907. doi:10.1111/j.1540-6520.2009.00356.x

Fernández-Viñé, M. B., Gómez-Navarro, T., & Capuz-Rizo, S. F. (2010, May). Eco-efficiency in the SMEs of Venezuela. Current status and future perspectives. *Journal of Cleaner Production*, *18*(8), 736–746. doi:10.1016/j.jclepro.2009.12.005

Giannopoulou, E., Yström, A., Ollila, S., Fredberg, T., & Elmquist, M. (2010). Implications of openness: A study into (all) the growing literature on open innovation. *Journal of Technology Management & Innovation*, *5*(3), 162–180. doi:10.4067/S0718-27242010000300012

Kaivanto, K., & Stoneman, P. (2007, June). Public provision of sales contingent claims backed finance to SMEs: A policy alternative. *Research Policy*, *36*(5), 637–651. doi:10.1016/j.respol.2007.01.001

Kollmer, H., & Dowling, M. (2004). Licensing as a commercialisation strategy for new technology-based firms. *Research Policy*, *33*(8), 1141–1151. doi:10.1016/j.respol.2004.04.005

Kornai, J. (2010) *Innovation and Dynamism: Interaction between Systems and Technical Progress*, Working Paper No. 2010/33, UNU-WIDER

Kühne, B., Vanhonacker, F., Gellynck, X., & Verbeke, W. (2010, September). Innovation in traditional food products in Europe: Do sector innovation activities match consumers' acceptance? *Food Quality and Preference*, *21*(6), 629–638. doi:10.1016/j.foodqual.2010.03.013

Jones, O. & Macpherson, A. (2006) Inter-organizational Learning and Strategic Renewal in SMEs: Extending the 4I Framework, *Long Range Planning*, 39:155-175 (33)

Kohn, S., & Hüsig, S. (2006, August). Original Research Article. *Technovation*, *26*(8), 988–998. doi:10.1016/j.technovation.2005.08.003

Raymond, L., & St-Pierre, J. (2010, January). Original Research Article. *Technovation*, *30*(1), 48–56. doi:10.1016/j.technovation.2009.05.005

Spithoven, A., Clarysse, B., & Knockaert, M. (2010). Building absorptive capacity to organise inbound open innovation in traditional industries. *Technovation*, *30*(2), 130–141. doi:10.1016/j.technovation.2009.08.004

Tilley, F., & Tonge, J. (2003). Introduction. In O. Jones & F. Tilley (Eds.), *Competitive Advantage in SME's: Organising for Innovation and Change* (pp. 1–12). John Wiley & Sons.

Vanhaverbeke, W. (2006). The interorganizational context of open innovation. *Open innovation: Researching a new paradigm*, 205-219.

Ziedonis, A. A. (2007). Real options in technology licensing. *Management Science*, *53*(10), 1618–1633. doi:10.1287/mnsc.1070.0705

KEY TERMS AND DEFINITIONS

Open Innovation Concepts: Concept that uses the purposive inflows and outflows of knowledge to accelerate internal innovation, as well as external innovation and expand the markets for increased use of innovation.

Open Innovation Practices: Practices carried out by the entrepreneurships incorporating open innovation researches and methods.

Open Innovation Researches: Researches utilizing open innovation concepts, methods and strategies.

Open Innovation Strategies: Strategies that incorporate fresh perspectives, knowledge and inspiration from the inside and outside of an entrepreneurship, thus allowing to go beyond day-to-day thinking and opens up the way to entirely new possibilities for any value addition.

SMEs Development: Development of the small and medium enterprises sector of the business arena; in terms of economic, knowledge, human skills and other forms of value additions.

ENDNOTES

1. Small and Medium scale Enterprises (SMEs) by definition are small firms with a small number of headcounts and the turnover fall below a certain limit. In terms of headcounts, they vary from country to country and regions to regions. In Europe, the headcounts are less than 250 and the turnover is less than 50 million Euros; https://ec.europa.eu/enterprise/policies/sme/facts-figures-analysis/sme-definition/index_en.htm. The study has adopted this definition.
2. https://www.innogetcloud.com/news/sale-al-mercado-la-primera-herramienta-de-open-innovation-en-formato-cloud/1
3. https://www.qmarkets.net/home
4. http://www.innovationfactory.eu/services/ideanet/
5. http://www.mindmatters.net/Software/INNOVATOREnterprise.aspx

Chapter 5
Open Innovation in SMEs:
Contexts of Developing and Transitional Countries

Hakikur Rahman
Institute of Computer Management and Science, Bangladesh

ABSTRACT

Despite the increased acceptance by corporate business houses around the globe, the adaptation of strategies and concepts belonging to the newly evolved dimension of entrepreneurships, the open innovation (OI), countries in the East, West, or South are yet to adapt appropriate strategies in their business practices, especially in order to reach out to the grass roots and marginal communities. By far, firms belonging to the small and medium sized enterprises (SMEs) sector, irrespective of their numbers and contributions towards their national economies, are lagging behind in accepting open innovation strategies for their business improvements. While talking about this newly emerged business dimension, it comprises of complex and dynamically developed concepts like, management of intellectual property aspects, administration of patents, copyright and trademark issues, or supervision of market trend for minute details related to knowledge attainment.

INTRODUCTION

Innovation is no more an experimentation, but a genuine reality within the business processes, given the circumstances of economic crisis, global competition and novelties of technologies. Perplexing further to face the reality and overcome crises, firms are day by day adopting newly developed ideas, concepts, and perceptions to fit into the business dimension from within and outside the boundaries of their own entities, thus channeling the entrepreneurships through the paradigm of open innovation (OI). By far, majority of the corporate business houses and multi-national enterprises are competing or collaborating with a common goal in promoting value added products, processes, or services. Notwithstanding, they are transforming the entire business development infrastructure to face the reality and move ahead (Van Hemert & Nijkamp, 2010).

DOI: 10.4018/978-1-7998-5849-2.ch005

Furthermore, it has been observed that over the past two decades, there has been a substantial shift in the global innovation landscape. Multinationals from developed economies are increasingly globalizing their research and development (R&D) activities and are developing an open innovation model to source innovations from outside the firm, including from emerging economies such as those in Africa and Asia. In addition, emerging economy enterprises, which traditionally have played a secondary role in the global innovation landscape, have now begun to catch up in developing their own innovative capabilities (Li and Kozhikode, 2009).

However, it has also been observed that a major portion of the business community that belongs to the small and medium enterprises (SMEs) sector, in spite of their justified contribution to economic growth and employment generation, is not always in advantageous situations in the arena of open innovation due to many factors, seen, unseen, attended, un-attended, researched, un-researched, and deserves further research (United Nations, 2006; World Business Council, 2007). In addition, towards the argument regarding the effect of firm size on the effectiveness of innovation continues, the particularities of open innovation from the perspective of emerging market small and medium enterprises (EM SMEs) need to be addressed (Xiaobao, Wei and Yuzhen, 2013).

Earlier studies suggest that open innovation activities positively influence innovation outcomes in large enterprises. However, few studies have investigated the implications of small and medium-sized enterprises' (SMEs) adoption of open innovation (Parida, Westerberg and Frishammar, 2012). A few studies on open innovation address OI practices in SMEs and how their use of OI and the resulting benefits differ from those of large enterprises. The lack of resources in SMEs to engage in looking outward is said to be a barrier to OI, but at the same time this shortage is mentioned as a motive for looking beyond organizational boundaries for technological acquaintance (Spithoven, Vanhaverbeke and Roijakkers, 2013).

In this context, Edwards, Delbridge and Munday (2005) argue that, in spite of increasing attention being given to the role of SMEs and innovation, there is a gap between what is understood by way of the general innovation literature and the extant literature on innovation for SMEs. They further argue that studies of innovation in SMEs have largely failed to reflect advances in the innovation literature. Supporting these arguments, this study has tried to find out relevance of open innovation among SMEs, and particularly the emergence of OI strategies in developing and transitional economies.

Brunswicker and Vanhaverbeke (2015) explore how SMEs engage in external knowledge sourcing, a form of inbound open innovation. They draw upon a sample of 1,411 SMEs and empirically conceptualize a typology of strategic types of external knowledge sourcing, namely minimal, supply-chain, technology-oriented, application-oriented, and full-scope sourcing. They find that each strategy reflects the nature of external interactions and is linked to a distinct mixture of four internal practices for managing innovation. Wynarczyk (2013) assesses the impact of open innovation practices on the innovation capability and export performance of UK SMEs. The overall results exhibits that the international competitiveness of SMEs is highly dependent on the cumulative effects and interrelationship between two key internal components, i.e. R&D capacity and managerial structure and competencies, coupled with two external factors, i.e. open innovation practices and the ability of the firm to attract government grants for R&D and technological development. Along this perspective, Hossain and Kauranen (2016) synthesize the extant literature on open innovation in SMEs. The study finds that adopting OI by SMEs improves their overall innovation performance. They also found that a larger number of studies are based on a quantitative approach.

There are a considerable number of literatures that are available about innovation, and in recent years various models have been suggested to describe its nature. Models have also be divided according to their innovation processes (such as, linear models, chain-linked models, triple-helix, Coventry University Enterprises Limited model, Rahman and Ramos model and others), or according to the fitness for developed or developing countries (Lee, Park, Yoonc and Park, 2010; Rahman and Ramos, 2013).

When one talks about cross-boundary issues on business concerns, innovation requires not only harvesting and transferring knowledge, but also an effective absorption capacity within the entity. However, the main components of this, in terms of organization, resources, and culture are in principal similar both in the domestic and international arenas. Furthermore, internationalization poses particular challenges in terms of openness and the tolerance or, encouragement of diversity if knowledge is to be absorbed successfully, and translated into innovation (Zahra and George, 2002; English-Lucek, Darrah and Saveri, 2002; Williams and Shaw, 2011).

To advance into the context of this trend of research this study has observed that, countries ranking as developed economies are ahead in the race adopting open innovation in their business development, while countries among the developing and transitional economies are struggling to fit into the race of the champions. This chapter based on a study, though not a specific case of one country, has tried to illustrate a few discrete scenarios from five developing countries through longitudinal literature review. The chapter has tried to provide a generic context of innovation (inclined to open innovation) in those randomly selected countries, and present challenges they are facing, including some recommendations, before concluding for further extensive research. Along this route, the chapter has tried to build a framework synthesizing the aspects of the findings. It is expected that this study will contribute to enhance knowledge of readers in refreshing the basic concept of open innovation and application of OI strategies among SMEs in developing countries. Furthermore, as majority of OI strategies nowadays are mainly dependent on utilization of information technologies, this study could form a start up literature or a guide towards future open innovation practices adopting strategic and pragmatic business processes.

BACKGROUND

Joseph Schumpeter (1883–1950), one of the first theorists who studied the economy through the innovative eye, stated that innovation is about new ways of doing things by combining existing elements into new products through a creative process (De Jong, Vanhaverbeke, Kalvet & Chesbrough, 2008). Along this route, innovation through the creation, dissemination and utilization of knowledge has become a key driver of economic growth. However, factors influencing innovation performance have changed in this globalized knowledge based economy, partly due to the advent of new information and communication technologies (ICTs), and partly due to the increased global competition. Innovation results from increasingly complex interactions at the local, national, regional and global levels among individuals, firms, industries and other knowledge institutions. Moreover, governments exert a strong influence on the innovation processes through financing and steering of public bodies which are directly responsible for knowledge creation and dissemination (universities, public and private labs, research houses or intermediaries), and through the provision of financial and regulatory inducements (Carayannis, Popescu, Sipp & Steward, 2006; Vanhaverbeke, 2017).

In this context, firstly, the new ICTs; secondly, the government and its politics; thirdly, universities and research houses; fourthly, entrepreneurs, suppliers, vendors, and partners; and finally, consumers have roles in forming environments pertaining to the launching of innovation within and among entrepreneurships. By far, all these actors need to collaborate and actively participate to create the environment, thus even, turning the innovation processes from traditional or closed ended towards rather non-traditional or open ended, terming it as open innovation.

The literature on open innovation (OI) is characterized by studies based on both large companies and SMEs. Among all, one of the less explored issues in SMEs is what inhibits them to adopt OI. Thus, Bigliardi and Galati (2016) look into the threefold objective to identify which factors hinder the adoption of OI in SMEs, to investigate if different behaviors exist among SMEs in relation to these factors, and to understand if the same factors effectively influence the level of acceptance of OI. Based on a survey on 157 Italian SMEs, we identified four main barriers (namely, 'knowledge', 'collaboration', 'organizational', and 'financial and strategic'). Drawing upon data from the fifth UK Innovation Survey, Freel and Robson (2017) shed light on how management choices on the nature of appropriation relate to management choices on the extent of openness within SMEs. To this end, their findings indicate a threshold effect of both informal and formal appropriation mechanisms on the likelihood of engaging in both coupled and inbound open innovation.

However, due to the close acquaintance and strong industry-university relationship, including familiarity with new ICTs and exploring their benefits, developed countries are much on the lead in creating and commercializing new knowledge. On the contrary, though developing nations are familiarizing their entrepreneurships through university spin-offs and increased intensification of industry-university relation to commoditize ready-made knowledge, but the situation is far behind to compete with the developed world. This applies both to the standardization of university-industry relationship and to the competency of the university, which need further investigation (Kroll and Liefner 2008). In this aspect, Savitskaya (2009) argues that the contribution to the understanding of open innovation practices in developing countries resides in demonstrated role of the government for creating favorable conditions for entrepreneurs to open up and integrate into innovation system in the country. She assumed that open innovation system needs a certain level of governmental support to emerge in developing economies.

Furthermore, when comes to the question of introducing open innovation in entrepreneurships, the focus directly or indirectly goes to developed countries, even so towards large and corporate business houses. But a prospective observation this study has made is that, with increased relationship between public funded research houses and entrepreneurs, including government initiatives, the sector of business entities that belongs to the SMEs are catching up in the run by adopting open innovation, mainly in developed countries, and very recently in a few developing countries. To set the benchmarking of a post doctoral research on assessing current scenario of open innovation dynamics in developing countries this study incorporates some specific observations along this context and this chapter is the result of a horizontal study on a few countries of that category who are trying in adopting open innovation (rather, trying to be innovative) in their businesses.

Apart from the entrepreneurship development, due to the very basic inheritance of the marginal societies in developing nations, a considerable interest of the entrepreneurs among SMEs has focused on their roles in the alleviation of widespread poverty. However, looking beyond the immediate, pressing concern of the poor, Andrew Warner (2001) has advanced the concept that SMEs are the building blocks of innovation and sustainable growth in developing countries, such as SMEs represent foci of technological creativity. Supporting Kowalski (2009) this study accepts that, these concepts are linked as sustained economic growth, which can alleviate real poverty. Hence, as SME development drives economic growth in a country, there is a concomitant reduction in poverty. Now the question appears as what could be an acceptable approach in establishing a sustained business environment in developing countries' perspective in the longer run? And, what could be the appropriate strategies they should adopt to enter into the open innovation paradigm? Moreover, as long as the developing countries are trying to adopt novel ideas and strategies as a booster of economic activity, especially by adopting open innovation strategies for the development of small and medium enterprises, the study has find that these concepts are relatively unfamiliar in developing countries.

Though Lee et al. (2010) mentioned, there is considerable literature about innovation, and various models have been suggested to describe its nature, such as product innovation and process innovation; radical innovation and incremental innovation; systemic innovation and component innovation; technology-push and market-pull; and more recently closed innovation, open innovation, collaborative or crowdsourcing innovation. Models can also be divided according to their innovation processes or according to the fitness for developed or developing countries, etc. but, this research argues that models as such on developing countries perspectives are scant.

Following these observations, this study explores the role and impact of SMEs in the developing and transitional economies, and discusses about a few countries' context focusing the emancipation of SMEs policies and practices accommodating open innovation (rather, innovation). The study observes that to roll out open innovation at the grass roots level of developing and transitional countries, it needs more additional input in addition to just being innovative. Next section looks into some details about innovation in business sector in developing countries.

INNOVATION AND DEVELOPING COUNTRIES

Based on the arguments made above, there arise several issues in terms of implementing open innovation for SMEs development in developing countries. Firstly, it must be understood that the term "developing countries" comprises a wide variety of nations that are at very different stages of economic development, have very heterogeneous levels of technological capabilities, and have very diversified cultural differences. Hence, the innovation appropriability dynamics will be very different, for example, in advanced developing countries such as some Latin American or Asian economies where industrial, export and innovation capabilities are more or less strong, *vis à vis* most least developed countries (LDCs), mainly rely on traditional agricultural activities and have poorer productive and technological capabilities. Predominantly, there is a reasonable innovation gap in between them.

Secondly, it is often thought that developing countries are mainly imitators or adopters of technologies and knowledge developed elsewhere. Hence, the debate on introducing OI strategies in developing countries is often focused on whether environments are more favorable for technological changes in those countries. While lax or strong intellectual property rights (IPRs) are thought to favor imitation, copy and reverse engineering; and hence are seen by some authors as a favorable factor for the deployment of learning processes that could lead in the medium and long run to the creation of genuine innovation capabilities in those countries; it is often stated that strong IPRs are a condition for developing countries to receive updated technology transfers by means of licenses and foreign direct investment (López, 2009).

Thirdly, reasonable policy update is desired at national and local contexts in transforming business environments in favor of open innovation. Developing countries are yet to be familiarized with the newly evolved OI strategies. To enhance OI adoption and to create a sustained platform of OI, developing countries should come up with policies at their national levels, emphasizing local businesses.

Fourthly, one has to recognize that, SMEs are critical to the economies of all countries, especially the developing ones (Payne, 2003), and encouraging innovation in SMEs remains at the heart of policy initiatives for stimulating economic development at the local, state, national and regional levels (Jones & Tilley, 2003; Edwards, Delbridge & Munday, 2005). According to Ernst, Mytelka and Ganiatsos (1994), innovation in developing countries is based on the continuous and incremental upgrading of existing technologies or on a new combination of them.

Fifthly, majority of the developing countries and LDCs (Least Developing Countries) are suffering from the weakness of the basic infrastructure necessary for economic activities. Activities for innovation and the use of intellectual property for promoting investment and R&D (research and development) in those countries are faced with additional difficulties caused by the lack of basic infrastructure and a "knowledge infrastructure" (Takagi and Czajkowski, 2012).

Realizing these issues, in recent years, there is a considerable interest among entrepreneurs in establishing SMEs in developing countries. There are probably two main reasons for this. One is the belief that SME development may prove to be an effective antipoverty initiation. The second is the belief that SME development is one of the building blocks of innovation and sustainable growth. These two reasons are of course inter-linked because most of the research evidence says that growth and real poverty reduction go hand in hand. If SMEs development helps growth, more than likely it helps reduce poverty as well (Warner, 2001).

Finally, organizational approaches (with patronage from the highest corners of the government) in the form of providing assistance in finding funds, knowledge and technologies are meant to be common practices at the beginning of the innovation cycle, till it matures to take over on its own both at the local level and national level.

Evidently, across Southeast and South Asia, the contribution of SMEs to the overall economic growth and the GDP is relatively high. Some examples include;

- Bangladesh where SMEs contribute 50% of industrial GDP and provide employment to 82% of the total industrial sector employment;
- India, where SMEs' contribution to GDP is 30%;
- Indonesia, where SMEs accounted to 99.985% of the total number of enterprises with their output contributed to 53.28% of GDP in 2006 and SMEs account for 96.18% of the total employment;
- Nepal, where SMEs constitute more than 98% of all establishments and contribute 63% of the value-added segment;
- Thailand, where SMEs account for more than 90% of the total number of establishments, 65% of employment and 47% of manufacturing value-added; and
- The Philippines, where SMEs comprise 99% of the total manufacturing establishments and contribute 45% of employment and 18% of value added in the manufacturing sector (Kowalski, 2009).

However, when comes the question of finding good cases or case studies or national initiatives on adoption of open innovation for SMEs development at the context of developing countries, they are rare. Although, the phenomenon on innovation of SMEs has captured the interest of many scholars, few studies have been found on studying the issue from the developing countries' perspective.

Literature on innovation indicates that over the last two decades, there has been a systematic and fundamental change in the way firms undertake innovating activities. Particularly, there has been a tremendous growth in the use of external networks by firms of all sizes. Innovation is seen as a process which results from various interactions among different actors. Inter-organizational and cross-sectoral networks, which facilitate the accelerated flows of information, resources and trust necessary to secure and diffuse innovation, have emerged as leaders. However, as SMEs with scarce resources, have less R&D, and generally face more uncertainties and barriers to innovation, networks represent a complementary response to insecurity arising from development and use of new technologies, while reducing uncertainties in innovation. Moreover, in the era of ''open innovation'', according to Chesbrough (2003), firms consistently rely on external sources of innovation by emphasizing the ideas, resources and individuals flowing in and out of organizations, searching for and using a wider range of external ideas, knowledge and resources, networks, which are becoming essential for the creation of successful innovations for SMEs (Zeng, Xie & Tam, 2010), but seems unfamiliar in developing countries.

Furthermore, in the perspective of innovation systems in developing countries, production and exchange of knowledge (mainly technical; internal or external or both) and information are not the only prerequisites for innovation; several additional factors play as key roles, such as policy, legislation, infrastructure, funding, and market developments (Klein-Woolthuis, Lankhuizen & Gilsing, 2005). In addition to these, the concept of knowledge absorption is often used related to intra- and inter- firm knowledge transfer and ability to implement the acquired knowledge, and the notion of absorptive capacity can be related to cross-region or cross-country knowledge exchange. This is most relevant to developing countries, who are believed to be imitators, rather than innovators, and their innovative development happens in terms of adaptation of existing technologies to satisfy local realities (Savitskaya, 2009).

This chapter likes to discuss a few SMEs development initiatives in five developing countries in terms of adopting innovative approaches. The study has tried to collect researches or examples based on policies and practices adopting open innovation. The selection criterion follows random sampling and availability of searched literature within accessible search engines.

CASE DESCRIPTIONS

Small and medium enterprises are being recognized in different ways in different countries .Most countries have adopted the benchmarks of employment. Some classify them in terms of assets, a few in terms of sales and others, in terms of fund. In a few countries, a hybrid definition is used, such as employment and assets or turnover. Although definition differ across countries, they have one thing in common; the vast majority of SMEs are relatively small and over 95% of SMEs in Asia employ less than 100 people . Based on this, broad comparison on the characteristics and role of SMEs is still possible even with differing definitions (Pandey & Shivesh, 2007).

This study has considered six countries from Africa and Asia. Among them the two countries in Africa, South African one is based on the Sekhukhune Living Labs experience and Ugandan one is showing the national contexts focusing SMEs development. Among the four Asian countries; from Bangladesh, India, Indonesia and China, the national policy perspectives have been illustrated, which show evolution of entrepreneurships towards innovation paradigm. Countries in this section have been selected at random basis, however, the intention is to find out the trend of doing any innovative (rather open innovative) entrepreneurships among these countries, any initiative taken by their governments to promote innovative entrepreneurships, and to find out any catalytic agents in this aspect. They are being described next following alphabetical order.

Bangladesh

Government of Bangladesh formulated the National Industrial Policy 2005 by giving emphasize for developing Small and Medium Enterprises[1] as a thrust sector for balanced and sustainable industrial development in the country with the vision for facing the challenges of free market economy and globalization. In the policy strategies, smooth and sustainable development of SMEs all over the country has been considered as one of the vehicles for accelerating national economic growth including poverty alleviation, and generation of employment. Most of the industrial enterprises in Bangladesh are typically SME in nature. Generally SMEs are found to be labor intensive with relatively low capital intensity. SMEs also possess a character of privilege as cost effective and comparative cost advantages by nature. In this aspect, the SME policy strategies have been formulated in line with the acknowledged principles for achieving the Millennium Development Goals (MDGs) by the Government (Govt. of Bangladesh, 2005a).

Furthermore, the provisions of facilities for attracting foreign investments have been envisaged in the Industrial Policy. The government has taken an initiative to formulate a separate SME policy to provide entrepreneurs with necessary guidance and strategic support in respect of the establishment of SME industries all over the country (Govt. of Bangladesh, 2005b).

A few of the broad objectives of the SME policy strategies are to:

- accept SMEs as an indispensable player in growth acceleration and poverty reduction, worthy of their total commitment in the requisite overall policy formulation and execution;
- SME policy strategies shall essentially be linked with broad based and integrated manner in line with the poverty reduction strategy paper (PRSP) of the Government of Bangladesh;
- encourage and induce private sector development and promote the growth of foreign direct investment (FDI), develop a code of ethics and establish good governance, ICT based knowledge management and customer supremacy in the market alliances;
- identify and establish the network of infrastructure and institutional delivery mechanisms that facilitate the promotion of SMEs;
- re-orient the existing fiscal and regulatory framework and government sponsored institutions supporting the goals of SME policy;
- have credible management teams in terms of the delivery of needed services, leadership, initiation, counseling, mentoring and tutoring;
- create innovative but rewarding arrangements so that deserving and especially enterprises with desired entrepreneurial qualifications and promise can be offered financial incentives within industries prescribed on some well-agreed bases;
- assist implement dispute settlement procedures that proactively shield small enterprises especially from high legal costs and insidious harassments;
- take measures to create avenues of mobilizing debt without collaterals to match (either using *debt-guarantee schemes* or mapping *intellectual-property* capital into pseudo-*venture capital)* in order to assist small enterprises in dealing with their pervasive lack of access to finance (Govt. of Bangladesh, 2005a).

This study notes that use of debt-guarantee scheme, mapping of intellectual property or concept of venture capital are very basic ingredients of open innovation strategies.

For promotional support the following booster sectors has been identified and the list has been set to be reviewed every three years:

- Electronics and Electrical
- Software Development
- Light Engineering
- Agro-processing and related business
- Leather and Leather goods
- Knitwear and Ready Made Garments
- Plastics and other synthetics
- Healthcare and Diagnostics
- Educational Services
- Pharmaceuticals/ Cosmetics/ Toiletries;
- Fashion-rich personal effects, wear and consumption goods (Govt. of Bangladesh, 2005a).

Moreover, the government has established an SME Foundation as a pivotal platform for the delivery of all planning, developmental, financing, awareness-raising, evaluation and advocacy services in the name of SME development as one of the crucially-important element of poverty alleviation. The Foundation suppose to provide a one-window delivery of all administrative facilities, including some resources needed for capacity-building in appropriate industry association(s), for SMEs in Bangladesh (Govt. of Bangladesh, 2005a; 2005b)

China

China is attempting to catch-up in terms of innovating their entrepreneurships, which is fundamentally different from earlier latecomers like Japan and Korea. The basic elements of Chinese catching up strategy are: market size, market-oriented innovation, global alliance and open innovation, spillover of FDI and role of government. Moreover, the core capability of Chinese company is an integration capability of market knowledge, outsourcing and learning (Liu, 2008).

Since the realization of the open policy in 1978, China has made great efforts to change from a highly centralized planned state to the near market economy. The role of SMEs has been expanding in the changing socio-political context. They not only play a greater role in the economies (accounting for more than 99% of all firms being SMEs), but also contribute in a large extent to the increased levels of business activity and employment (Siu, 2005). Zeng, Xie & Tam (2010) argue that, the manufacturing industry is the main driving force of social development and economic growth in developing countries. In this context, Zeng, Xie & Tam (2010) mention that China, with more than two decades of market oriented reform, there has been a rapid growth in the manufacturing industry. Hence, it is necessary to explore the external cooperation network of manufacturing SMEs in order to help them improve their industrial competitiveness. However, there is a paucity of studies on the impact of external cooperation network on the innovation of Chinese manufacturing SMEs. *This study notes that collaborative networking is one of the most effective preconditions for adopting OI strategies.*

Using a structured questionnaire survey, Zeng, Xie & Tam (2010) examine the innovation networking activities of some surveyed SMEs in Shanghai, the largest city and economic center in China. Their study aims to explore the relationships between different cooperation networks and innovation performance of SME. Based on a survey of 137 Chinese manufacturing SMEs, they empirically explore the relationships between different cooperation networks and innovation performance of SME using the technique of structural equation modeling (SEM). Their study finds that there are significant positive relationships between inter firm cooperation, cooperation with intermediary institutions, cooperation with research organizations and innovation performance of SMEs, of which inter firm cooperation has the most significant positive impact on the innovation performance of SMEs.

This study supports the above mentioned parameters as the basic building block in establishing a platform of open innovation. However, the result of Zeng, Xie & Tam (2010) reveals that the linkage and cooperation with government agencies do not demonstrate any significant impact on the innovation performance of SMEs. Moreover, their findings confirm that the vertical and horizontal cooperation with customers, suppliers and other firms plays a more distinct role in the innovation process of SMEs than horizontal cooperation with research institutions, universities or colleges, and government agencies, which is quite opposite to the context of developed countries. This study suggests that further studies need to be carried out to re-confirm this hypothesis or find out any future diversions.

India

In India, the term small scale industries (SSIs[2]), is used far more often than SMEs and is based upon investment in assets[3] (Saini & Budhwar, 2008). However, despite various liberalizations and schematic changes to meet the emerging requirements of the business sector, availability of finance continues to be a major problem for small enterprises in India. Realizing this fact, some of the development financial institutions (DFIs) and forward looking commercial banks have put in operation a number of innovative schemes, and among them the Small Industries Development Bank of India (SIDBI) has taken the lead. The majority of the experiments have started showing good results. The SSI sector plays a significant role in the Indian economy. For the past one decade, it has been consistently registering about three per cent higher real growth rate in terms of GDP (8.9 per cent during 1999–2000) compared to the growth recorded by the industrial sector as a whole. The SSI sector contributes over 41 per cent of the total industrial production, 31 per cent of the country's total exports, and jointly with traditional industries (for example Khadi, handloom, handicrafts, sericulture, and coir) the relative percentage goes up to 58 per cent (Narain, 2001).

In terms of finance, transaction lending such as asset based lending, factoring and leasing have been in use to fund SMEs for some time, and there is some evidence of relationship lending in India. Moreover, in developing countries, the private economy would comprise largely of family businesses. It is estimated that in India, family businesses account for 70% of the total sales and net profits of the biggest 250 private-sector companies (Economist, 1996), and almost all the micro-small-and-medium-enterprise (MSME) would be family firms. Inter-family relationships and family succession play an important role in the performance of family firms, and financial institutes would need to take this into account in their credit decisions. A study by Marisetty, Ramachandran & Jha (2008) finds that family businesses in India where succession takes place without fights and splits show higher profitability (Thampy, 2010). *This study notes that India is accepting several strategies towards open innovation, such as providing financial supports; liberalizing market conditions; adopting lending, factoring and leasing; and foremost promoting networking.*

Indonesia

Based on Indonesian Presidential Decree no. 99/1998, Small Enterprise is being defined as "Small scale people economical activities with major business category in small business activities and need to protect from unhealthy business competition". However, the government of Indonesia defines small enterprises as firms with total asset up to RP. 200 million (1USD is equivalent to 9887 Indonesian Rupee as on June 17, 2013) excluding land and building, the total annual sales are not more than Rp. 1 billion owned by Indonesian citizens, not subsidiary or branch of medium or large enterprise, personal firm. While medium enterprises are firms with total asset more than Rp. 200 million but not exceed Rp. 10 billion excluding land and buildings. Apart from these definitions, Biro Pusat Statistik (Statistic Center Body) defined SMEs based on number of employee. Small enterprises employ 5 to 19 people, while medium enterprises employ 20-99 people (Hamdani and Wirawan, 2012).

Similar to other countries, SMEs in Indonesia also play an important role in social and economic growth of the country, due to great number of industry, GDP contribution, and total employment. SME's characteristics have been found to be more agile and adaptable with capability to survive and raise their performance during economical crisis than larger firm. But with the increase of business competition, in particular against large and modern competitor, SMEs are being placed in a vulnerable situation. Hence, the development of sustainable SMEs has become an important step to strengthen and sustain Indonesian economy. According to the statistics 43.22 millions SMEs accounted to 99.985% of the total number of enterprises in Indonesia in 2006. Their output contributed 53.28% of GDP in 2006, and, with over 85.42 million workers, SMEs account for 96.18% of the total employment in Indonesia. The role of SMEs has become more important, because according to research by AKATIGA, the Center for Micro and Small Enterprise Dynamic (CEMSED), and the Center for Economic and Social Studies (CESS) 2000, SMEs have the unique ability to survive and raise performance during economical crisis, due to their flexibility in adapting production process, ability to develop with their own capital, capacity to pay high interest loan and only a little get involve with the bureaucracy. It is expected that with this vital stance in the economy, the development of SMEs would contribute to economic and social development through economic diversification and accelerated structural changes that promote stable and sustainable long-term economic growth (Padmadinata, 2007; Hamdani and Wirawan, 2012)

It has been observed that the success factors of Indonesian SMEs are capital access, marketing and technology, while legality was a challenge to business success. Education and source of capital were related significantly to business accomplishment. However, these later two factors seemed to need moderating variables since poor operational explanations were needed to link these two with business achievement. Increasing business competition, in particular against large and modern competitors, put SMEs in a vulnerable position. In Indonesia, most SMEs operate along traditional lines in production and marketing. They also have several challenges such as lack of knowledge, lack of qualified human resource as well as quantity, non conducive atmosphere, lack of facility, limitation of market and information access, and bureaucracy (Indarti and Langenber, 2004; Hamdani and Wirawan, 2012).

South Africa

In European context, supporting open innovation among SMEs, Living Labs are providing significant input in terms of co-creation, exploration, experimentation and evaluation[4]. As a knowledge centre of the European Network of Living Labs (ENoLL), the Sekhukhune Living Lab focuses on small, medium and micro-enterprises (SMMEs) which are regarded as important growth engines in South Africa. However, several barriers are inhibiting rural entrepreneurship and access to mainstream or global supply chains and markets. Schaffers et al. (2007), in their research mentions that, long distances, high transport/transaction costs and low economies of scale are the consequences of typical rural conditions such as physical remoteness and low economic activity levels there. Furthermore, the problems associated with these barriers worsen dramatically if roads are poor, telecommunications bandwidth is limited or expensive, and many rural entrepreneurs have limited computer literacy and do not own a truck, motorcar or computer. These are the typical complexities faced by rural entrepreneurs in most of the developing countries, and in South Africa's "deep rural areas" such as Sekhukhune.

Through ENoLL, Sekhukhune Living Lab introduces a range of services through the facilitation of so called Infopreneurs, which are micro, self-sustainable service enterprises that channel and deliver services for local SMMEs and citizens into the community. These Infopreneurs are the 1st tier target SMME group of the work and interventions of the C@R Living Lab. They provide knowledge-based services such as cross-organizational business process enabling, SWOT analysis and logistics brokerage to assist start-up, grow and cluster other SMME's in various sectors (for example, health, mining, construction).

These Infopreneurs are being deployed in existing infrastructure and getting benefit from ongoing local initiatives supported by the South-African government. Franchise-like agreements are shaping the collaboration among partners. However, the focus of Living Lab development is on establishing collaboration tools and processes, particularly addressing the accessibility of knowledge-based services that are relevant to local SMME businesses, in harnessing increased mobile connectivity and enabling rural service channels that enhance effective collaboration amongst SMMEs in communities and between first and second economy enterprises.

The ubiquitous infrastructure shortcomings of South-Africa (such as, constricted bandwidth) are being taken into account when setting up these knowledge service agents. By forming clustered enterprises via Infopreneur services, consolidation of supply chain volumes is achieved with lower transaction and transportation costs. The strategy is to create Infopreneur service bundles to enhance local business and geo-economic intelligence that helps SMMEs to seamlessly interoperate among each other and first economy enterprises (Schaffers et al., 2007). *Intermediaries are an essential element of promoting open innovation dynamics in diverse and difficult environments, as such this study notes.*

Uganda

In Uganda, SMEs[5] are increasingly taking the role of the primary vehicles for the creation of employment and income generation through self-employment, and treated as tools for poverty alleviation. SMEs also provide the economy with a continuous supply of ideas, skills and innovation necessary to promote competition and at the same time, efficient allocation of scarce resources.

Furthermore, mentioned by Kasekende (2001), a few strong SMEs in Uganda, like Capital Radio, Kabira International School, Masaba Cotton Co. Ltd and Africa Basic Foods were formed through *joint venture* arrangements with foreign partners from the United Kingdom and the United States. These and other SMEs have provided domestic linkages, comprising link between agriculture and industry and between SMEs and large-scale industries. This has created opportunities for employment and income generation both in rural and urban areas at relatively low cost, thus ensuring a more equitable income distribution. In turn, the stimulation of activities in both rural and urban areas has mitigated some of the problems that unplanned urbanization tends to create, thus offering an efficient and progressive decentralization of the economy. In this aspect, SMEs play a crucial role in creating opportunities to achieve equitable and sustainable growth. SMEs in Uganda are providing employment and income generation opportunities to low income sectors of the economy.

However, due to their characteristics and nature, SMEs in Uganda suffer from constraints that lower their resilience to risk and prevent them from growing and attaining economies of scale. The challenges are not only in the areas of financial investment and working capital, but also in *human resource development*, *market access*, and *access to modern ICTs*. Furthermore, *access to financial resources* is constrained by both internal and external factors. Internally, most SMEs lack creditworthiness and management capacity, so they have trouble securing funds for their business activities, for example procuring raw materials and products, and investing in plant and equipment. From the external viewpoint, SMEs are regarded as insecure and costly businesses to deal with because they lack required collateral and have the capacity to absorb only small amount of funds from financial institutions. Foremost, due to high intermediate costs, including the cost of monitoring, they are rationed in their access to credits and having difficulties in enforcing loan contracts (Kasenkende, 2001).

To overcome such constraints, the government and other players such as the Bank of Uganda (BOU) have designed programmes and policies to support SMEs that are market driven and non-market distorting. The government has created stable macroeconomic conditions, liberalized the economy, and encouraged the growth of the micro-financing business. In conjunction with donors, the government has designed a medium-term competitive strategy and a Rural Financial Services Programme to benefit SMEs. However, the challenge to SMEs in accessing financial services will remain dependent on how they themselves increase their creditworthiness (Kasekende, 2001). *This study observes that to widen OI strategies at the national and local level, Uganda has been moving in the appropriate direction.*

SYNTHESIS AND THE FRAMEWORK

Synthesizing the countries of this study provides different dimensions of business growth in their countries, accommodating innovation. Ranging from policy initiation to networking, to liberization, to institutionalization are significantly observable there. The table below shows various aspects of the synthesis in the form of a framework, however, this does not mean that any country is superseding another.

Table 1. Observed Tendencies on SMEs development

Country	Pattern at National Context	Observed Tendency
Bangladesh	Policy initiation	Awareness development, acceptance of policy, initiation of policy, and patronization from the government
China	Action through vertical and horizontal integration	Cooperation among customers, suppliers, other firms, partners, research institutions, universities and government agencies; Market driven initiation accommodating global competition, dependency on FDI (foreign direct investment), and patronization from the government
India	Identification	Identify the potential business sector where thrust should be given
Indonesia	Translation	Conceptualization, awareness development, adoption of appropriate policy, recognizing challenges in the formulation process
South Africa	Clusterization and Institutionalization	Build a sustainable infrastructure serving local community at local context
Uganda	Utilization	Application of appropriate strategies at designated levels of enterprises

CURRENT CHALLENGES

This section starts using a quote of Saini and Budhwar about the understanding on SMEs, saying "The concept of SME itself is quite problematic" (2008: 417). This study finds another important quote from their paper, where Storey notes, ''there is no single, uniformly acceptable, definition of a small firm. There are differences as to size, shape and capital employed. In the USA there is no standard definition of small business. Even a firm employing up to 1500 employees is considered as small by American Small Business Administration. The concept in USA is industry-specific; mostly income and persons employed will determine whether a firm falls in the category of small business or not'' (1994: 8).

The European Commission classifies firms according to the number of employees as: micro (0–9), small (10–99) and medium (100–499). However, in Oslo Manual (OECD, 2005) the EC has incorporated turn over, in addition to the number of employees. In China, it includes companies employing less than 200 persons; and in Japan those employing less than 300 persons are considered to be SMEs (Srivastava, 2005: 166). Even, sometimes the definition of SMEs depends on the stage of national economic development and the broad policy purposes for which the definition is required. But, the essential fact is that, whatever may be the definitional problems, SMEs occupy an important place in the economy of most countries; especially they are favored in developing countries due to their employment potential (Saini and Budhwar, 2008).

Furthermore, access to finance has been identified as a key element for SMEs to succeed in their drive to build productive capacity, compete, create job opportunities and contribute to poverty alleviation in developing countries. Without finance, SMEs cannot acquire or absorb new technologies nor they can expand to compete in global markets or even establish business linkages with larger firms. Finance has been identified in many business surveys as the most important factor determining the survival and growth of SMEs in both developing and developed countries. Access to finance allows SMEs to undertake productive investments to expand their businesses and acquire the latest technologies, thus ensuring their competitiveness. Poorly functioning financial systems can seriously undermine the microeconomic environment of a country, resulting in lower growth in income and employment (UNCTAD Secretariat, 2001)

Despite their dominant numbers and importance in job creation, SMEs face difficulty in obtaining formal credit or equity. For example, maturities of commercial bank loans made available to SMEs are often limited to a period far too short to pay off any sizeable investment. Meanwhile, access to competitive interest rates is reserved for only a few selected blue-chip companies while loan interest rates offered to SMEs always remain high. Moreover, banks in many developing countries traditionally lent overwhelmingly to the government, which are less risky and offer higher returns. Such practices have congested most private sector borrowers and increased the cost of capital for them. Governments cannot expect to have a dynamic private sector as long as they absorb the bulk of private savings. In the case of venture capital funds (an essential ingredient of open innovation entrepreneurship), governments have been concentrated in high technology sectors. Similarly, the international financial institutions have ignored the plight of SMEs. These preferences and tendencies have aggravated the lack of financing for SMEs (UNCTAD Secretariat, 2001).

Technological advancements have contributed to remarkable changes to the nature of current production systems. This has also created impact on the nature of work, workers and skills involved. SMEs may take benefit from these advancements in their operations, but they do not recognize the critical role of effective human resource policies for their success. Furthermore, the need for a skilled workforce in SMEs certainly becomes apparent during periods of such technological changes. Particularly, SMEs have to undergo some changes when they compete with global companies and other large buyers, as they are dependent on supply contracts from the same. This puts substantial pressure on SMEs to control both their costs and quality and meet the different legal requirements. Moreover, this poises a serious challenge for SMEs, especially for those operating in developing countries with labor-intensive technologies, where labor cost is a major concern. Many of them resort to disputed practices, such as employment of child labor to reduced labor costs and violation of labor standards including denial of minimum wage, and other minimum-work conditions. Majority of them also lack access to relevant data and information about new markets, legal provisions regulating their working, and product innovations, which hinders their survival. In addition to these, it has been found that their accessibility to professional management tools is almost absent (Zeng, Xie and Tam, 2010).

In terms of innovation, not all countries have the opportunity or ability to capitalize on the opportunity to catch up. For a developing country, it is not easy to proceed from stage of imitation to stage of innovation (Zeng, et. al., 2010). Bell and Pavitt (1993) pointed out, just installing large plants with foreign technology and foreign assistance will not assist in the building of technological capability. The prevailing fact is that the relation between competition patterns, productive structures and innovation in developing countries are very different from that in developed countries, and hence one should also expect to find differences in the pattern of use of intellectual property rights (IPRs) and other innovation mechanisms. Furthermore, there are differences when comparing developing countries at different stages of industrial and technological development (López, 2009). Hence, researching into open innovation focusing SMEs development in developing countries requires further intensive study and research.

This section concludes with a final sentence that, among these five nations, being driven by geographical, cultural, economical and most of all economical aspects, are very different from each other in achieving innovation in their entrepreneurships, which is a challenge to develop a generic framework for developing countries. However, as this research continues, efforts will be given to include a few more countries of similar socio-economic-cultural contexts and in-depth study will be carried out.

RECOMMENDATIONS AND FUTURE RESEARCH

From this study on a few country specific aspects of SMEs development, if one likes to interpret them towards the dimension of open innovation, the question will arrive, as how important open innovation thinking should be at the national level to guide the policy makers and other decision support systems in policymaking. In terms of adopting open innovation, especially in developing countries, there may be other priorities in policymaking due to the relatively modest absorptive capacity of incumbent enterprises and under-developed innovation institutions. In such countries it would probably easier to start with the relatively simple guidelines with simpler framework, for example developing basic innovation and interaction skills, rather than starting with more sophisticated interventions to enhance technology markets or stimulate corporate entrepreneurship (matured stages of technology exploration or technology exploitation). Future work may explore if there is an optimal sequence in the innovation system as how to adopt various open innovation policy guidelines, and if the developed framework needs to be refined for this purpose (De Jong, Vanhaverbeke, Kalvet and Chesbrough, 2008; Rahman & Ramos, 2010; Gassmann, Enkel and Chesbrough, 2010; Vanhaverbeke, Chesbrough and West, 2014). West, et. al. (2014) link to three key trends in open innovation research: better measurement, resolving the role of appropriability and linking that research to the management and economics literature.

In the recent years banks in developed countries have launched a number of initiatives that both improve the profitability of lending to SMEs, and provides SMEs with better access to finance and financial products that are better tailored to their needs. A number of leading banks have demonstrated that providing financial services to SMEs can be turned into a profitable business. Although the business environments in developing countries and developed countries differ in many respects, the problems of servicing SME customers remain similar, such as high perceived risk, problems with information asymmetry and high administrative costs. Hence, recent innovations in developed countries to improve SMEs access to credit may provide valuable insights for developing country banks to become more SME-oriented and increase the volume and the quality of their services (Warner, 2001).

Davidsson (2006) forwarded the idea of the Small Business Innovation Program, and suggested that, perhaps in one way to adjust the conditions and challenges of a developing country one can pursue the following focus areas;

- Education, training and skill development programmes for entrepreneurship;
- Routines, initiations and contacts for initiating start-ups;
- Communication with government officials to better understand legislation and regulation in the area of entrepreneurships, including marketing environments;
- Availability of skills those are useful for potential consumer markets;
- Improving online access by skills and resources;
- Access to financial resource and contacts for foreign direct investments;
- Strengthen the technological capacity;
- Successful e-business models; and
- Establish stronger, more effective representation of small enterprises' interests at local, and national government and international level.

Foremost, there is an urgent need to make the best out of the public and private resources invested in fundamental and applied research. Both budget pressures and the need to solve crucial challenges, such as transitioning to an environmentally sustainable economy and supporting the equitable growth of developing countries mean that science will be required to generate technology at an ever-increased rate to maintain the continuous stream of social and market driven innovations (Ruiz, 2010).

CONCLUSION

The emerging global market has become the focus of sustained research in the past two decades due to several reasons. Firstly, the emerging markets comprise the majority of the world's population and land, and they continue to grow faster than the developed world. Secondly, the emerging markets are increasingly being recognized as a diverse set of business, cultural, economic, financial, institutional, legal, political and social environments within which to test, reassess and renew acquired wisdoms about how the business works, to gain deeper insights into prevailing theories and their supporting evidence, and to make new discoveries that enhances human welfare in all environments including the world's poorest countries, the developing economies, the transitional countries and the developed world (Kearney, 2012).

SMEs are in general initiated by a single entrepreneur or a small group of people, and are often managed by owner–managers. Their organizational structures are typically flat. SMEs do not have many layers (mainly due to small number of both employees/supervisors and specializations in human skills) because the owner/s is/are mostly at the top of decision making affairs (which still keeps them bureaucratic as most of the times employees do not dare to challenge the supervisors/owner/s). However, the good thing come from this nature is that it adds to their flexibility. Many researchers argue that entrepreneurs mostly seek to derive several advantages by undertaking operations at a smaller level in terms of flexibility, informality, sustainability, and structural adaptability (Zeng, Xie & Tam, 2010).

However, this study argues that, to observe the rolling out of innovation processes in developing and transitional economies, a multi-facet research has to be carried out, including broader aspects of the entire context of open innovation dynamics and incorporating larger sample size. The discussions presented so far, is an attempt to overview the open innovation paradigm and relevant public policy context in a developing country. The indicative remarks may offer insights for future research in the fields of open innovation and innovation policy initiation not only in the developing economies, but also in the transitional economies. This study had its limitations. Scant literature and lack of necessary tools, such as survey or interview or other instruments are among them. Nevertheless, introducing these tools is expected to bring along opportunities for further research.

REFERENCES

Bell, M., & Pavitt, K. L. R. (1993). Technological accumulation and industrial growth: Contrasts between developed and developing countries. *Industrial and Corporate Change*, *2*(2), 157–210. doi:10.1093/icc/2.2.157

Bigliardi, B., & Galati, F. (2016). Which factors hinder the adoption of open innovation in SMEs? *Technology Analysis and Strategic Management*, *28*(8), 869–885. doi:10.1080/09537325.2016.1180353

Brunswicker, S., & Vanhaverbeke, W. (2015). Open innovation in small and medium-sized enterprises (SMEs): External knowledge sourcing strategies and internal organizational facilitators. *Journal of Small Business Management*, *53*(4), 1241–1263. doi:10.1111/jsbm.12120

Carayannis, E. G., Popescu, D., Sipp, C., & Steward, M. (2006). Technological learning for entrepreneurial development (TL4ED) in the knowledge economy (KE): Case studies and lessons learned. *Technovation*, *26*(4), 419–443. doi:10.1016/j.technovation.2005.04.003

Chesbrough, H. W. (2003). *Open innovation: The new imperative for creating and profiting from technology*. Harvard Business School Press.

Davidsson, J. (2006). *Small Business Innovation Program: Business development and entrepreneurial training with intellectual property in developing countries*. B-Open Nordic AB.

De Jong, J. P. J., Vanhaverbeke, W., Kalvet, T., & Chesbrough, H. (2008). Policies for Open Innovation: Theory, Framework and Cases. Research project funded by VISION Era-Net.

Economist. (1996). The family connection. *The Economist, 341*(7986), 62.

Edwards, T., Delbridge, R., & Munday, M. (2005). Understanding innovation in small and medium-sized enterprises: A process manifest. *Technovation*, *25*(10), 1119–1127. doi:10.1016/j.technovation.2004.04.005

English-Lucek, J. A., Darrah, C. N., & Saveri, A. (2002). Trusting strangers: Work relationships in four high-tech communities. *Information Communication and Society*, *5*(1), 90–108. doi:10.1080/13691180110117677

Ernst, D., Mytelka, L., & Ganiatsos, T. (1994). Technological Capabilities: A Conceptual Framework. UNCTAD.

Freel, M., & Robson, P. J. (2017). Appropriation strategies and open innovation in SMEs. *International Small Business Journal*, *35*(5), 578–596.

Gassmann, O., Enkel, E., & Chesbrough, H. (2010). The future of open innovation. *R & D Management*, *40*(3), 213–221. doi:10.1111/j.1467-9310.2010.00605.x

Govt. of Bangladesh. (2005a). *Policy Strategies for Small & Medium Enterprises (SME) Development in Bangladesh*. Ministry of Industries, Government of the People's Republic of Bangladesh.

Govt. of Bangladesh. (2005b). *Bangladesh Industrial Policy 2005*. Ministry of Industries, Government of the People's Republic of Bangladesh.

Hamdani, J., & Wirawan, C. (2012). Open Innovation Implementation to Sustain Indonesian SMEs. *Procedia Economics and Finance*, *4*, 223–233. doi:10.1016/S2212-5671(12)00337-1

Hossain, M., & Kauranen, I. (2016). Open innovation in SMEs: A systematic literature review. *Journal of Strategy and Management*, *9*(1), 58–73. doi:10.1108/JSMA-08-2014-0072

Indarti, N., & Langenberg, M. (2004). Factors Affecting Business Success Among SMEs: Empericical Evidences from Indonesia. In *Proceedings of the Second Bi-annual European Summer University*. University of Twente.

Jones, O., & Tilley, F. (Eds.). (2003). *Competitive Advantage in SMEs: organizing for innovation and change*. Wiley.

Kasekende, L. (2001). Financing SMEs: Uganda's Experience. In *Improving the Competitiveness of SMEs in Developing Countries: The Role of Finance to Enhance Enterprise Development* (pp. 97–107). United Nations.

Kearney, C. (2012). Emerging markets research: Trends, issues and future directions. *Emerging Markets Review*, *13*(2), 159–183. doi:10.1016/j.ememar.2012.01.003

Klein-Woolthuis, R., Lankhuizen, M., & Gilsing, V. (2005). A system failure framework for innovation policy design. *Technovation*, *25*(6), 609–619. doi:10.1016/j.technovation.2003.11.002

Kowalski, S. P. (2009). SMES, Open Innovation and IP Management: Advancing Global Development, A presentation paper on the Theme 2. In *The Challenge of Open Innovation for MSMEs - SMEs, Open Innovation and IP Management - Advancing Global Development*. WIPO-Italy International Convention on Intellectual Property and Competitiveness of Micro, Small and Medium-Sized Enterprises (MSMEs).

Kroll, H., & Liefner, I. (2008). Spin-off enterprises as a mean of technology commercialisation in a transforming economy – evidence from three universities in China. *Technovation*, *28*(5), 298–313. doi:10.1016/j.technovation.2007.05.002

Lee, S., Park, G., Yoonc, B., & Park, J. (2010). Open innovation in SMEs—An intermediated network model. *Research Policy*, *39*(2), 290–300. doi:10.1016/j.respol.2009.12.009

Li, J., & Kozhikode, R. K. (2009). Developing new innovation models: Shifts in the innovation landscapes in emerging economies and implications for global R&D management. *Journal of International Management*, *15*(3), 328–339. doi:10.1016/j.intman.2008.12.005

Liu, X. (2008). *China's Development Model: An Alternative Strategy for Technological Catch-Up*. SLPTMD Working Paper Series No. 020, University of Oxford.

López, A. (2009). Innovation and Appropriability, Empirical Evidence and Research Agenda. In The Economics of Intellectual Property: Suggestions for Further Research in Developing Countries and Countries with Economies in Transition. World Intellectual Property Organization.

Marisetty, V., Ramachandran, K., & Jha, R. (2008). *Wealth effects of family succession: A case of Indian family business groups*. Working Paper. Indian School of Business.

Narain, S. (2001). Development Financial Institutions' and Commercial banks' Innovation Schemes for Assisting SMEs in India. In *Improving the Competitiveness of SMEs in Developing Countries: The Role of Finance to Enhance Enterprise Development*. United Nations.

OECD. (2005). *Oslo Manual: Guidelines for Collecting and Interpreting Innovation Data* (3rd ed.). Organization for Economic Co-operation and Development.

Padmadinata, F. Z. S. (2007). Quality Management System and Product Certification Process and Practice for SME in Indonesia. *Proceedings of the National Workshop on Subnational Innovation Systems and Technology Capaicity Building Policies to Enhance Competitiveness of SMEs*. UN-ESCAP and Indonesian Institute of Science (LIPI).

Pandey, A.P. & Shivesh. (2007). *Indian SMEs and their uniqueness in the country*. Munich Personal RePEc Archive, MPRA Paper No. 6086. Available online at https://mpra.ub.uni-muenchen.de/6086/

Parida, V., Westerberg, M., & Frishammar, J. (2012). Inbound open innovation activities in high-tech SMEs: The impact on innovation performance. *Journal of Small Business Management*, *50*(2), 283–309. doi:10.1111/j.1540-627X.2012.00354.x

Payne, J. E. (2003). *E-Commerce Readiness for SMEs in Developing Countries: a Guide for Development Professionals*. Available at URL: http://learnlink.aed.org/Publications/Concept_Papers/ecommerce_readiness.pdf

Rahman, H., & Ramos, I. (2010). Open Innovation in SMEs: From Closed Boundaries to Networked Paradigm. *Issues in Informing Science and Information Technology*, *7*, 471–487. doi:10.28945/1221

Rahman, H., & Ramos, I. (2013). Open Innovation Strategies in SMEs: Development of a Business Model. In *Small and Medium Enterprises: Concepts, Methodologies, Tools, and Applications* (Vols. 1–4; pp. 281–293). Information Resources Management Association. doi:10.4018/978-1-4666-3886-0.ch015

Ruiz, P. P. (2010).*Technology & Knowledge Transfer Under the Open Innovation Paradigm: a model and tool proposal to understand and enhance collaboration-based innovations integrating C-K Design Theory, TRIZ and Information Technologies* (Master's dissertation). Management School of Management, University of Bath, UK.

Saini, D. S., & Budhwar, P. S. (2008). Managing the human resource in Indian SMEs: The role of indigenous realities. *Journal of World Business*, *43*(4), 417–434. doi:10.1016/j.jwb.2008.03.004

Savitskaya, I. (2009). Towards open innovation in regional Innovation system: case St. Petersburg. Research Report 214, Lappeenranta University of Technology, Lappeenranta.

Schaffers, H., Cordoba, M. G., Hongisto, P., Kallai, T., Merz, C., & van Rensburg, J. (2007). *Exploring business models for open innovation in rural living labs.* Paper from the *13th International Conference on Concurrent Enterprising*, Sophia-Antipolis, France.

Secretariat, U. N. C. T. A. D. (2001). Best Practices in Financial Innovations for SMEs. In *Improving the Competitiveness of SMEs in Developing Countries: The Role of Finance to Enhance Enterprise Development* (pp. 3–58). United Nations.

Siu, W. S. (2005). An institutional analysis of marketing practices of small and medium-sized enterprises (SMEs) in China, Hong Kong and Taiwan. *Entrepreneurship and Regional Development*, *17*(1), 65–88. doi:10.1080/0898562052000330306

Spithoven, A., Vanhaverbeke, W., & Roijakkers, N. (2013). Open innovation practices in SMEs and large enterprises. *Small Business Economics*, *41*(3), 537–562. doi:10.100711187-012-9453-9

Srivastava, D. K. (2005). Human resource management in Indian mid size operations. In Indian mid-size manufacturing enterprises: Opportunities and challenges in a global economy. Gurgaon: Management Development Institute.

Storey, D. (1994). *Understanding the small business sector.* International Thomson Business Press.

Takagi, Y., & Czajkowski, A. (2012). WIPO services for access to patent information - Building patent information infrastructure and capacity in LDCs and developing countries. *World Patent Information*, *34*(1), 30–36. doi:10.1016/j.wpi.2011.08.002

Thampy, A. (2010). *Financing of SME firms in India Interview with Ranjana Kumar, Former CMD, Indian Bank; Vigilance Commissioner, Central Vigilance Commission.* doi:10.1016/j.iimb.2010.04.011

United Nations. (2006). *Globalization of R&D and Developing Countries, United Nations Conference on Trade and Development.* UNCTAD/ITE/IIA/2005/6, United Nations.

Van Hemert, P., & Nijkamp, P. (2010). Knowledge investments, business R&D and innovativeness of countries: A qualitative meta-analytic comparison. *Technological Forecasting and Social Change*, *77*(3), 369–384. doi:10.1016/j.techfore.2009.08.007

Vanhaverbeke, W. (2017). *Managing open innovation in SMEs.* Cambridge University Press. doi:10.1017/9781139680981

Vanhaverbeke, W., Chesbrough, H., & West, J. (2014). Surfing the new wave of open innovation research. *New Frontiers in Open Innovation, 281*, 287-288.

Warner, A. (2001). *Small and Medium Sized Enterprises and Economic Creativity*. A paper presented at UNCTAD's intergovernmental Expert Meeting on "Improving the Competitiveness of SMEs in Developing Countries: the Role of Finance, Including E-finance, to Enhance Enterprise Development".

West, J., Salter, A., Vanhaverbeke, W., & Chesbrough, H. (2014). *Open innovation: The next decade*. Academic Press.

Williams, A. M., & Shaw, G. (2011). Internationalization and innovation in Tourism. *Annals of Tourism Research, 38*(1), 27–51. doi:10.1016/j.annals.2010.09.006

World Business Council. (2007). Promoting Small and Medium Enterprises for Sustainable Development. World Business Council for Sustainable Development.

Wynarczyk, P. (2013). Open innovation in SMEs: A dynamic approach to modern entrepreneurship in the twenty-first century. *Journal of Small Business and Enterprise Development, 20*(2), 258–278. doi:10.1108/14626001311326725

Xiaobao, P., Wei, S., & Yuzhen, D. (2013). Framework of open innovation in SMEs in an emerging economy: firm characteristics, network openness, and network information. *International Journal of Technology Management, 62*(2/3/4), 223-250.

Zahra, A. Z., & George, G. (2002). Absorptive capacity: A review, reconceptualization, and extension. *Academy of Management Review, 27*(2), 185–203. doi:10.5465/amr.2002.6587995

Zeng, S. X., Xie, X. M., & Tam, C. M. (2010). Relationship between cooperation networks and innovation performance of SMEs. *Technovation, 30*(3), 181–194. doi:10.1016/j.technovation.2009.08.003

ADDITIONAL READING

Albuquerque, E. M. (2007). Inadequacy of technology and innovation systems at the periphery. *Cambridge Journal of Economics, 31*(5), 669–690. doi:10.1093/cje/bel045

Alcorta, L., & Peres, W. (1998). Innovation systems and technological specialization in Latin America and Caribbean. *Research Policy, 26*(7–8), 857–881. doi:10.1016/S0048-7333(97)00067-X

Alexander, N., & Dawson, J. (1994). Internationalisation of retailing operations. *Journal of Marketing Management, 10*(4), 267–282. doi:10.1080/0267257X.1994.9964274

Bell, R. M., & Albu, M. (1999). Knowledge systems and technological dynamism in industrial clusters in developing countries. *World Development, 27*(9), 1715–1734. doi:10.1016/S0305-750X(99)00073-X

Bougrain, F., & Haudeville, B. (2002). Innovation, collaboration and SMEs internal research capacities. *Research Policy, 31*(5), 735–747. doi:10.1016/S0048-7333(01)00144-5

Brunswicker, S., & Ehrenmann, F. (2013). Managing open innovation in SMEs: A good practice example of a German software firm. *International Journal of Industrial Engineering and Management*, *4*(1), 33–41.

Buckley, P. J., & Casson, M. C. (1985). *The economic theory of multinational enterprise*. Macmillan. doi:10.1007/978-1-349-05242-4

Cook, I., & Horobin, G. (2006). Implementing eGovernment without promoting dependence: Open source software in developing countries in Southeast Asia. *Public Administration and Development*, *26*(4), 279–289. doi:10.1002/pad.403

Dodgson, M., Gann, D., & Salter, A. (2008). *The management of technological innovation: Strategy and practice*. Oxford UP.

Edwards, T., Delbridge, R., & Munday, M. (2005). Understanding innovation in small and medium-sized enterprises: A process manifest. *Technovation*, *25*(10), 1119–1120. doi:10.1016/j.technovation.2004.04.005

Ernst, D. (2006). Innovation offshoring: Asia's emerging role in global innovation networks. *East-West Center Special Reports*, *10*, 1–50.

Ili, S., Albers, A., & Miller, S. (2010). Open innovation in the automotive industry. *Research Management*, *40*(3), 246–255.

Jacob, M., & Groizard, J. L. (2007). Technology transfer and multinationals: The case of Balearic hotel chains' investments in two developing economies. *Tourism Management*, *28*(4), 976–992. doi:10.1016/j.tourman.2006.08.013

Jaeger, A. (1990). The applicability of western management techniques in developing countries: a cultural perspective. In A. Jaeger & R. Kanungo (Eds.), *Management in Developing Countries* (pp. 131–145). Routledge.

James, J. (2002). Low-cost information technology in developing countries: Current opportunities and emerging possibilities. *Habitat International*, *26*(1), 21–31. doi:10.1016/S0197-3975(01)00030-3

Jones, O., & Tilley, F. (Eds.). (2003). *Competitive Advantage in SMEs: organizing for Innovation and Change*. Wiley.

Lall, S. (2000). The technological structure and performance of developing country manufactured exports, 1985–1998. *Oxford Development Studies*, *28*(3), 337–369. doi:10.1080/713688318

Lall, S., & Teubal, M. (1998). Market stimulating technology policies in developing countries: A framework with examples from East Asia. *World Development*, *26*(8), 1369–1385. doi:10.1016/S0305-750X(98)00071-0

Lundvall, B.-A. (Ed.). (1992). *National Systems of Innovation: Towards a Theory of Innovation and Interactive Learning*. Pinter.

Maranto-Vargas, D., & Gómez-Tagle Rangel, R. (2007). Development of internal resources and capabilities as sources of differentiation of SME under increased global competition: A field study in Mexico. *Technological Forecasting and Social Change*, *74*(1), 90–99. doi:10.1016/j.techfore.2005.09.007

Narula, R. (2004). R&D collaboration by SMEs: New opportunities and limitations in the face of globalisation. *Technovation, 25*(2), 153–161. doi:10.1016/S0166-4972(02)00045-7

NESTA. (2008). *UK global innovation: Engaging with new countries, regions and people*. NESTA.

Niosi, J. (2008). Technology, development and innovation systems: An introduction. *The Journal of Development Studies, 44*(5), 613–621. doi:10.1080/00220380802009084

Popa, S., Soto-Acosta, P., & Martinez-Conesa, I. (2017). Antecedents, moderators, and outcomes of innovation climate and open innovation: An empirical study in SMEs. *Technological Forecasting and Social Change, 118*, 134–142. doi:10.1016/j.techfore.2017.02.014

Prashantham, S. (2008). *The internationalization of small firms: A strategic entrepreneurship perspective*. Routledge.

Rusinovic, K. (2008). Transnational embeddedness: Transnational activities and networks among first- and second-generation immigrant entrepreneurs in the Netherlands. *Journal of Ethnic and Migration Studies, 34*(3), 431–451. doi:10.1080/13691830701880285

KEY TERMS AND DEFINITIONS

Developed Economies: While there is no one set definition, but typically a developed economy refers to a country with a relatively high level of economic growth and security. Some of the most common criteria for evaluating a country's degree of development are its per capita income or gross domestic product (GDP), the level of industrialization, general standard of living and the amount of widespread infrastructure. Increasingly other non-economic factors are included in evaluating an economy or country's degree of development, such as the Human Development Index (HDI) which reflects relative degrees of advancement in education, literacy, and health.

Developing Economies: Comprise low- and middle-income countries where most people have lower standard of living with access to fewer goods and services than most people in high-income countries. Developing countries are broadly split into two categories, the middle-income and the low-income groups.

Emerging Economies: Are the most economically progresses of developing countries. In terms of GNP per capita, they correspond to the medium-low and medium-high country groups but are characterized by a regulated and functioning securities exchange, or in the process of developing one, and the fact that shares traded on the stock exchanges must be available for purchase by foreign investors, even if subject to certain restrictions.

Entrepreneurs: An entrepreneur is a person who has possession of a new enterprise, venture or idea organizes, operates a business or businesses and assumes significant accountability for the inherent risks and the outcome.

Entrepreneurships: It is the process of discovering new ways of blending resources. When the market value generated by this new blending of resources is greater than the market value these resources can generate elsewhere individually or in some other combination, then the entrepreneur makes a profit.

First Economy Enterprises: These are the enterprises that are comprised of established businesses in sustained form.

Second Economy Enterprises: These are the form of enterprises that are mainly belong to the working poor, or marginalized communities, and working in the informal economy.

ENDNOTES

[1] Enterprises shall be categorized using the following definition (fixed investment implies exclusion of land and building, and valuation on the basis of current replacement cost only): Small enterprise: an enterprise should be treated as *small* if, in today's market prices, the replacement cost of plant, machinery and other parts/components, fixtures, support utility, and associated technical services by way of capitalized costs (of turn-key consultancy services, for example), etc, excluding land and building, were to be up to Tk. 15 million; Medium enterprise: an enterprise would be treated as *medium* if, in today's market prices, the replacement cost of plant, machinery, and other parts/components, fixtures, support utility, and associated technical services (such as turn-key consultancy), etc, excluding land and building, were to be up to Tk. 100 million; a. For non-manufacturing activities (such as trading or other services), the Taskforce defines: Small enterprise: an enterprise should be treated as *small* if it has less than 25 workers, in full-time equivalents; Medium enterprise: an enterprise would be treated as *medium* if it has between 25 and 100 employees.

[2] In India, the industrial sector has two broad segments viz., (a) Small Scale Industries (SSI) and (b) Others (i.e. medium and large industries). The Government of India notifies the definition of small-scale industry from time to time based on the investment ceiling. The present definition is, "an industry in the small scale sector shall have investment in plant and machinery not exceeding INR 10 million" (approx. US$22,000). A sub-component of micro enterprises, known as the "Tiny Sector" forms part of the overall SSI sector. Medium sized industries are out of the purview. India, thus, follows the concept of SSIs and not SMEs.

[3] In India, until recently there has been no formal concept of SME or medium enterprises. However, the term small scale industry (SSI) is well known; this is different from the SME sector in other countries. The Government of India had a policy of providing assistance of different types to SSIs through various state agencies. Lately, Indian Parliament has enacted the Micro, Small and Medium Enterprises Development Act, 2006.1 As per this Act, medium manufacturing or production enterprises are those which have an investment in plant and machinery between Rs. 50 million and 100 million (1$ US = Rupees 40.10 approximately in July 2007). The investment referred to in this definition is that in ''initial fixed assets'' i.e., the plant and machinery (which excludes land & building). Under this Act, a micro enterprise has been defined as one where the investment in plant and machinery does not exceed Rs. 2.5 million and a small enterprise as one where such investment is more than Rs. 2.5 million but does not exceed Rs. 50 million. Whereas, a medium enterprise is one in which the investment limit is between Rs. 50 million and Rs. 100 million. In this Act there is no reference to the term SME. One may, however, combine the definitions of small and medium enterprises to derive a concept of SME. This would mean that an SME in the Indian context is an enterprise in which the investment in plant and machinery is between 2.5 million and 100 million.2 The definition of the terms ''small'' and ''medium'' enterprise in India is investment specific, while in the rest of the world it reflects a combination of factors including terms of employment, assets or sales or combination of these factors (Saini & Budhwar, 2008).

[4] http://www.openlivinglabs.eu

5 SMEs are widely defined in terms of their characteristics, which include the size of capital investment, the number of employees, the turnover, the management style, the location, and the market share. Country context plays a major role in determining the nature of these characteristics, especially, the size of investment in capital accumulation and the number of employees. For developing countries, small-scale generally means enterprises with less than 50 workers and medium-size enterprises would usually mean those that have 50–99 workers. In Uganda, a small-scale enterprise is an enterprise or a firm employing less than 5 but with a maximum of 50 employees, with the value of assets, excluding land, building and working capital of less than Ugshs. 50 million (USD 30,000), and an annual income turnover of between Ugshs. 10–50 million (USD 6,000–30,000). A medium-size enterprise is considered a firm, which employs between 50–100 workers. Other characteristics have not been fully developed.

Chapter 6
Visualising Ethics Through Hypertext

Domenic Garcia
Department of Education, Malta

ABSTRACT

Ethics has always been a main domain of philosophical study. Hypertextual discursive expression has never been considered as the technology that can revolutionise the way we interpret ethical events and actions. This chapter intends to show that hypertextual discursive expression is not only useful, but also vital in any undertaking that seeks to reframe our ethical deliberations and judgements. The final objective is to convey the message that technology is becoming ever more essential in transforming philosophical thinking.

METHODOLOGY

The methodology is qualitative, focusing on the hermeneutical aspect, validity and effect of hypertextual writing in contrast with unilinear writing. The reason for choosing a qualitative method arises from 1) the fact that the work, which is proposing a new software that can redeem the individual from the confinement of judging actions as being totally 'right' or 'wrong' without considering the grey areas, is at its preliminary a stage and 2) the software has not, as yet, been fully implemented and tested. At this stage in this research it is best to first embark upon documentary analysis. The data is here divided into two subjects – the philosophical and the technological topics. The first data is directed towards the philosophy of interpretation and reference is here being made to Gadamer 2004; Linge 2008; and Grondin 1994. In the second part, the analysis of data shifts to the pragmatic/technological aspect, focusing mainly on the application of a software (Garcia, 2020) that would render ethical judgement more emancipative. The objective of the data analysis is not directly involved with the new software that is introduced here but contains, at the same, features that are either related to, or can be identified with, features of the software (Eyman, 2015; Bell, 2014; Macdonald 2018; Amadieu et. al., 2015). Reference is also made to data which could be distinct from the software proposed (AlAgha, 2012; Baetens &Truyen, 2013; Koehler, 2013; Marcante, 2011).

DOI: 10.4018/978-1-7998-5849-2.ch006

INTRODUCTION

This chapter[1] aims to forge a pragmatic approach to interpreting and judging actions as 'right' or 'wrong' through a visual description of hypertextual writing. Embracing such a practice will have a positive consequential effect over the prevailing and deeply rooted philosophical and jurisprudential traditions of hermeneutical understanding and ethics, potentially encouraging an innovative way of achieving social justice - a pressing issue in contemporary times. However, before elaborating on this visual description, a brief explain of the Gadamerian notion of philosophical hermeneutics with reference to his notion of 'fusion of horizons will be given.' The purpose of this overview is to identify fresh avenues of discussion and reflection. One of these compellingly new paths will propose re-thinking hermeneutical understanding of ethical deliberation and judgement within the framework of a specific type of hypertextual writing. The present aspiration is to fine-tune our perception and adopt a broader frame of mind when judging actions in the near future.

THE PHILOSOPHICAL BASES OF HYPERTEXT – *ITS HERMENEUTICAL IMPLICATION*

Contemporary studies of philosophical hermeneutics are predominantly centred on the Gadamerian conception, giving due importance to the dialogue that should take place between what has been written in the past and the present situation of the interpreter. This Gadamerian notion contributed greatly to understanding interpretation. This chapter shall move a step further by shifting attention to another discussion, the aim of which is to encourage the reader to actively consider a profound change in the approach to hermeneutic study. This shall be done by entertaining the possibility of understanding and reshaping future hermeneutics within a different form of discursive expression - hypertextual writing. The idea is not to start and end the discussion with hermeneutics but to enrich it even further by considering, for the most part, ethical deliberation and judgement. It is not the intention to approach this discussion in a manner that is fraught with presumption. The intention is, rather, to broach the subject in the form of a plea addressed to the reader of this work to come up with constructive thoughts on what might be the future of interpreting and judging actions as 'right' or 'wrong.' It is hoped that the reader, is thus encouraged to adopt such a positive frame of mind, and will also be inclined to contribute to the setting up a platform of discussions to create a new framework that can be better placed to deal with such an important subject as the future of judgement and interpretation across the globe. This chapter, crude as it may be, will lay some foundations.

In order to do this, it is pertinent to first reconsider the Gadamerian notion of hermeneutics itself, and thereby foreground the important contribution of this eminent thinker to the subject matter of this work. It shall here be demonstrated that the Gadamerian notion of hermeneutics is in crises but it shall not do so by departing from his understanding of the 'fusion of horizons' but by showing, rather, that this 'fusion of horizons' needs to be revisited and reconsidered under the wealth of possibility offered by a working knowledge of hypertextual writing. This chapter will, within the practicality of a computer application, rethink the Gadamerian notion of hermeneutics through the hypertextual framework. The need to revisit Gadamer's philosophical hermeneutics arises for two main reasons. The first is considered to be intrinsic to Gadamer's considerations of hermeneutics and the second is to be considered as extrinsic. The intrinsicality lies in what is call the *metaphysics of discourse*, a term which highlights the fact that the way words are written down, following each other to form a structured whole, will create an infrastructure of absolutizing discourse, giving a teleological spirit to discourse itself. This form of discourse is, in fact, the house of rationality and is the sole host of the setting in which discursive rationality is displayed. Gadamer never directly challenged this scientificization that lies hidden in *architecture of language* (Garcia, 2020, p.97) itself and it is, partially, the purpose of this chapter to do so in order to understand the validity of hypertextual writing when it comes to interpretation and judgement. The extrinsicality lies in the fact that, since one lives in a multidimensional society, understanding ought to be supplied with a technology that offers the possibility to express discourse in multiple ways. It will be argued that, when compared to a discursive infrastructure that is unilinear and conformative, the technology of hypertextuality is markedly closer and more sensitive to both human nature and a multidimensional society. It becomes essential, therefore, to keep the idea of multiplicity constantly present. The importance of having the possibility of discoursing in a multiplicity of horizons, rather than focusing on just one horizon, becomes a prerequisite. It will, therefore, be more appropriate to think of *fusions of horizons* rather than a 'fusion of horizons.'

Apart from the proposed aim of this chapter, it would be impossible to give an exhaustive exposition of the Gadamerian notion of hermeneutics. The purpose is, rather, to demonstrate the metamorphoses Gadamer's concepts undergo when envisaged within the framework of hypertextual writing which, as will be argued, offers a better form of justice, thus vindicating its seekers. As already pointed out, this will be done in a very pragmatic way - conjoining hermeneutic understanding with the digital technology of writing, and taking stock of the implications this has on the handling of ethical deliberations of actions deemed 'right' or 'wrong' as well as the issuance of justice (Garcia 2020, p. 7). This shall be done by introducing a computer application that will be vital in transforming the way one will handle and interpret justice in the future. This chapter offers a blueprint of what this computer application would look like. The figures presented here are solely based on an imagination of how this software could appear.[2] Attention should be taken to the fact that the issue at stake lies not in this hypothetical technology as such, but in the far-reaching effects of what is, clearly, a simple approach. In other words, the proposed hypertextual technology could reveal new ways in which one could better interpret and handle justice in the future.

Essentially, the phrase *metaphysics of discourse* borrows its meaning from Heidegger's critique of metaphysics. Heidegger points to the metaphysical error (Linge, 2008, p. xlviii) that established a universal understanding of Being instead of focusing on individual entities – beings (Woodward, 2009, p. 57). In its simplified form, this "ontological difference" imposed certain standards for who beings are, by universalising a Being that can fit all beings. This was patently untenable for Heidegger as it made us forget "being itself as the event of disclosure or openness that allows beings to come forward into unconcealedness" (Linge, 2008, p. xlix). Linge explains how Heidegger found fault with the idea of envisaging beings as static and permanent, unchanged by time or any form of culture or tradition. According to Heidegger, this way of understanding beings started with Plato and eventually came to be defined as humanism in the Western world (Linge, 2008, p. xlix). This metaphysical representation resulted in a misrepresentation or what can be described as an artificialisation of who man, as a thinking being, is. This permanent and stable structure stands for the main concept of truth up to the modern period (Garcia, 2020, p. 123).

This kind of thinking, however, came under serious attack, especially in the twentieth century, as scientific ways of understanding human beings started to become less popular among social scientists and philosophers. The metaphysical idea of acquiring an absolute truth was inherited by science as it provided the methods of acquiring a stable foundation by which it could provide an absolute and universal truth. According to Grondin (1994), Heidegger alluded to his dissatisfaction with this "fixed Archimedean point by exposing its metaphysical presuppositions" (Grondin, 1994 p. 107). Heidegger wanted to show the importance of the 'hermeneutic-existential' as over the 'apophantic as' when studying humanity (Garcia, 2020, p. 137). But as long as one remains stuck in "Aristotle's phenomenological approach to the analysis of the logos, our understanding of Dasein remains a "superficial 'theory of judgement,' in which judgement becomes the binding or separating of representations and concepts" (Heidegger, 1962, para 159). Heidegger wanted to 'transgress the taken-for-granted assumption of being and wanted, instead, to study Dasein in its lifeworld' (Garcia, 2020, p. 138). He wanted, therefore, to divest Dasein from the analysis of the as-structure – in its domain of neutral things, in its uniform plane. He aimed, rather, to study being as an existential-hermeneutic *as* (Garcia, 2020, p. 138). This understanding of the lifeworld was taken up by Gadamer in *Truth and Method*. According to Gadamer, truth follows when the scientificization of the method is overtaken by the study of the lifeworld. Linge maintains that Gadamer concurs with the idea voiced by Heidegger that Dasein should not be understood in the "idealist sense of the objectification of finite spirit or in Husserl's implied sense of the lifeworld as disclosive of the constitutive accomplishments of the transcendental ego." (Linge, 2008, p. xlvi)

Heidegger provides us with a humane understanding by giving priority to the situated character of the individual rather than imposing a method by means of which the individual could be analysed and judged. Heidegger's intention is to understand Dasein by considering temporal possibilities rather than entertaining a timeless analysis which makes one forget oneself. According to Gadamer:

The question of being, as Heidegger poses it, breaks into an entirely different dimension by focusing on the being of Dasein which understands itself....Being and Time already begins to counteract the forgetfulnessofbeing which Heidegger was later to designate as the essence of metaphysics. What he calls the 'turn' is only his recognition that it is impossible to overcome the forgetfulness of being within the framework of transcendental reflection.

Heidegger's aim was to understand Dasein by considering temporal possibilities (Grondin 1994 p. 107). He wants to undo historicism by rendering invalid the use of scientific methodology in history (Garcia, 2020, p.138). According to Heidegger, 1962:

In the end, the emergence of a problem of 'historicism' is the clearest symptom that historiology endeavours to alienate Dasein from its authentic historicality. Such historicality does not necessarily require historiology. (para: 448)

Like Heidegger, Gadamer (2004) maintains that:

Everything that makes possible and limits Dasein's projection ineluctably precedes it. This existential structure of Dasein must be expressed in the understanding of historical tradition as well, and so we will start by following Heidegger (p. 254).

Gadamer wanted to do away with the established universal method of doing hermeneutics, and thus go against precepts held by, among others, "Dilthey, Droysen and neo-Kantians" (Grondin, 1994, p. 108). According to Grondin, Gadamer is more in line with Helmholtz's understanding of human science. For Gadamer, therefore, human science does not follow the strict rules of natural science and, therefore, logical induction (Garcia, 2020, p. 139). For Gadamer, human sciences are more akin to the kind of "artistic induction stemming from an instinctive sense or tact, for which there are no definable rules" (Grondin, 1994. P. 108). Like Heidegger, Gadamer wants to uncover our "own foreprojections" (Grondin, 1994, p.111).

Both Heidegger and Gadamer set the stage for a new understanding of hermeneutics by removing any form of calculability in understanding, including the way one comes to understand truth and the meaning of texts. It is with this in mind that one can introduce an understanding of what is here called the *metaphysics of discourse* or the *metaphysics of text* or, rather, the *metaphysics of the architecture of discourse or text*. These terms refer to that unilinear mode of writing that automatically compels the reader or its user to arrive at a definite and absolute conclusion. Texts are typically characterized by a metaphysical feature, namely that of setting up teleological ways of understanding and coming up with a tidy formula of how 'truth' can be obtained. This feature has long been present in writing, and goes back, in fact, to its early stages. In *How language Works*, (2005) Crystal explains that expression in writing works by having words follow each other to form a structured whole. Text may be written from left-to-right or right-to-left as in Arabic, or from top-to-bottom as in traditional Japanese or from bottom-to- top as in forms of Ancient Greek. This is our method of writing and it is the one passed on to early learners. Children learn this mode of writing at school and there is no known way of expressing ourselves in writing (Garcia, 2020, p. 2). It can be said that this mode of writing is very effective, so much so that it has become the standard form of writing which, given its uncompromising nature, admits nothing but a definite and absolute truth. Macdonald (2018) emphasizes the idea that there cannot be just one side to a story and that there is always more than one perspective. These various facets tend to remain hidden from view since the reader invariably succumbs to the unilinear format of the story.

Without delving into the intricacies of the history of writing, this area of study merits, nevertheless, a brief overview. The purpose is to draw the reader's attention to the importance of the architecture of writing and how it might transform linguistic expressions in such a way as to conceal thought rather than provide the openness that would allow beings to come forward into "unconcealedness" (Linge, 2008, p. xlix). Some perspectives will, therefore, be mentioned, borrowing from scholars who, while not necessarily involved with philosophy or hermeneutics in particular, may still offer significant insights into the subject matter of this chapter. In her article, *Multivariant Narratives*, Marie-Laure Ryan describes how writing "froze the free ordering of plots into a fixed sequence" (Ryan, 2004, p. 415). In their introduction to *History of the Book* (2009), editors Simon Eliot and Jonathan Rose write about the nature of their role as book historians, describing how:

Book historians do not claim that books explain everything, but they do recognize that books are the primary tools that people use to transmit ideas, record memories, create narratives, exercise power, and distribute wealth (p. 1).

One may also include other important factors that force linguistic expressibility to confine itself to a method that limits, rather than offers, the possibility of what Heidegger may have been aiming for when describing the unconcealedness of Dasein and the de-scientificization of the human sciences. As the brief description above attests, Heidegger and Gadamer wanted to emancipate the individual from the metaphysical, and from the scientificization and methodological way of treating Dasein. However, it should equally be noted that it is the unilinearisation of the text and the physicality of the book, which should also be marked as the root cause of scientificization and of the metaphysicization of Dasein. This will be the focus of this work, intended to bring together a critique of the unilinear architecture of writing and my theorizations regarding hypertextual writing. However, it is important to mention that such a critique of unilinear architecture was left unexplored not only by both Heidegger and Gadamer but also by traditional mainstream philosophers, especially those specializing in philosophical hermeneutics. It will, therefore, be argue that the metaphysics of discourse, or the metaphysics of the text, concealed the individual from view, confining communication to rigid linguistic expression and hindering the free flow of thoughts, rather than encouraging emancipation.

By observing how the physical aspects of the book evolved, it becomes clear that authentic linguistic expression necessarily undergoes a number of changes in order to assume the required characteristics of a book's contents. It must, therefore, be suitably refashioned if it is to be in line with the dictates imposed by the regimentation of the book. In the southern Iraqi marshes, as early as the late millennium BC, writing in linear order became essential to depict the sequence of words in the form of a sentence (Robson, 2009, p.69). In order to maintain this form, scribes developed the cuneiform – the pressing of a length of reed into a clay stylus in order to create linear strokes or wedges (Robson, 2009, p. 69). In Egypt, classical Greece and Rome, papyrus was rolled rather than folded so as to prevent breakage. It was eventually replaced by the codex in the second century AD. To change the Greek or Latin roll into a book, the scribe would copy, starting from the left, "distributing the text over the entire roll into columns of the same width and height, little dots in the upper margin may have guided him too in keeping the intercolumnia, the open space between columns, uniform." (Roemer, 2009, p. 93) According to Roemer, the popularity of the codex increased mainly due to its practicality, the codex obviously being less fragile than its predecessor, the roll. The pages of the book could lie "flat on top of each other." (Roemer, 2009, p. 93) The book also had greater capacity as both sides of its pages could be written on and it could easily be opened and closed whereas the roll, by contrast, had to be rolled completely back after use (Roemer, 2009, p.93).

These scholars give reason to question whether the fluidity and the variability of expression could be made to fit into the scientificization, the practicality, the physicality and the politics of the codex or the book, in general (Garcia, 2020, p. 12). Moreover, the architecture of the text also contains, in its very nature, a number of seductive and deceptive features that lure the author towards a writing path the nature of which may automatically coerce the utterer to come to a premature or hasty, unexhaustive conclusion (Garcia, 2020, p. 2). 'The unilinear infrastructure's lack of potential for a fluid and variable linguistic expression gives the speaker no choice but to opt for an either/or situation' (Garcia, 2020, p. 13). In cases like, for example, ethical deliberations and judgements, that which is supposed to be a redemptive tool proves instead to be a castigative tool, potentially giving the impression, albeit false, that it is a redemptive medium. Without realizing, a setback is inevitably experienced the moment a violent unilinear discursive path is used. This kind of violence is on a par with the violence inherent in the flawed assumption that 'metaphysical idioms of the platonic style are the sole providers of an absolute 'truth'' (Garcia, 2020, p. 8). The dialectic process that results in an absolute, universal truth is consistent with the chronological process of linguistic expression in that it is designed to engender nothing but a rigid and absolute answer. It can be said, in fact, that it is only by means of this medium, namely this architectural style of discourse, that dialectics becomes possible, thereby giving the impression that a single and unparalleled finality does indeed exist. Insisting on an inviolable conclusion as the end result of our utterances is in itself a subtle act of violence which, while appearing to emancipate us as individuals, effectively does the opposite to forcefully undermine each act of verbal communication.

This unilinearisation of writing sets the stage for a restricted mode of communicating since it provides the only means by which truth is acquired. This makes unilinear discursive expression metaphysical by nature, a feature which has long passed unnoticed. While unilinear communication has evidently been embraced as the only tool available to emancipate Dasein, it has, instead, contributed to the forgetfulness of Dasein by not allowing thought to be fluid. This architectural style has dominated the form of all thought processes, forcibly shaping each thought to accommodate the confines of the book.

With this background knowledge, one may start to question whether the Gadamerian notion of fusion of horizons does effectively take place. Gadamer rightly maintains that one cannot, in an attempt to understand and interpret, step out of one's own culture or tradition. Gadamer uses the term 'prejudice' (Vorurteile) not in a pejorative sense but in a positive one by taking it to mean 'prior understanding' – a precondition of an interpreter's own form of life. It stands for, in Gadamer's view, the initial situation from which the interpreter automatically proceeds. This prejudice or historicity should not impede one from understanding. Gadamer takes his lead, in fact, from Heidegger's concept of fore-understanding, the idea of which is to help one better understand any phenomenon. According to Gadamer (2004) understanding a tradition:

Undoubtedly requires an historical horizon. But it is not the case that we acquire this horizon by placing ourselves within a historical situation. Rather, we must always already have a horizon in order to be able to place ourselves within a situation. For what do we mean by 'placing ourselves' in a situation? Certainly not just disregarding ourselves. This is necessary, of course, in what we must imagine the other situation. But into this situation we must also bring ourselves...To acquire a horizon means that one learns to look beyond what is close at hand – not in order to look away from it, but to see it better within a larger whole and in truer proportion...hence the horizon of the present cannot be formed without the past. There is no more an isolated horizon of the present than there are historical horizons. Understanding, rather is always the fusion of horizons we imagine to exist by themselves. (p. 273)

All this appears to make perfectly good sense. What if, however, this idea of seeing better by grasping the 'larger whole and in truer proportion' is, in reality, severely limited? The act of communicating requires that all thought be narrowed down in order to meet the demands imposed by the castigative unilinearity of the text dominated, as it is, by a teleological force. By considering this issue, readers are invited to address whether or not the truth that lies within the fusion of horizons is, in fact, limited by the unilinearisation of the text (Garcia, 2020, p. 134). Individuals have been taught how to navigate and plot an entrenched discursive path and the more this path becomes chronological, the more unlikely the potential for knowingness becomes. In spite of all the good intentions to achieve better understanding, the individual isin fact slated to fail because he or she is doing the very opposite by digging deeper into unknowingness and forgetting the potential of who he or she really is. The result is that he or she believes that a comprehensive understanding has been reached by fusing the past with the present situation. But this fusion is only an artificial fusion as it narrows down our supposed understanding to the bare minimum, offering nothing but an oddly single strand of understanding, the singularity of which is built upon the idea of a metaphysical absolute 'truth'. With this in mind, it is essential to note that the unilinear architecture of writing has resulted in the rarely challenged idea of one beginning and one end which allows the individual no freedom to express himself in as exhaustive a manner as he might deem proper.

As it has been pointed out already in the book, *Rethinking Ethics Through Hypertext*, all traditional discursive expressions – whether of a scientific nature or an ethical nature - typically lure us into what is called an energising force that inexorably pushes those of us seeking to express ourselves towards a point of finality (Garcia, 2020, p.65). This is particularly concerning when it comes to expression of an ethical nature, the very issue on which this work is focused.

The intended proposal is an idea of a richer alternative the concept of *understanding better*. A cue from Gadamer's perspective shall be considered and proceed to suggest a more authentic form of *fusion of horizons*. This shall be done by introducing a type of hypertext that accommodates not simply one type of fusion of horizons but is so designed as to take into account what is referred to as *fusions of horizons* (Garcia, 2020, p.95). The reason for this is that what is said at the moment of utterance leaves many things unsaid. What ends up being uttered is, in fact, only a partial rendering of what the individual would have liked to convey (Garcia, 2020, p.117). The unilinear architecture of writing words, assembled after each other to form a structured whole, inevitably distances itself from the threedimensional character of reality as well as actual events as they will have unfolded in real time (Garcia, 2020, p.58). It seems that Gadamer had hinted at a kind of unfulfillment when a genuine conversation takes place. This is what Gadamer had to say in the beginning of part three of his *Truth and Method*:

We say that we "conduct" a conversation, but the more genuine a conversation is, the less its conduct lies within the will of either partner. Thus a genuine conversation is never the one we wanted to conduct. Rather, it is generally more correct to say that we fall into conversation, or even that we become involved in it. The way one word follows another, with the conversation taking its own twist and reaching its own conclusion, may well be conducted in some way, but the partners conversation are far less the leaders of it than the led. No one knows in advance what will "come out" of a conversation. Understanding or its failure is like an event that happens to us. Thus we can say that something was a good conversation or that it was ill fated. All this shows that a conversation has a spirit of its own, and that the language in which it is conducted bears its own truth within it – i.e., that it allows something to "emerge" which henceforth exists (p. 385)

This may give reason to believe that the architecture of language is intrinsically at fault when it comes to an exhaustive understanding. This work is therefore meant to challenge the veracity of a fusion of horizons as the mode of understanding and interpreting texts and people. This is not to say that Gadamer's approach should be dismissed. On the contrary, it is thought that his idea of a fusion of horizons deserves more credit particularly, given its potential to incorporate difference, in contexts of multiculturalism. Although unilinearity works efficiently in all modes of discourse, it will only lead to one final conclusion thus emulating the metaphysical idiom of a universal understanding with its insistence on one finite truth. The efficiency of unilinear writing is convenient but it comes at a cost, for in promising to deliver one finite truth, it might consequently not be exhaustive of all façades of truthfulness, especially in issues of an ethical nature.

As pointed out by Gadamer, a conversation or a text might not have successfully communicated thoughts in their entirety. The individual might feel that a lot has been left unsaid. He or she may occasionally believe that re-writing or re-telling an event will set the record straight and vindicate him or her (Garcia, 2020, p.66). However, a unilinear expression is incapable of exhausting all the relevant discursive expression that will shed more light on the description of an action or event. This is especially the case when it comes to discourse ethics. What is being suggested is not an erasure of the original narrative expressed by the speaker, but an amalgamation of his or her different narratives (Garcia, 2020, p.66). Such an amalgamation makes it likelier for justice to be done, particularly in view of the limitations of a single absolute narrative. This amalgamation of different narratives could become available through a multiplicity of discursive paths. It is here proposed that such a robust architecture augurs well for the realization of an emancipative discourse. It is useless, or even perhaps dangerous, to scientificize the complexity of ethical discourse through a unilinear mode of expression. Ethical deliberations need a technology that can handle such complexity.[3] Discourse ethics should therefore be equipped with a mode of expression that can expose all valid discourse, thus exhausting all perspectives. The proposed project seeks to give a better understanding of reality by creating a type of hypertextual writing that tends to visualize events which are then classified as 'right' or 'wrong' in terms of ethical deliberations. While this project's focus is markedly distinct from the processes involved in the production of stories, poems or fiction (Koehler, 2013, p. 383, Marcante, 2011), it nevertheless tries to emulate and adopt the pedagogical skills of creative writing as proposed by Koehler. Like Shackelford (2014a), it would here be important to explore further that which may be left 'unexplored' (Shackelford, 2014a, p. 30) using 'emerging digital writing practices' (Shackelford, 2014a, p. 29) to do so. However, the idea is not simply to explore and examine techniques alternating between the print medium and digital writing – hypertext. This work will also explore another aspect of this form of digital writing: its use as a source for deliberation in the context of the courtroom rather than as a construct of the narrative of fiction.

What is crucial in this attempt to rework discursive expression is the opportunity that is provided to the individual to validate all the things that he or she wants to say. Discourse ethics should not create an annihilation of expression but an amalgamation of discourse – a democratisation of expression and an equality of expressive discourse. It is only then that the truth of the *fusionsof horizons*is experienced. This is of course not an easy undertaking, particularly because it makes hermeneutics increasingly more complex as the opportunity to bring a multiplicity of expressions to the fore becomes a real possibility. The difficulty does not, however, mean that one should give up this responsibility. On the contrary, one should find ways to navigate the various discourses in order to achieve multiple understandings of justice rather than simply one final understanding. This is particularly important in cases where the person charged of a criminal offence is to face prosecution. It is therefore the responsibility and also in the interest of all to see the manifestation of a multiplicity of discursive paths when it comes to deliberation and the issuance of justice.

By acknowledging that different patchworks of discourse do indeed exist and by exposing all of them in order to achieve a more thorough level of understanding, one is significantly better informed. This contrasts sharply with situations in which a person is exposed to a linearly expressed story, the style of which is typically rigid and artificial. Conversation is broadened when it does not rely on one single consensus that is routinely established by the formal discourse of an institutional setup, in the style proposed by Habermas. A non-teleological discourse is closer to a paralogical style of Lyotard. But this, in turn, does not mean doing away with normative discourse. As argued in Rethinking Ethics Through Hypertext (2020), the non-teleological way of 'seeing' discourse gives the reader the power to understand without, at the same time, conditioning him to accept a particular normative, discursive conclusion. This is particularly important in our post-truth era where giving attention to one particular conclusion becomes unrealistic for a multicultural mentality. A post-truth era cannot be sold while simultaneously resorting to a metaphysical way of expression. A post-truth era demands a change in ways of discursive expressions. The rationale of ethical deliberations lies in their potential to offer each participant the opportunity to oversee the entire landscape of an event, story or testimony. As explained, this expansive range of information is even more crucial when it comes to conducting interrogations and describing certain events, particularly those which take place in legalistic institutions. It can be said that Gadamer (2008) is already hinting at the limitations of language in such contexts when he notes that:

A person who has something to say seeks and finds the words to make himself intelligible to the other person. This does not mean that he makes 'statements.' Anyone who has experienced an interrogation – even if only as a witness – knows what it is to make a statement and how little it is a statement of what one means. In a statement the horizon of meaning of what is to be said is concealed by methodical exactness; what remains is 'pure' sense of the statements. That is what goes on record. But meaning thus reduced to what is stated is always distorted meaning.

To say what one means, on the other hand – to make oneself understood – means to hold what is said together with an infinity of what is not said in one unified meaning and to ensure that it is understood in this way. (p. 464)

A lot of what is said by Gadamer in the above quotation is echoed in this chapter. A person pleading for an excuse, especially in a formal setup, may possibly realize that he is unable to redeem himself as the statements made in his type of predicament tend to, as Gadamer observes, say little of "what one means" (p. 464). It is believed that the "methodological exactness" that Gadamer refers to is also typical of the architecture of language as it is too rigid, confining its user to a teleological finality and limiting, in turn, what one wants to say. It becomes evident that the scientification of unilinearity and its focus on the exactness required to establish a clear conclusion inevitably conceal meaningfulness. The unfortunate thing is that 'what goes on record is just one unilinear narrative' (Garcia, 2020, p.95) – one horizon reductively sourced from one part of a substantially larger whole. This *onehorizon* is, therefore, not exhaustive enough, offering us a partial and a half-baked idea of culture and tradition. This limits Gadamer's notion of hermeneutics through the exact methodology of unilinearity.

THE HYPERTEXTUAL REDEMPTION - *RE-THINKING HERMENEUTICAL UNDERSTANDING OF ETHICAL DELIBERATION AND JUDGEMENT WITHIN THE FRAMEWORK OF A TYPE OF HYPERTEXTUAL WRITING*

An alternative to this mode of writing is here suggested in a bid to redeem discursive expression from its limitation by exposing all discursive intentions. Reference is here being made to the use of *hypertextual writing* or what is also call the *multiplicity of discursive paths*. By taking advantage of this option, the horizons concerned become more exhaustive, offering the possibility of a better outcome for truth by way of a more appropriate form of expression. The program that is proposed here provides an exhaustive space. This technological expressive device gives the opportunity for the writer to express himself or herself exhaustively. According to Van Zundert (2016):

Digital editions often trumpet the ability to represent text exhaustively, celebrating the fact that there is no need to make decisions on what to leave out. Indeed, it is an asset that digital scholarly editions may be capacious almost without limit (p. 102).

This type of hypertext can be thought of as being part of a fragmentation, that is, as a text that is part of a whole but which can also, at the same time, be considered as a standalone text in its own right. Contrary to the position that text 'resists integration into a larger whole and whose incompleteness is an essential feature,' (Baetens &Truyen, 2013, p. 480), the type of hypertext proposed here can only be appreciated if seen in its exhaustive totality. In line with Baetens & Truyen (2013), this type of hypertext consists of fragments that make up a text that are not "linked" but "juxtaposed"' (Baetens & Truyen, 2013, p. 480). This type of hypertext is not to be considered as linked because the idea of linking, in this case, tends to bring back the notion that there should be a chronology. Juxtaposing, in the sense of placing texts without having to directly link text, tends to be more in line with the reality of events happening in an unchronological way. While events do not necessarily have to be related to one another, they may nevertheless be, at the same time, remotely connected, with each event playing its essential part in the building of a totality. Thus, the hypertextual writing that is considered here is mainly paratactic (Baetens & Truyen, 2013, p. 480) without avoiding the syntactical (Baetens & Truyen, 2013, p. 480) aspect in the sense that such a type of hypertext cannot avoid all aspects of events and actions when deliberating on the ethical aspects.

The kind of hypertextual writing suggested here will, as a result of its pragmatic nature, offer a better chance for meaningfulness. It will also deal, in turn, with the impasse created by the sole use of unilinear writing which tends to both narrow meaning and limit its transmission. 'Meaningfulness' is used instead of the word 'meaning' (Garcia, 2020, p. 101) to refer to the reality that actual events and reactions characterised by an ethical slant may indeed house a multiplicity of outcomes. The idea of 'meaningfulness' is used to indicate the deliberate absence of a secure meaning and an objective structure and the simultaneous consideration of a number of equally valid reasons that explain why an action occurred the way it did (Garcia, 2020, p. 67). The medium of unilinearity gives the impression that there should be one absolute conclusion or truth or meaning. This rigorous insistence on a single and supposedly indisputable result epitomizes the exact understanding of the metaphysics of discursive expressions. One may observe, in line with Foucault's notions, (Garcia, 2020, p. 2) that having been normalized into this form of expression, one is unable to wholly express himself or herself in such a unilinear manner.

This work suggests the use of technology in order to achieve, on a practical terrain, the aims which have been hereabove described in a philosophical way. The overall intention is to make ethical deliberation and judgement more pragmatic as well as offer the reader the hope that the act of expressing oneself through writing may be all the richer as a result. This hope will undoubtedly leave a mark on hermeneutics especially when it comes to the interpretation of justice, the authentic understanding of freedom of speech and empathy, and the opportunity to fairly represent the unrepresented. According to Shackelford 2014b, while digital cultures:

continue to evolve in response to complex social, cultural, technological, and material relations, it is worthwhile to identify a broader system of thinking in order to account for the range of contemporary sciences, technological practices, social relations, and expressive forms that reconceive themselves as, or in terms of, dynamic systems. (p. 64).

It is with this in mind that this project evolves, aiming to improve the technology of hypertext by taking into consideration the deep-seatedness of human rationality and, therefore, ethical deliberation and judgement.

Admittedly, the technology of hypertext has evolved from the traditional mode of unilinear writing that was invented mainly to record financial transactions and keep track of food stocks. It can be observed, however, that the technology of hypertext, by comparison, comes closer to resembling the nature of *Dasein*. It does so by providing a writing space where the individual can express thoughts without futilely struggling to reconcile them with a chronology that regiments an authentic freedom of discursive expression (Garcia, 2020, p. 1). Hypertext therefore offers us the possibility to be true to ourselves and others. It should not be seen as a distant and abstract device but should be perceived, rather, as part of what a human is (Stiegler, 1998). It furnishes humans with a tool that closely corresponds to a constantly evolving epistemology. Hypertextuality will be envisaged as a writing tool that facilitates our changing epistemology. As Landow (1992) explains, "writers in these areas offer evidence that provides us with a way into the contemporary episteme in the midst of major changes" (p. 2). Hypertext presents us with a multiple and associative infrastructure (Bolter, 2001, p. 42) that can offer us the possibility to express both the discourse ethics of the normative – selling out the principle of universalization – and the correspondingly deserving notion of particular narratives. As a concept, hypertextual writing deconstructs the philosophy of hierarchy in the metaphysical sense and becomes the genesis of a new meaningfulness which results in re-thinking and even, perhaps, revolutionizing ethics and justice (Garcia, 2020, p. 99).

So how would such a technology look like and how would it work? What is proposed here is the creation of a computer application that will eventually become a handy tool for all to use. What makes this software innovative is its application. This hypertextual application, will give the user the possibility to write in a multiplicity of paths. This possibility will create new avenues for the writer, transforming his or her speech into an authentic discursive expression, uncastigated by a unilinear writing that restricts, rather than encourages, communication. The unilinear architecture of discourse makes it easier for the speaker to be untruthful as the track of discourse will condition one to follow a uni-directional path of expression and thus leave out other valid possibilities of expression that might be equally significant when giving judgement on that which is considered to be ethical. In contrast to this, hypertextual technology is tailored to give the individual the power to guide his or her own expression without the fear of being accused of contradictions, as this software will allow text to be juxtaposed (Baetens & Truyen, 2013) rather than present discourse chronologically. This may come about by offering subjects the chance to write inside spheres in a set-up where each sphere is only remotely connected to other textual spheres. The hypertextual spheres can be considered as a special type of hypertext that contains soft links. Soft links, in contrast to hard links, allow the reader to follow and move on to other textual spheres with ease. Furthermore, soft links are not created and marked out by visible lines but are represented by the void space that appears and which is, in other words, the void background behind the spheres.

In this case, the idea is, to borrow from Bjork's study in Digital Humanities and the First-year Writing Course (2012), that the user, who might be an early learner or an adult learner, is in a position to start navigating through the spheres without following any paths or links. The links are not present in the program itself but are nevertheless at the user's disposal to create his or her own mental linking. This will create a new concept of hypertextuality as it eschews a hard form of linking. Other users reading the same text will, undoubtedly, opt to take a different path, randomly selecting (refer to figures 2 and 3 below) the hypertextual text. It is up to the user to decide how to acquire knowledge and how to decipher ethical actions and events. This type of software will therefore give the reader the power to select his or her own knowledge paths, providing the means for an authentic freedom of expression. This kind of software is close to the notion of concept mapping which 'requires that learners structure themselves a spatial overview of the semantic organization' (Amadieu et. al., 2015, p. 100). This type of concept mapping will be created by the user himself or herself thus creating his or her own impression of the story. It is hoped that the reader of this chapter will start to sense the importance of hypertext as a technology that is not alien to human nature but is, rather, an intricate part of the human condition.

The software works in the following way: As one opens the program, a coloured sphere moving in a black void (background) appears. As shown in figure 1, the spheres are writing spaces that allow the individual the room to express a thought or describe a particular event without being constrained by other texts as shown. Thismay potentially make redundant the stipulation that one must perforce confine oneself to a chronological form of writing. The philosophy behind the moving sphere/s is that they are meant to reflect events and random actions as well as any thoughts that may come to mind. One should remember that actions and events of an ethical nature do not tidily unfold in a chronological way and cannot be reflected textually in a simplistic manner. Thoughts that are meant to reflect reality will therefore have to correspond with such reality. This will encourage the feeling that writing does not have to be chronological and thus liberate its user from the violence made manifest through the unilinear use of discourse (Garcia, 2020, p. 91).

Figure 1.

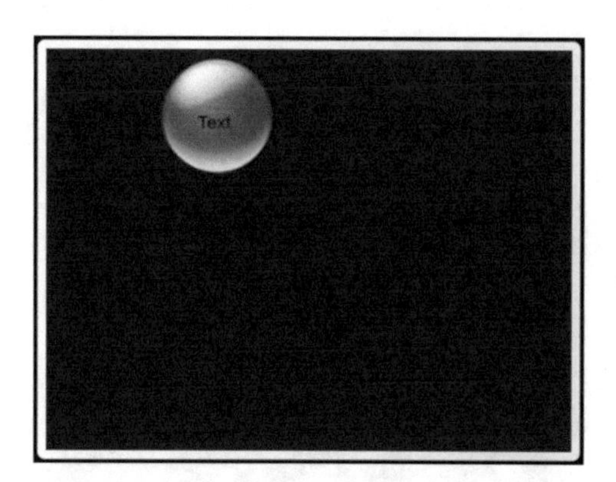

Figure 1 is a visual description of two important features with a philosophical intent. The first significant feature is the manifestation of possible writing spaces that are not confined to the two-dimensionality of the paper or to other forms of two-dimensional writing devices. In the case of a two-dimensional writing device, what happens in the real world is inevitably transformed into a two-dimensional description, distorting the transmission of authentic and truthful discourse. The void, on the other hand, allows the user more freedom for thoughts to become dechronologized and to better reflect the horizons of ethical events and actions. This encourages the freedom to write without the often-associated fear of contradiction. In addition to the text being written inside the sphere, the other significant feature is the movement of the sphere itself (which is here representing the actual software). The moving sphere emphasizes the fact that texts cannot claim dominance over other texts. This corresponds with the notion that events do not, in reality, follow the narrow chronology of writing in a two-dimensional way.

Whatever is written appears inside a sphere. However, this does not mean that the writer is forced to remain within the same sphere. If this were to be the case, there would be an undesired element of chronology. In order to eliminate chronology, the user is provided with the opportunity to write on other spheres that will appear when clicking on the 'ENTER Button'.

Each time the user clicks on this button more spheres materialise. By moving randomly, these spheres serve to deconstruct any chronology whatsoever of discursive expressions. An important feature in this set-up is that there is no chance of any one text taking precedence by interfering in such a way as to dominate the other texts. In this way each text will be independent of all the other texts. As suggested, this technique is simple but its implications are so important that it has the potential to significantly deconstruct the metaphysics of writing by removing any form of finality and hierarchy (Garcia, 2020, p. 68).

In such a favourable context, texts are not meant to aim for an absolute 'truth' which embeds itself in one finite conclusion. This computer software will effectively have a wide-ranging impact on understanding the horizons of others and the way judgements are reached. Differently put, it will remove the calculability in writing which has always tried to emulate natural science and the mathematics of deduction (Garcia, 2020, p. 121). Figure 1 (as with all the other figures that will follow) becomes the technology that approximates human nature by liberating its users from the violence made manifest through unilinear writing. As is further shown in figure 2, the user of this software is in the position to emerge from the forgetfulness that is automatically imposed by unilinearity (Garcia, 2020, p. 136). It is in fact possible for the reader to do away with the instructive method of unilinear writing. Figure 2 demonstrates how the teleological feature of writing is deconstructed by eliminating not only a starting point but also the teleological inclination to reach one finite conclusion.

Figure 2.

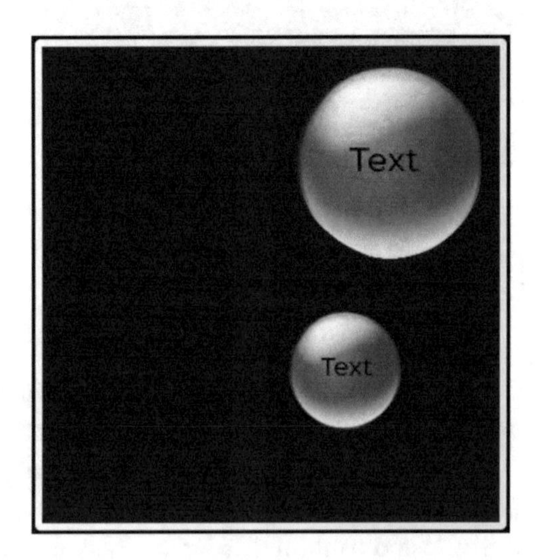

Figure 2 shows how the screen will appear after the 'ENTER button' is clicked (followed from figure 1). A new writing sphere will appear. As seen in figure 2, the new hypertextual sphere is indistinguishable from the other sphere (that is, the sphere from figure 1). As in the description of figure 1, the second hypertextual sphere is not more or less important or valid than the first sphere in figure 1. This does not happen in a unilinear architecture of writing where what is said in the first paragraph, inevitably conditions the second paragraph. Figure 2 is also a visual description of a type of hypertext that does not contain visible links, as described in the main text above. The links are created in the mind of the reader. The writing void is, in itself, a link, as all hypertextual spheres are made whole in the void of the screen. It is within the power of the reader to select the direction taken in this multiplicity of writing paths.

Another important feature with a philosophical intent is the possibility that spheres might also be moving out of the screen as shown in figure 3.

Figure 3.

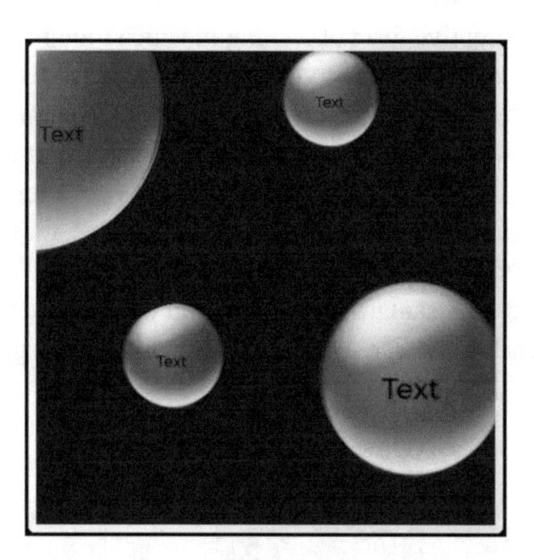

Figure 3 purposely shows the fourth screen in order to give the understanding that each time the 'ENTER button' is clicked, a new sphere is displayed and every time the 'ENTER button' is clicked another hypertextual sphere appears. What is important here is that, as the writer adds more hypertextual spheres, the spheres start to move out of the screen and disappear. The idea behind this feature is to show the reader that no one text is better than the other - all texts are equally important and may leave the screen at any time. Spheres are automatically moving unless controlled by the writer or the reader. This will be explained in figure 5. This screen will start to give the reader a holistic understanding of the text and its correspondence with reality.

Such a feature is particularly meant to raise, in an empirical and visible way, the awareness that there is no discourse that is more important than others. Textual equality takes centre stage and may further encourage an authentic way to treat other social equalities. In order to acquire such textual equality, the algorithm's design allows the user to rotate and to distance the sphere away from the screen.

This effectively shows how hypertextuality can also be used in areas that may tentatively be described as more socially-charged and considerably more consequential than hypertextual fiction. As pointed out above, this work does not dismiss the important of fictional narratives. On the contrary, it is only through the gains made in hypertextual fictional narratives and the experimentation carried out in such areas that one may adapt this knowledge in order to improve the practice of interpretation and devise new ways of dealing with judgements of an ethical nature.

Figure 4 illustrates another important feature of hypertext. Hypertextual writing gives the user the opportunity to present his or her writing without having to, for example, obfuscate and violate his or her own thoughts by placing certain thoughts at a higher rung of the hierarchy, thus favouring them over others.

In chronological writing, one falls prey to linearity, pandering to its demand to hierarchically organize information and assign a more prominent place to certain thoughts. In a hierarchy of paragraphs, for example, the further down a thought is placed, the more it is bound to appear subordinate to the thoughts that have preceded it. This alienating way of writing in a chronological manner undoubtedly creates a path of discourse from which it becomes exceedingly difficult to liberate oneself. A chronology of writing also creates, for both writer and reader, the sensation of time. As can be shown in figure 4, such a chronological arrangement together with the sensation of time are deconstructed so that one start to experience, instead, a more redemptive form of writing in which thoughts are presented without having to actively undermine other thoughts that, in actual fact, happened at the same time. It is only through the unilinear mode of writing that thoughts are coerced into a single chronological file. In addition, such a form of hypertextual writing will create a medium of timeless space in which one may air the present-ness of thoughts. The reader should here be reminded that this is not a return to a metaphysical timeless analysis that makes one forget oneself. This should instead make one aware that one's realities of ethical actions and events are composed of a multiplicity of happenings and the resulting representation would be unrealistic and un-authentic representation if other valid perspectives of the reality were to be cast aside and ignored. What is needed is a 'large pool of information' (AlAgha, 2012, p. 275)

Figure 4.

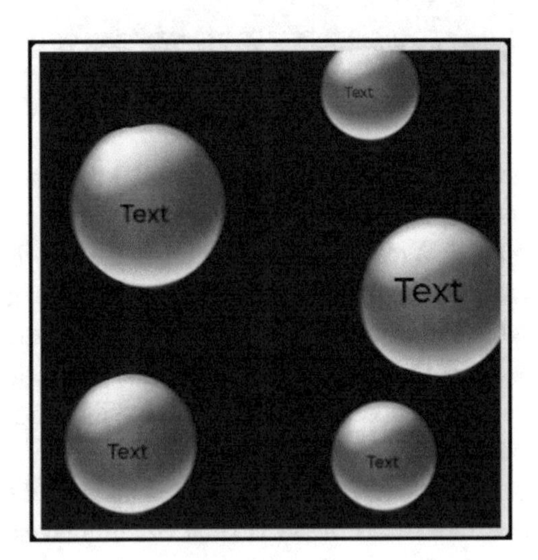

It is difficult to read what is inside the moving spheres and it is impossible to read small texts inside these moving spheres. Figure 4 indicates a feature of the software that could a) become static and b) be enlarged. The reason is obvious as the hypertextual spheres will have to be made legible. The reader of this type of hypertextual software will be able to control the direction of the spheres. The reader may select a particular sphere and enlarge it so as to make it legible. After reading a particular hypertextual sphere the reader may 'push' the read sphere back to select another hypertextual text. This will create juxtaposition – placing texts in proximity to each other and offering the possibility to contrast them with one another. This would allow contradictory texts to form part of the whole understanding of actions and events with ethical intent. The final product, although paratactic, does not avoid the syntactical.

Spheres can take a rotational movement for the reader to select any particular text to be read. This can be done by using the up or down button on the keyboard. In this way texts can be selected and read by zooming in on the type of sphere that the reader wants to pore over. Although unilinearity to write inside the spheres, is still used (something which cannot be avoided at the present moment), this unilinearity is subverted by the presence of other equally valid texts.

This computer software will be an optimal device for those situations in which an accused or a person who wants to give testimony or express herself or himself in discourse cannot really do so. This is not due to an inability to express herself or himself in writing, but due to the fact that what he or she says in a chronological way risks having a negative impact on what he or she might want to say later on. It is here reiterated, even at the cost of sounding redundant, that events that happen in reality cannot correspond with the chronological manner in which he or she writes. There is no correspondence whatsoever between the method of unilinearity and the behaviour of actions and events as it plays out in the real world. When giving a testimony, one automatically panders to this alienating mode of unilinear writing unwittingly becoming the violent agent of his or her own being. Differently put, one acts as if one is suffering from acute memory loss because the moment one begins to express himself or herself, one instantly falls victim to his or her unknowingness of unilinearity. What is more alienating is the fact that the individual is made to embrace unilinear writing in the belief that its role as a redemptive tool is beyond reproach. What is being suggested, therefore, is a redemptive solution to this alienating and violent role that one perforce take on unknowingly, having always been taught since childhood to think in such a manner.

The proposed computer software can be described as a technological device that offers an alternative to the unilinear mode of writing since it allows the user to write in a multiplicity of paths. It is important to begin to teach young individuals about this potentially game-changing option from an early age at school (Garcia, 2020, p. 92). It is only when one learns to engage with such a form of expression that one may really start to experience the emancipation and freedom of expression that one so hopes for. This is what is here called a new understanding of the *fusions of horizons* in the Gadamerian sense (Garcia, 2020, p.95). The philosophy of hypertextual writing will therefore build upon the important work initiated by Gadamer, giving it a fresh dimension in an increasingly digital age. However, in addition to the Gadamerian approach to hermeneutics, it is here believed that interactivity is paramount when it comes to less felicitous encounters that involve, for example, filing a charge against an individual and being part of the process to convict someone. It is my view that this hypertextual software offers a practical alternative to the individual who yearns to be truthful to an event but is unable to, namely because the reality of happenings does not correspond with the two-dimensionality of paper as well as linearity's chronological nature.

In cases where a person accused of wrongdoing is to face charges, hypertext writing will offer him or her a tangible possibility to bring to light all the happenings of the event (Garcia, 2020, p. 91). It might be the case that the small events that might be considered as unimportant can, in fact, be very helpful in the reconstruction the whole story of the event in its entirety. This can be compared to the fragments found in an archaeological site where all the fragments are important for in the quest for a more comprehensive understanding of the past. The same can be said with the reconstruction of the whole story, so vital for the fusions of horizons in future hermeneutical understanding. In hypertextual writing, the individual is given the possibility to lay down all the horizons, all the fore-structure of oneself.

With such hypertextual software, the individual is granted the proper writing space, a space which does not stifle his acts of expression and does not force him to conform to a rigid narrative that is disembodied from the context of related happenings. The accused will therefore have the possibility to write by associativity.

Now, if the judge, jury, and all those involved, were given the facility to actually use the technological devices available and be able to travel around the 'space of events,' interactivity would be greatly enhanced. In this way, hypertextuality could help in rethinking, the way, interpretation and judgement are perceived.

There are many types of hypertext. The one described in this work was chosen because it is the one that comprehensively caters for the issues under consideration. The type of hypertext that is here referring to is a *read-only, non-hierarchical, exploratory, network hypertext* (Garcia, 2020, p. 101). *Heterarchy* is a system that best fits this type of hypertext. This system consists of fragments of texts that are written in an unranked way (Garcia, 2020, p.68). In this case, the ranking of texts could differ according to the reader in question. Such a feature could be integrated with networked hypertext, a type of hypertext consisting of nodes that are interconnected. The node does not have any dominant axis to orient the reader and it has no beginning.

Figure 5.

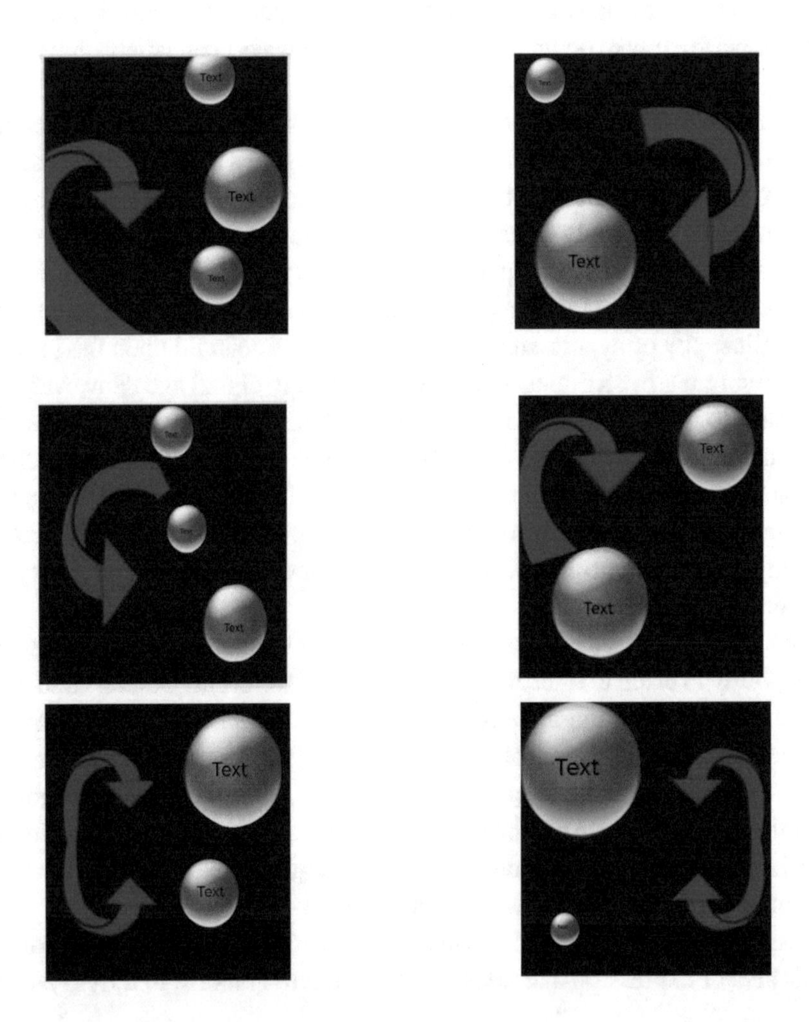

The above figures clearly depict that the reader is in total control of the moving spheres. Although the spheres are moving randomly, the reader of the text can take control and redirect the spheres in a clockwise or anticlockwise direction. Text may also be directed in an upward or downward rotation. The figures above demonstrate that the hypertextual spheres are fashioned in such a way that the user is in full control of them at each and every juncture.

As seen in the figures of rotations above, the ability to rotate networked hypertext further removes the hierarchical characteristic, allowing the reader to 'virtually travel through the texts' in such a way that one text does not gain 'precedence over the other' (Garcia, 2020, p.68). With rotational hypertext, one gets the impression that one is moving within a cycle of events (Garcia, 2020, p. 68). This is however counteracted by giving the user the possibility to move the text the way he wants. Therefore, in addition to rotational hypertext, the reader will have the possibility to change the directions the way he likes – moving the text up or down (Garcia, 2020, p. 68). This will give the user, who could be, for instance, reading testimonies or other kinds of legal and formal texts, an increased freedom to reflect on the information at hand. The feature known as 'exploratory hypertext' (Kolb, 1996, p. 21) will make it possible for the reader to explore the text under investigation. The reader will have to learn to integrate information rather than read in a chronological manner – 'electronic writing seems in some ways to be more like hieroglyphics than it is like alphabetic writing. The computer welcomes elements that we in the West have long come to regard as inappropriate to writing.' (Bolter, 2001, p. 37) Finally and as already pointed out, since one is here dealing not with fictional narratives but with texts that address testimonies to provide evidence and the issuance of judgements about an individual's culpability or innocence, the nature of such text calls for *read-only hypertext.* As Landow explains, readers of such texts are 'limited to choosing their own reading paths.' (P. 9) As the figures above suggest, the subject of linking is not tackled in this work even though it is a major feature of hypertext – 'linking is the most important fact about the hypertext, particularly as it contrasts to the world of print technology.' (Landow, 1994, p. 6) The brief reference to links focuses solely on their relevance 'to the kinds of speech acts performed rather than to the logical steps made' (Kolb, 1996, p. 22) Linking's illocutionary force is of our concern insofar as it connects speech acts and thus prevents any type of linking that forces the user into an either/or situation. This project also allows the reader to interact with the text by contributing (Eyman, 2015, p32) other spheres of texts. Small links could be added here to show the reader that the text relates to a particular sphere.

FUTURE RESEARCH

Before concluding it would be of interest to outline various avenues for future research and, as was pointed out at the start of this chapter, to show these possibilities in this software's pragmatic applications. This research has the potential to address a number of sectors and involve individuals across all age groups. What follows is an overview of the diverse groups of individuals that may benefit from future research and how the research itself may consequently evolve

The Young Learner

This form of hypertextual writing should primarily be introduced to the young learner at an early stage in his or her education. In this way the learner would not be mentally conditioned to write in a unilinear way but would, rather, be provided with a new mode of discursive expression that could make manifest a number of different paths. This research could be carried out at school during writing classes. Such a project would ideally involve research into two main areas: the activity of role play, first of all, which is meant to represent and reflect actual events that occur in reality, followed by the activity of writing that is meant to describe these very events. The idea is that a group of young learners are encouraged to think about an everyday event that involves making a judgement, small as it may be, about certain actions. This may range from preparing breakfast or preparing for school or the activity of arriving at school and meeting friends. The events under investigation, in other words, may include all that is common and that occurs on an everyday basis, including the small actions, deliberations and judgements that may occur. A number of research actions may follow after this role play. This would represent the second stage of the research: the stage of discursive expression. The number of spectators of the role playing would be divided into two. The first group would be asked to describe the role play in a unilinear manner while the other group would be introduced to the program described above, thereby offering them the possibility to describe the event in hypertextual form. Results would then be compared, with the researcher thus being in a position to see the amount of detail given by reading through the result of both unilinear texts and hypertextual texts. How would the researcher know who gave the best description of the event? Apart from the obvious fact that reading both versions would yield a sense of which one describes the event more faithfully, the researcher would also ask the actors of the role play itself for their opinion as to which text best represents the event. This kind of research contains many variables that have to be accounted for such as, for example, the writing skills of the young learners and the ability to perform. Research would, however, start to give a clear indication on the potentiality of the software described above.

The Adult Learner

While the research task of the adult learner would be more difficult to carry out, it might also prove to be more ground-breaking. The difficulty obviously stems from the fact that this software is being introduced to adult learners who, having been taught to write and express themselves in a unilinear manner, would instinctively prefer to write unilinearly. If, however, they accept to express themselves in a hypertextual way, they may come to realise that such a form of writing is designed to reveal more, especially when it involves expressing the reality of idividuals. It is better tailored, therefore, to do justice to the individual by welcoming a more robust freedom insofar as expression is concerned. Such research would underscore the far-reaching implications of this software application as the truth of what might easily remain hidden when communicating in a unilinear way, might be revealed without the fear of being somehow misconstrued. The research would focus mainly on the notion of judgement or of, rather, being judgemental. One effective way for the researcher would be to ask the adult learner to write about a situation or a topic such as, for instance, sexual orientation or the value of spirituality, in a unilinear way. After doing so, the researcher would ask the adult learner about the degree of apprehension experienced. The learner would then be asked to write about the same topic, this time using hypertextual software. The researcher would once again investigate the accompanying level of apprehension. In this way, the researcher would be in a position to understand whether exposing one's horizons in a multiplicity of paths, as the software suggests, would make the learner feel more emancipated. This kind of research focuses on a very deep notion; the notion of freedom of speech or, rather, what the notion of freedom of speech might mean in a hypertextual setup.

The Accused

Another potential area of research might focus on people who are either charged with a criminal offence or people who are convicted of a criminal offence. The researcher would have to investigate whether unilinear techniques of formalistic writing are in fact, redemptive of those accused in a court of law. In such a formalistic set-up, each utterance carries tremendous weight. In other words, the illocutionary act has the power to condemn the accused to serve a period of time behind bars. In these difficult situations, the research could be divided into two areas. One are could focus on people who are accused and not convicted, with the researcher inviting the accused to go through the report of accusation. Having done so, the accused could then be asked to examine the accusation and re-write his or her own accusation by using hypertextual software rather than the unilinear manner typical of official reports. The researcher may identify expressions which are more revealing and better suited to the issuance of justice than the narrative provided in the unilinear version, the latter potentially leaving many things unsaid due its emphasis on formality. The other area of research involves the convicted. In this case, the researcher could ask (after due permission from the authorities) the convicted to give his or her own version of what actually happened and what in fact was said inside the courtroom. The convicted may feel at liberty to express himself or herself without inadvertently having important facts screened for the narrative. These facts may prove unquestionably significant in that they may give a new and different light to the case.

The Autistic

People suffering from autism might also benefit from this hypertextual software. Research could be carried out with people who might find it hard to express themselves in a unilinear and chronological way. Research might open up new avenues as to the form of expression that would provide an opportunity for integration and emancipation for this sector of the population.

AVOIDING CONCLUSIONS

Although this conclusion shall steer clear from drawing a conclusion, It is here appropriate to reflect on the use of hypertextual technology and make a few remarks in the process. The proposed hypertextual technology teaches individuals how to resist being judgmental by providing a multiple vision of occurrences. With regard to Heidegger and Gadamer, it deconstructs the metaphysical concept of *being* and comes closer to the authentic *being*. As has been pointed out, the technology's rationale takes its cue from two major philosophical principles, namely the fundamental doctrine of the philosophy of hermeneutics as expounded by Gadamer, and the understanding of *being* developed by Heidegger. The computer software provides a space in which individuals are free to express a kind of discourse that eliminates any hierarchy and chronology in writing, thereby acquiring an authentic freedom of speech and the ability to properly exercise their human rights in expressive discourse. The intention of this work was to introduce a computer application as better means by which one can interpret and treat deliberations of an ethical nature in the future.

The idea behind this technology is to deconstruct the notions of absolutization and radicalisation when interpreting ethical actions. It tries to deliver the message that one cannot classify actions as either 'right' or 'wrong'. It can be said that this work does not exhaust all the details and more research is needed to refine this hypertextual technology. As such, my plea is to find a constructive audience that appreciates the implications of developing such a project and thus makes a collective effort to improve upon it. It is hoped that this software will provide the right technology to make us more humane - conserving the individual's rights to authenticity without the fear of self-expression. This technology is therefore intended to cater for all true liberal expression. It is meant to make its user see through, instead of reading actions and events of ethical importance as if these very actions and events actually happened chronologically. This would be in line with Gadamer's theorizations, as it would provide us with the opportunity to better understand all the horizons of the individual.

This computer software should encourage the individual to come forth in unconcealedness. While unilinearity routinely imposes forgetfulness on its user, the same cannot be said for a multiplicity of discursive paths which is so designed to avoid this crippling pitfall. This tactic of seeing through justice and its consequential interpretation is vital to transform the way one shall handle and interpret justice in the future, offering a fresh vision for meaningfulness and hermeneutic understanding. Such a technology eliminates any form of categorization and any form of calculability. It avoids selection as is proposed through the metaphysics of writing and prevents any form of economisation insofar as intentions and desires are concerned. It is a technology which allows the individual to regain his lost strength of discursive expression, becoming a redemptive form of expression - a truly democratic form of expression. One cannot deny the reality that 'truth' is rendered fake through the medium of calculability given that it excludes, rather than embraces, the different perspectives that an individual might have.

This technology will only become possible if one teaches this mode of discursive expression to students from an early stage. A truly emancipative way of expression is one which offers authentic freedom. The opportunity for this freedom can only arise if children learn to be masters of their own discourse. It is hoped that by means of this project, the learner's exposure to hypertext will encourage its use as an effective mode of expression (Bell, 2014, p. 143) to the extent that it will give rise to a new way in which justice and freedom of speech are expressed. In order for this to happen, the learner must not be forced to distort self-expression, categorizing thoughts according to the rigid infrastructure of language. In addition, it is here proposed that this mode of expression be made available to those who desire to be authentic with themselves and others and would like to reveal their authenticity unhindered. This hypertextual software offers a voice to those who are unrepresented and who might have different and alternative perspectives (Garcia, 2020, p.124).

It is also hoped that the proposed technology will assist individuals who, having been charged with a criminal offence, find themselves unjustly forced to express actions in a chronological manner. By making use of hypertext, they would be able to express themselves without forcibly altering that which they would like to communicate. As was pointed out, actions and events do not just happen in a chronological way - there might be a number of events and actions that are avoided in order to make the formal text appear clear and sound. The resulting unilinear text that is presented in formal juridical institutions is the one considered as having the correct format. A key feature of this format is that of making a plea for an excuse and it must always be communicated in conformity with the established formalities of juridical institutions.

In spite of the fact that these elements of clarity, chronologizing, and scientificization are embedded in the architecture of unilinear writing to create a sense of order, the written statement which results is not necessarily redemptive of the individual being charged. Since it must perforce conform to the formality and the clarity of the unilinear infrastructure of language it will not reflect this person's freedom to express himself or herself in authentic ways. This is the violence embedded in unilinear discourse that is supposed to be redemptive. The individual, once charged, will have to be castigated, instantly becoming a victim of discourse before having even barely experienced the formal rigmarole of juridical institutions.

The hope is that if juridical institutions acknowledge the shortcomings of linearity and learn to duly adapt while familiarising themselves with reading and writing through the medium of hypertext, a more emancipative justice will ensue. All that this requires is the good will to accept such software and the intention to educate oneself in this new way of expression. If law students, the police, parole personnel and all those who are involved in juridical institutions, become acquainted with this mode of discursive expression and accept its emancipative potential, individuals will start to be the authentic architects of truth, unphased by the rigidness of chronological writing. Hermeneutics will undoubtedly be positively affected as one will start to experience not a *fusion of horizons* but *fusions of horizons*.

REFERENCES

AlAghi, I. (2012). KnowledgePuzzle: A Browsing Tool to Adapt the Web Navigation Process to the Learner's Mental Modle. *Journal of Educational Technology & Society*, *15*(3), 275–287.

Amadieu, F., Salmerón, L., Cegarra, J., Paubel, P.-V., Lemarié, J., & Chevalier, A. (2015). Learning from Concept Mapping and Hypertext: An Eye Tracking Study. *Journal of Educational Technology & Society*, *18*(4), 100–112.

Baetens, J., & Truyen, F. (2013). Hypertext Revisited. *Leonardo*, *46*(5), 477–480. doi:10.1162/LEON_a_00644

Bell, A. (2014). Schema Theory, Hypertext Fiction and Links. *Style (Fayetteville)*, *48*(2), 140–161.

Bjork, O. (2012). Digital Humanities and the First-Year Writing Course. In *Digital Humanities Pedagogy: Practices, Principles and Politics* (pp. 97–119). Open Book Publishers. doi:10.2307/j.ctt5vjtt3.9

Bolter, J. D. (2001). *Writing Space: Computers, Hypertext, and the Remediation of Print* (2nd ed.). Lawrence Erlbaum Associate Publishers. doi:10.4324/9781410600110

Eliot, S., & Rose, J. (2009). Introduction. In S. Eliot & J. Rose (Eds.), *A Companion to the History of the Book* (pp. 1–6). Wiley-Blackwell.

Eyman, D. (2015). Defining and Locating Digital Rhetoric. In Digital Rhetoric: Theory, Method, Practice (pp. 12-60). Open Book Publishers. doi:10.2307/j.ctv65swm2.5

Gadamer, H. G. (2004). *Truth and Method.* Continuum International Publishing Group. (Original work published 1960)

Garcia, D. (2020). *Rethinking Ethics Through Hypertext.* Emerald Publishing Limited.

Grondin, J. (1994). *Introduction to Philosophical Hermeneutics* (J. Weinsheimer, Trans.). Yale University Press. (Original work published 1991)

Heidegger, M. (1962). *Being and Time* (J. Macquarrie & E. Robinson, Trans.). Blackwell Publishing. (Original work published 1927)

Koehler, A. (2013). Digitizing Craft: Creative Writing Studies and New Media: A Proposal. *College English*, *75*(4), 279–397.

Kolb, D. (1996). Discourse Across Links. In C. Ess (Ed.), *Philosophical Perspectives on ComputerMediated Communication* (pp. 15–26). State University of New York Press.

Landow, G. P. (1992). *Hypertext 2.0. The Convergence of Contemporary Critical Theory and Technology*. The John Hopkins University Press.

Landow, G. P. (1994). *Hyper/Text/Theory*. The Johns Hopkins University Press.

Linge, D. E. (Ed.). (2008). Hans-Georg Gadamer Philosophical Hermeneutics (D. E. Linge, Trans.). Berkeley, CA: University of California Press. (Original work published in 1976)

Odin, J. K. (2010). *Hypertext and the Female Imaginary*. University of Minnesota Press.

Robson, E. (2009). The Clay Tablet Book in Sumer, Assyria, and Babylonia. In S. Eliot & J. Rose (Eds.), *A Companion to the History of the Book* (pp. 67–83). Wiley-Blackwell.

Roemer, C. (2009). The Papyrus Roll in Egypt, Greece, and Rome. In S. Eliot & J. Rose (Eds.), *A Companion to the History of the Book* (pp. 84–94). Wiley-Blackwell.

Ryan, M.-L. (2004). Multivariant narratives. In S. Schreibman, R. S. Siemens, & J. Unsworth (Eds.), *A Companion to Digital Humanities* (pp. 415–430). Blackwell Publishing. doi:10.1002/9780470999875.ch28

Shackelford, L. (2014a). Literary Turns at the Scene of Digital Writing. In Tactics of the Human: Experimental Technics in American Fiction (pp. 28-56). University of Michigan Press, Digitalculturebooks.

Shackelford, L. (2014b). Tracing the Human through Media Difference. In LTactics of the Human: Experimental Technics in American Fiction (pp. 57-94). University of Michigan Press, Digitalculturebooks.

Stiegler, B. (1998). *Technics and time, 1: The fault of epimetheus* (R. Beardsworth & G. Collins, Trans.). Stanford University Press. (Original work published 1994)

Van Zundert, J. (2016). Barely Beyond the Book? In M. J. Driscoll & E. Pierazzo (Eds.), *Digital Scholarly Editing* (pp. 83–106). Open Book Publishers.

Woodward, A. (2009). *Nihilism in Postmodernity*. The Davies Group, Publishers.

ADDITIONAL READING

Bynum, T. W., & Moor, H. J. (Eds.). (1998). *The Digital Phoenix: How Computers Are Changing Philosophy*. Whitstable Litho Ltd.

Carusi, A. (2006). Textual Practitioners: A Comparison of hypertext theory and Phenomenology of Reading. *Arts and Humanities in Higher Education*, *5*(2), 163–180. doi:10.1177/1474022206063652

Chartier, R. (1995). *Forms and Meanings: Text, Performers, and Audiences from Codex to Computer*. University of Pennsylvania Press. doi:10.9783/9780812200362

Crystal, D. (2005). *How Language Works*. Penguin Books.

Drucker, J., & Nowviskie, B. (2004). Speculative Computing: Aesthetics Provocations in Humanities Computing. In S. Schreibman, R. S. Siemens, & J. Unsworth (Eds.), *A Companion to Digital Humanities* (pp. 431–447). Blackwell Publishing. doi:10.1002/9780470999875.ch29

Ess, C. (1994). The Political Computer: Hypertext, Democracy and hypertext. In G. P. Landow (Ed.), *Hyper/Text/Writing* (pp. 225–267). The Johns Hopkins University Press.

Ess, C. (Ed.). (1996). *Philosophical Perspectives on Computer-mediated Communication*. SUNY Press.

Ess, C. (2004). Revolution? What Revolution? Success and Limits of Computing Technologies in Philosophy and Religion. In S. Schreibman, R. S. Siemens, & J. Unsworth (Eds.), *A Companion to Digital Humanities* (pp. 132–142). Blackwell Publishing. doi:10.1002/9780470999875.ch12

Flusser, V. (2011). *Into the Universe of Technical Images* (N. A. Roth, Trans.). University of Minnesota Press. (Original work published 1985) doi:10.5749/minnesota/9780816670208.001.0001

Joyce, M. (1990). Afternoon, a story [computer program]. Watertown, MA: Eastgate Press. (first published 1987).

Kolb, D. (1994). Socrates in the. Labyrinth. In G. P. Landow (Ed.), *Hyper/Text/Writing* (pp. 232–344). The Johns Hopkins University Press.

Lancashire, I. (2004). Cognitive Stylistics and the Literary Imagination. In S. Schreibman, R. S. Siemens, & J. Unsworth (Eds.), *A Companion to Digital Humanities* (pp. 397–414). Blackwell Publishing. doi:10.1002/9780470999875.ch27

Moulthrop, S. (1991). Victory Garden [computer program]. Watertown, MA: Eastgate Systems.

Moulthrop, S. (1994). Rhizome and Resistance Hypertext and the Dreams of a New Culture. In G. P. Landow (Ed.), *Hyper/Text/Theory* (pp. 299–320). The Johns Hopkins University Press.

Rosenberg, M. E. (1994). Physics and Hypertext: Liberation and Complicity in Art and Pedagogy. In G. P. Landow (Ed.), *Hyper/Text/Theory* (pp. 268–298). The Johns Hopkins University Press.

Ryan, M.-L. (2004). Multivariant Narratives. In S. Schreibman, R. S. Siemens, & J. Unsworth (Eds.), *A Companion to Digital Humanities* (pp. 415–430). Blackwell Publishing. doi:10.1002/9780470999875.ch28

Sandbothe, M. (2000). Interactive - Hypertextuality - Transversality A media-philosophical analysis of the Internet. *Hermes. Journal of Linguistics*, (24), 81–108. https://tidsskrift.dk/her/article/download/25570/22483/

Schreibman, S., Siemens, R. S., & Unsworth, J. (Eds.). (2004). *A Companion to Digital Humanities*. Blackwell Publishing.

Von Foerster, H. (1984). Disorder/Order Discovery or Invention? 1-6. Retrieved from http://ada.evergreen.edu/~arunc/texts/cybernetics/heinz/disorder.pdf

Wachtel, E. (Ed.). (1993). *Writers and Company*. Alfred A. Knopf Canada.

Warwick, C. (2004). Print Scholarship and Digital Resources. In S. Schreibman, R. S. Siemens, & J. Unsworth (Eds.), *A Companion to Digital Humanities* (pp. 366–382). Blackwell Publishing. doi:10.1002/9780470999875.ch25

KEY TERMS AND DEFINITIONS

Judgement: As against the perceived notion of coming to a conclusion about certain ethical actions, judgement is here taken to be the state of sustained perception of all related events.

Metaphysics of Writing (the Metaphysics of the Unilinear Infrastructure of Language): Compared to the traditional metaphysical idiom of the Platonic style as the mode providing the only means to 'truth,' the metaphysics of writing can also be here applied to traditional unilinear writing. The reason for this is that unilinear writing dominates all thought processes: thinking must be shaped in such a way as to accommodate the confines of the unilinear writing.

Multiplicity of Discursive Paths: The amalgamation of different narratives. Such an amalgamation makes it likelier for justice to be done, particularly in view of the limitations of a single absolute narrative.

Pragmatic Hypertextual Ethics: The possibility of having a constellation of text that contains both normative and paralogical discourse.

Teleological Aspect: Unilinear writing that gives, both the reader and the writer, a sense of a definite conclusion.

ENDNOTES

1. This chapter is a practical version, the philosophy of which is extracted from my doctorate thesis and the book published by Emeralds Publishing Limited. See: Garcia, D. (2020). *Rethinking Ethics Through Hypertext.* UK: Emerald Publishing Limited.
2. A big thank you to Matthias Garcia who tried to understand and patiently transformed my imagination into the figures below.
3. See: Odin, J.K. (2010). *Hypertext and the Female Imaginary*, Minnesota, University of Minnesota Press.

Chapter 7
Citizen Participation in Community Surveillance:
Mapping the Dynamics of WhatsApp Neighbourhood Crime Prevention Practices

Anouk Mols
https://orcid.org/0000-0003-0355-9849
Erasmus University, Rotterdam, The Netherlands

ABSTRACT

Despite their recent emergence, WhatsApp neighbourhood crime prevention (WNCP) groups are an already pervasive phenomenon in the Netherlands. This study draws on interviews and focus groups to provide an in-depth analysis of the watchfulness and surveillance activities within these groups. The conceptualisation of WNCP through the lens of practice theory shows that the use of ICTs in the form of WhatsApp amplified all three dimensions of neighbourhood watchfulness practices. It examines how friction at the intersections of materialities, competencies, and meanings affect neighbourhood dynamics as well as the personal lives and experiences of people involved. While voluntary citizen participation in crime prevention leads to an increase in social support, feelings of safety, and active prevention of break-ins, it also defaults to forms of lateral surveillance which transcend digital monitoring practices. Pressing issues related to social media use, participatory policing, surveillance, and the normalisation of distrust and intolerance have an impact beyond its localised Dutch context.

INTRODUCTION

"When there are cars in the neighbourhood we're not familiar with, or when we are not sure about people we have never seen before, we'll take a picture and send it: Do we know anything about this?" (Pauline, moderator of a WNCP group in City C)

DOI: 10.4018/978-1-7998-5849-2.ch007

Pauline is the moderator of a WhatsApp neighbourhood crime prevention (WNCP) group in a city in the Netherlands. This study examines surveillance practices within WNCP groups which have gained popularity in the Netherlands since 2013. Neighbours are connected via a WhatsApp group in order to exchange warnings, concerns, information about incidents, and suspicious situations in their street. As illustrated above, Pauline and her neighbours immediately materialise their suspicions into pictures of unfamiliar vehicles or persons. Her quote provides a preview into how voluntary citizen participation in crime prevention practices has inherently ambivalent consequences. This study explores how ICTs in the form of WhatsApp-equipped smartphones amplify neighbourhood watchfulness practices and how this defaults to precarious forms of surveillance.

WNCP groups have a low participation threshold because citizens can easily join a WhatsApp group (Bervoets, 2014) in order to participate in safeguarding practices. This often creates a positive feeling about being aware of neighbourhood activities, as well as feelings of safety (Lub & De Leeuw, 2017; Pridmore et al., 2018; Smeets et al., 2019). Moreover, the existence of WNCP groups can increase social cohesion in the neighbourhood (Van der Land et al., 2014). However, WNCP groups can also cause feelings of unsafety and distrust, discriminatory practices, risky vigilant behaviour, and privacy infringement (de Vries, 2016; Lub, 2018; Lub & De Leeuw, 2017, 2019; Mehlbaum & van Steden, 2018; Mols & Pridmore, 2019; Pridmore et al., 2018). Needless to say, these WNCP practices impact the experiences of neighbours as well as passers-by. For neighbours, participants and non-participants, an active WNCP group can change the neighbourhood dynamic into a watchful, and at times distrustful, atmosphere. And even if there are street signs signalling the existence of a WNCP group, passers-by are often unaware of the fact that they are actively being monitored by citizens. It is important to note that surveillance practices existed in neighbourhoods long before WNCP initiatives emerged. Alert neighbours who keep an eye out on the street and who contact neighbours or police in case of trouble are not a new phenomenon. However, the use of WhatsApp, a cross-platform smartphone-based instant messaging application (Church & de Oliveira, 2013) or a similar messaging app has changed neighbourhood interactions and practices. ICTs are known for creating new forms of interaction (Hampton, 2007) and the emerging use of WhatsApp groups within existing surveillance practices is currently changing neighbourhood dynamics and personal experiences, which makes this a pressing issue.

WNCP practices are a form of informal surveillance, the "casual, but vigilant, observation of activity occurring on the street and active safeguarding of property"(Bellair, 2000, p. 140). Notably, these practices can also be seen as forms of lateral, or interpersonal surveillance (Andrejevic, 2005; Trottier, 2012), whereby people actively monitor their peers, or in this case, their neighbours. The goal of this study is to understand mobile technology-driven informal surveillance activities through the lens of practice theory. Practice theory aims to explain society, culture, and social life through practices, practices are bundles of activities existing of material elements, competences, and meaning. Practices are not static because they change over time (Reckwitz, 2002; Schatzki, 2002; Shove et al., 2012). This makes practice theory particularly suitable to study societal developments such as the impact of the emergence of WNCP groups.

This study maps the consequences of a citizen initiative aimed at improving safety, and shows that surveillance practices in WNCP groups have emerged in different forms and shapes. By highlighting the diversity in WNCP practices and the fact that this phenomenon is still developing, we found that neighbours are improvising on a daily basis in their self-organised WNCP groups. Interviews and focus groups revealed how friction in the conjunction of dimensions in WhatsApp neighbourhood watchfulness practices affects the personal lives and experiences of people (often unknowingly) involved. This study offers an in-depth account of the socio-material elements assembled together in everyday neighbourhood watchfulness and surveillance practices. Pressing issues related to lateral surveillance, participatory policing, and the normalisation of distrustful and intolerant attitudes are highlighted in the results section of this chapter. The use of practice theory shows a co-constructed amplification of watchfulness practices revolving around a technical layer (WhatsApp) that brought qualitative change to existing neighbourhood practices and experiences. Despite a more or less unified articulation of purpose – protecting neighbourhoods – differentiations became visible across enacted neighbourhood watchfulness practices and their commonalities. Rather than avoiding these complexities, these tendencies are explored in clusters of actions and materialities. This study not only addresses the pressing nature of the issues arising in emergent WNCP activities in a specific local context, it also offers an exploration of lateral surveillance practices which go beyond digital forms of monitoring.

THE DUTCH CONTEXT OF NEIGHBOURHOOD SURVEILLANCE PRACTICES

WNCP practices can be seen as a form of (lateral) surveillance taking place in a Dutch context. Before exploring these watchfulness practices in the light of surveillance studies, it is important to grasp how WNCP groups relate to neighbourhood block watch groups and to provide an overview of their personal and communal benefits and drawbacks.

Neighbourhood Watchfulness Practices in the Netherlands

In the previous three decades, the Dutch government's approach shifted from active governance to regulations wherein citizens are co-producers of safety (Van der Land et al., 2014). Safety has become a shared responsibility of police and citizens. Similar to neighbourhood watch initiatives in other countries, neighbourhood block watch groups are active in the Netherlands. A recent inventory found 661 block watch groups in the Netherlands (Lub, 2018). Block watch revolves around small groups of participating citizens devoting their time to neighbourhood safety. The employment of WhatsApp for neighbourhood watch purposes radically increased the number of actively monitoring citizens in the Netherlands. There are more than 8000 registered WNCP groups in the Netherlands (see the overviews on https://wapb.nl), this shows that WNCP groups not only supplement but more often supplant physical neighbourhood watch groups. A key aspect of these groups is that citizens make themselves responsible for public safety (Mols & Pridmore, 2019; van der Land, 2017)

The sudden omnipresence of WNCP groups throughout the Netherlands caught the attention of policy makers and researchers who address its benefits and drawbacks in a growing body of 'grey' literature. The popularity of the groups can be explained by the low participation threshold and easy accessibility (Bervoets et al., 2016). According to Akkermans and Vollaard (2015), the introduction of WNCP to neighbourhoods in Tilburg considerably decreased the number of break-ins. Their study measured break-in rates before and after the introduction of WNCP and neighbourhood block watch. While these findings are often repeated by news media, the question rises if WNCP groups actually decrease property crime or if it is displaced to surrounding neighbourhoods? Tilburg's WNCP groups might have instigated a 'water bed effect', i.e. a temporary relocation of criminality or break-ins to other neighbourhoods (Van der Land et al., 2014). While the main purpose of WNCP groups is to prevent break-ins, increased social cohesion is also mentioned as a positive effect (Mehlbaum & van Steden, 2018; Van der Land et al., 2014). Moreover, some participants feel good about being aware of activity in their neighbourhood, or feel safer when they know about neighbourhood safeguarding practices (Pridmore et al., 2018; Smeets et al., 2019).

In contrast, an increased anxiety about safety can also be caused by WNCP group participation (Lub & De Leeuw, 2017) as well as an unwanted feeling of being 'monitored' (Mols & Pridmore, 2019). Apart from these personal drawbacks, the neighbourhood culture can be effected by stereotyping, racist behaviour, and privacy-infringing practices (de Vries, 2016; Lub & De Leeuw, 2019). Moreover, police are often not involved in neighbourhood safeguarding and surveillance practices (Mehlbaum & van Steden, 2018). Citizens engage in participatory policing practices when they assist law enforcement and engage in monitoring, information sharing, reporting, and preventative practices (Larsson, 2017). This inevitably leads to issues of accountability and responsibility (Mols & Pridmore, 2019). In order to make sense of the inherently conflicting nature of WNCP practices with their beneficial as well as detrimental consequences for neighbourhood dynamics and personal experiences, this study zooms in on how these practices can be seen as citizen-initiated surveillance activities.

Interpersonal Surveillance Activities

Ubiquitous and accessible mobile and interconnected technologies and devices increase the potential for pervasive forms of digitally mediated surveillance. In fact, people's day-to-day life can be seen as being under constant surveillance, and according to Lyon "humans are surrounded, immersed, in computing and networked technologies from dawn to dusk in every conceivable location." (Lyon, 2007, p. 1). Surveillance pertains to the collecting and processing of personal data for influencing or managing purposes (Lyon, 2007). It is important to note that the collection and processing of personal data is done by a great variety of actors (governmental, commercial or societal) for an even greater number of reasons. "Surveillance is not directed by one centralised entity, but is polycentric and networked" (Niculescu Dinca, 2016, p. 62). This has to do with the multiplicity of (interconnected) surveilling actors and with the diversity of data collected via different channels.

WNCP practices entail multiple forms of surveillance with each their own privacy-issues. First, the group conversations take place on WhatsApp, a commercial platform owned by Facebook, and whereas conversations might be protected by end-to-end encryption, unencrypted metadata still enables commercial surveillance of locations, connections, patterns, and personal information (Rastogi & Hendler, 2017). Second, the potential involvement of police actors in WNCP groups can lead to surveillance by law enforcement, often without knowledge or consent of participants (Mols & Pridmore, 2019). Finally, as indicated in the introduction, WNCP practices can be seen as a form of lateral surveillance (Andrejevic, 2002, 2007), also known as co-veillance (Mann, 2016), or interpersonal surveillance (Trottier, 2012). This form of social monitoring is based on citizens using digital services to find information about people in their network.

Existing literature is limited to interactive technologies, such as instant messaging, cell phones, Google, home surveillance products and services (Andrejevic, 2007), and social media (Lee et al., 2017; Trottier, 2012). WNCP practices also entail a digital counterpart of lateral surveillance when participants view the information neighbours openly share on WhatsApp (phone number, profile picture, status update), or when they use WhatsApp to check if their neighbours are or have been online, and if they've read their messages. Moreover, moderators also often screen new participants on Facebook to check if they are trustworthy. However, most interestingly, WNCP practices offer a different layer to lateral surveillance practices when neighbours actively watch one-another in person. The combination of digital and physical interpersonal surveillance practices is further explored in the results section.

RESEARCH APPROACH: A PRACTICE THEORY LENS

WNCP surveillance activities can be seen as practices; routinised types of behaviour which entail configurations of interconnected dimensions, such as: "forms of bodily activities, forms of mental activities, 'things' and their use, a background knowledge in the form of understanding, know-how, states of emotion and motivational knowledge."(Reckwitz, 2002, p. 249). Shove, Pantzar and Watson (2012) conceptualise these dimensions as competence, meaning and material. When studying practices, the conjunction of all three interacting dimensions should be taken into account (Shove et al., 2012) and hence, they form the lens through which WNCP group practices are analysed in this study.

The material dimension refers to objects as constitutive components of practices (Reckwitz, 2002). Tools, objects, hardware, infrastructures and bodies are paramount to practices, as most actions cannot take place without objects (Schatzki, 2002). Within neighbourhood watchfulness practices, the material dimension consists of physical streets and houses. Moreover, bodies are also material dimensions, in the form of citizens watching their street and passers-by who become (unknowingly) subject of watchfulness practices. Other material dimensions are the tools used by citizens to enhance their practice (binoculars), to collect proof of suspicious activities (cameras) and to contact police, neighbours or others (phones). To successfully engage in neighbourhood watchfulness practices, specific knowledge is required. Competence refers to "understanding and practical knowledgeability" in the broadest sense, including know-how, background information, practical skills (Shove et al., 2012, p. 23) as well as rules, principles, and explicit instructions (Schatzki, 2002). The particular knowledge needed for watchfulness practices includes the capability to assess suspicious persons and activities, the knowledge of what is normal and what is deviant behaviour, and the skills to protect the street. Yet, the material and the competence dimensions would not be put to practice without a particular purpose. This is conceptualised in the meaning dimension which entails "the social and symbolic significance of participation at any moment"(Shove et al., 2012, p. 23). For neighbourhood watchfulness practices, this motivational knowledge is based on the desire for a sense of security and the purpose of safeguarding the neighbourhood. Moreover, it signifies the protection of private space and alertness.

Neighbourhood watchfulness practices are the result of particular configurations of material, competence and meaning dimensions. Neighbours are the practitioners producing, carrying out, and reproducing these activities, whereby their shared practices form collective accomplishments (Barnes, 2001, p. 31). Neighbourhood watchfulness entails a bundle of practices including watching the neighbourhood, informing the police, taking action and interacting with neighbours. These practices are not limited to one location, they can be seen as co-located practice bundles (Shove et al., 2012), which emerged in different neighbourhoods. Dimensions of practice shape each other and change over time. Shove et al.'s approach to practice theory (2012) focuses on the evolving nature of practices as the researchers aim to analyse change in everyday life. In order to provide an in-depth account of watchfulness practices, a qualitative research design is used.

METHODS: A QUALITATIVE ANALYSIS OF WNCP PRACTICES

The diversity and the novelty of WNCP groups requires an in-depth qualitative understanding of the range of practices. Therefore, this study follows a constructivist grounded theory approach (Charmaz, 2014). Semi-structured in-depth interviews and focus groups have been conducted, transcribed, coded and analysed. An iterative three-stage inductive process (with open, axial and selective coding) forms the basis of the analysis (Corbin & Strauss, 2007; Lincoln & Guba, 1985).

Data Sample

Because WNCP groups exist in all types of neighbourhoods, sampling was aimed at maximised diversity on the basis of Statistic Netherlands' degree of urbanisation[1]. The Netherlands is one of the most densely populated countries in Europe and this scale identifies categories ranging from a rural 5 (fewer than 500 addresses/km^2) to an urbanised category 1 (2,500 or more addresses/km^2). Table 1 includes an overview of the respondents and indicates the urbanity level of their neighbourhoods. The respondents were recruited in several ways; via snowball sampling, Twitter and LinkedIn private and public messages, and the https://wapb.nl online inventory of WNCP groups. Moreover, the personal network of the researcher was addressed to engage low-profile and less prominent WNCP groups.

WNCP Interviews and Focus Groups Procedures

The empirical research for this chapter started with 14 interviews with WNCP group moderators. In addition, two focus groups were organised with citizen members in order to complement the moderators' views. The research design was approved by the department ethics committee. Before the start of the interviews and focus groups, all respondents signed informed consent forms which informed the respondents about the research procedures, data collection, and provided information about how to retract their participation and where to direct questions or complaints. The interviews were of semi-structured nature, the interview guide covered a diverse range of topics including the start and development of the WNCP groups, guidelines, examples of successful events and failures, monitoring practices and administrative efforts. On average, the interviews lasted 75 minutes. Most interviews were one on one, yet, two interviews included two respondents. The focus groups, which took approx. 95 minutes each, focused on personal experiences of WNCP participants. The interviews and focus groups provided in-depth accounts of the daily practices in and around WNCP groups. All conversations were conducted in Dutch (relevant quotes were translated into English for this chapter), pseudonyms are used to protect the privacy of respondents. The in-depth interviews and focus groups feed into the analysis of WhatsApp neighbourhood watchfulness practices which is presented in the next section.

Data Analysis Procedure

The in-depth interviews and focus groups have been conducted, transcribed, coded and analysed. The constructivist grounded theory approach (Charmaz, 2014) included an iterative three-stage inductive process (with open, axial and selective coding) (Corbin & Strauss, 2007; Lincoln & Guba, 1985). In the open coding stage, a large collection of verbatim codes was created which literally reflect what the respondents said (for example "keeping a distance" and "social control unwanted". Subsequently, these

Table 1. Respondent overview

Respondent	Neighbourhood	Urbanity Level 2016
Pauline	City C	1
Bas	City E	1
Dave		
Marion	Suburb M	2
Marc	Suburb D	2
Arnold	Town G	3
Lenny		
Kai	Town S	3
Saskia	Suburb H	3
Rick (Focus Group)		
Henry (FG)		
Jessica (FG)		
Bram (FG)		
Daniel (FG)		
Emma (FG)		
John	Village Z	4
Sven	Town B	4
Klara	Town L	4
Harold	Village H	4
Theo (FG)		
Chrissy (FG)		
Vera (FG)		
Betty (FG)		
Lucia (FG)		
Bert	Village W	5
Ron	Village N	5
Louise	Village S	5

open codes were clustered in broader conceptual categories (such as 'lateral surveillance practices'). Finally, these clusters formed the basis of the four distinct types of WNCP groups which are presented in the results section whereby overarching characteristics such as 'community co-veillance' were combined with practice theory dimensions.

RESULTS: NEIGHBOURS' EXPERIENCES OF WNCP PRACTICES

As mentioned earlier, watchfulness practices existed in neighbourhoods long before WNCP initiatives first emerged. However, WNCP groups amplified existing surveillance practices whereby WhatsApp use changed the configuration of the three dimensions (material, competence and meaning). Monitoring practices have occurred before in assemblages of houses, streets, bodies, binoculars, and the physical watching practices of neighbours. When citizens wanted to communicate about activities on the street, they had to establish (one-to-one) phone connections or engage in physical conversations. In WNCP practices, these assemblages are amplified by internet connections, (smartphone) cameras, and WhatsApp groups which enable neighbours to reach all members of their group instantaneously in order to inform or activate them. The seemingly invisible internet infrastructures manifest itself in connectivity icons and the WhatsApp interface on mobile phones and other devices. With cameras embedded in mobile phones, pictures can immediately be shared with neighbours, police, and others. Thus, the expanded material dimension directly enhances the competence dimension as WNCP participants have gained the facilities to activate and inform neighbours and police more effectively than before the use of WhatsApp. Moreover, the competence dimension also changed because participating in WNCP practices requires skills that were not needed before. Namely, citizens need to know how to use their smartphone, to install WhatsApp, and to communicate with their neighbours. Furthermore, the meaning component now includes connectivity, which is indispensable for the continuous communication with neighbours.

Distinctions in Competence and Meaning Dimensions

When WhatsApp groups were integrated in neighbourhood watchfulness and surveillance practices, a reconfiguration of the three dimensions took place.During the interviews it became clear that, while the material dimension of WNCP practices has similar aspects across groups, WNCP groups were different in their competencies and in the meaning they attributed to their practices. The competence dimension varies across groups because moderators differ in their access to professional knowledge and support. In some of the neighbourhoods, police and municipalities inform and advise the WhatsApp moderators

Table 2. Differences in Meaning and Competence levels across groups

WNCP Group	Meaning	Competencies
City C	Broad	Low
Town L	Broad	Low
Town S	Broad	Low
Village W	Broad	Low
Village Z	Broad	High
Suburb H	Broad	High
Suburb D	Broad	Medium
Town B	Narrow	Low
Village S	Narrow	Low
Village H	Narrow	Low
Suburb M	Narrow	High
Village N	Narrow	High
City E	Narrow	High
Town G	Narrow	High

actively (*high competencies*), whereas such knowledge is absent in other neighbourhoods (*low competencies*). In the interviews, the meaning or purpose of the groups was discussed, and it became clear that the meaning that members and moderators attribute to the practices they perform can be understood as ranging from narrow to broad. In neighbourhoods where the WNCP group has a *narrow meaning*, the focus is solely on safety; on preventing break-ins and burglaries. In contrast, other moderators want their groups to serve a *broad purpose* and also allow social support practices (such as watching neighbourhood children and supporting neighbours in need).

Table 2 lists the differences in meaning and competence, and based on these differences, four clusters can be identified with similar levels of competence and range of meaning. The matrix in Figure 1 visualises these clusters and displays tendencies which are used as analytical tools in the analysis. In order to grasp how these WhatsApp groups function and how their different configurations impact parties that are (sometimes unknowingly) involved in surveillance practices, the characteristics and particular issues of each cluster are examined next.

Figure 1. Clusters of WNCP groups with similar competencies and meaning

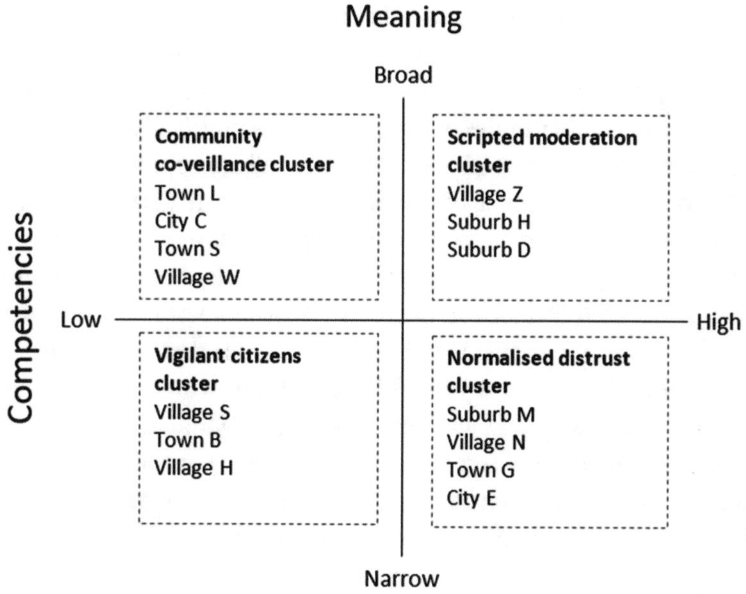

Meaning
Broad
Community co-veillance cluster
Town L
City C
Town S
Village W
Scripted moderation cluster
Village Z
Suburb H
Suburb D
Competencies
Low
High
Vigilant citizens cluster
Village S
Town B
Village H
Normalised distrust cluster
Suburb M
Village N
Town G
City E
Narrow

1: Community Co-veillance Cluster

The first cluster includes neighbourhoods in both rural village environments and urban contexts and is characterised by broad meaning and low competencies. More specifically, the WNCP groups in this cluster serve a broad purpose with room for social support, and they are initiated and sustained without police or municipal involvement. Group members have knowledge about their community but lack strongly enforced guidelines or access to professional (police) knowledge. As Klara (moderator Town L) notes: "I don't even know who our community police officer is." Consequently, group members use intuition and common-sense beliefs to make decisions about what is to be considered suspicious or deviant behaviour and how to react. There is no formalised hierarchy present as the low-profile set-up often includes multiple moderators (ranging from two to six moderators) and power structures remain invisible.

The groups in this Community co-veillance cluster are largely self-governing and have a relatively small group size (30-150 members). Their informal nature is emphasised by the broad meaning members attribute to the groups. First and foremost, the WhatsApp groups are targeted towards increasing safety in the neighbourhood. Kai, moderator in Town S, states: "I believe that it also provides sort of a sense of security". While preventing burglaries and break-ins is the primary purpose of the groups, they leave room for social exchanges and social support. For moderator Pauline (City C), their WNCP practices are characterised by communality: "Our sense of community is stronger now." This communality is accompanied by a sense of increased social support, neighbours help each other. Moderators and members shared examples of practical, functional and supportive content; e.g. a keychain found on the sidewalk (Pauline, moderator City C) and two parakeets that broke loose (Bert, moderator Village W). Moderator Kai (Town S) adds: "Last week someone lost his cat, that's fine you know, just post that in the group." More tangible forms of social support can also be activated via WNCP groups, as moderator Klara (Town L) illustrates: "We said that [points out of the window] there, someone had knee surgery, [we said] if your husband is working, you can send a message in the WNCP group, like: Hey, who can help me out?"

However, social support easily turns into social control and WNCP group members experience the watchfulness of others on a daily basis. Lateral surveillance, or co-veillance practices lead neighbours to not only monitor the streets but also their neighbours. These practices are guided by a physical dimension, the act of watching one another in the street. Neighbours physically keep an eye out on the street while simultaneously monitoring their neighbours digitally, WhatsApp offers digital settings that allow users to check when a contact was last seen online and if they have read their message. In Klara's Town L, neighbours are alert and keep an eye on the neighbourhood children, but this also leads to: "everyone knows at what time everyone goes to work. "Notably, not all neighbours feel comfortable with being monitored by their neighbours, Moderator Pauline (City C) illustrates: "It is not always pleasant, because it can really feel as if you are being watched, and to what extent is that good?" Inevitably, Pauline tries to protect her personal space: "Sometimes you will keep a bit of a distance, they are your neighbours, but eh... you do not have to be walking in and out all the time. "Similarly, Bert (moderator Village W) describes how he feels ambivalent about social support and social control: "Look, it is good that people care about each other and look out for each other... But, well, it is good that people look around anyway, but I'm like, social control, keep it out of my house, they don't have to know everything about me."

The Community co-veillance cluster shows how multi-dimensional informal surveillance practices impact personal lives and daily experiences of neighbours in their own house in their own neighbourhood. This is experienced as a benefit when neighbours keep an eye out for each other and offer social support, yet some respondents feel watched by their neighbours and prefer less social control. Lateral surveillance is both a curse and a blessing in these small and informal WNCP groups.

2: Scripted Moderation Cluster

As opposed to the Community co-veillance cluster with its informal moderation and small groups, the Scripted moderation cluster is characterised by intense moderation and relatively large group sizes. Founders of these groups appointed co-moderators to share administrative and monitoring responsibilities in order to effectively moderate the large populations (the groups typically include 250-350 members and cover a whole village or suburban area). Similar to the groups in the Community co-veillance cluster, the meaning of the watchfulness practices is not solely limited to burglary or break-ins, moderators also welcome notifications of missing children and other pressing issues. Yet, social support is not provided by these groups, since the scope and size of the groups makes them less personal and more detached from everyday issues.

Guidelines and rules play an important role in these groups. Most groups base their guidelines on the SAAR method, an abbreviation that stands for Signaleer (see), Alarmeer (alert the police), App (inform the neighbours), Reageer (react)(*Huisregels*, 2015). The SAAR method, devised in City E and adopted by many WNCP groups, can be seen as 'abstracted knowledge' that circulates between sites of enactment (Shove et al., 2012). In other words, competence transcends localised WNCP practices. Village E moderators translated their knowledge into the abstract SAAR method which was shared online and subsequently decoded and adopted in other neighbourhoods. In the Scripted moderation cluster, the use of the SAAR method and additional rules is informed and complemented by professional knowledge, as these groups are supported by community police officers who provide day-to-day advice. Community police officers streamline the (lateral) surveillance practices in these neighbourhoods by actively guiding the WNCP moderators and members.

What should be evident in this is that WNCP practices are piecemeal configurations, bringing together different types of (professional) knowledge and networks. Instead of using technologies specifically designed for watchfulness practices (such as the application Nextdoor), moderators and members make do with the resources at hand. This bricolage approach (Ciborra, 1992) includes tinkering with the messaging app itself as improvisation and adjustments occur on the part of both moderators and members. The piecemeal character of WNCP groups often leads to misunderstandings and tensions in the groups – tensions caused by members that share content deemed irrelevant or inappropriate by other members or moderators. John (Village Z) provides examples of irrelevant content shared in his group: "The sidewalk for example. If it's wonky and uneven, that has to do with safety, but you should go to the municipality. Or a damaged street light, you shouldn't post that in the WhatsApp group." Likewise, it is deemed irrelevant when members unnecessarily react to questions or notifications: "Someone asks: 'Does someone know about…?', and then everyone replies with: 'No'. While if they would wait until one person says 'Yes', no-one else needs to respond." (Lars, member Suburb H). Irrelevant content is often caused by people accidently sending messages in the wrong WhatsApp conversation. Member Rick (Suburb H) also made a mistake: "I wanted to WhatsApp my daughter that she had to come down for diner – she was in the attic. So, I type: 'dinner's ready'. But then WhatsApp restarted because of an update, and I thought it was ready to go, so I pressed 'Send'. And I looked… Oops, it's in the WNCP group."

Misunderstanding and tension arise in WNCP groups because certain key scripts are at play in the Scripted moderation cluster. These scripts – particular visions of the world inscribed in an object or artefact implying a specific relationship between the object and surrounding actors (Akrich, 1992) – combine a variety of actors. These range from an internet connection, a smartphone, multiple human users, and text-, photo-, video-, sound-, emoticon-based messages. The inscribed user (Latour, 1992) of WhatsApp is expected to be capable of using a smartphone, installing the application, and connecting with contacts. The high penetration rate of WhatsApp in the Dutch market suggests that many people follow this script successfully. Research shows that WhatsApp is primarily used as a private platform to interact with close ties (Karapanos et al., 2016; Waterloo et al., 2017). The informal nature of these interactions in combination with the speed and ease of the platform are not compatible with the serious nature of WhatsApp neighbourhood crime prevention. The SAAR method and other WNCP scripts clash with the script of WhatsApp. As the examples above show, this easily leads to friction in the functional environment of the WNCP group.

The effectiveness of WNCP can be thwarted by misguided reactions and mistakes, which lead to frustrations and conflicts among members and moderators. Failed scripts of WhatsApp as well as the tailor-made guidelines hinder everyday experiences in the WNCP groups. Moderators assemble dispersed network of neighbours, mobile phones, WhatsApp software, streets, and houses in order to protect their neighbourhood and private space against actors with bad intentions. In order to make neighbours uniformly participate in neighbourhood surveillance practices, the moderators in the Scripted moderation cluster wrote (sometimes extensive) manuals. However, in almost all WNCP groups, frustrations and conflicts emerge when members fail to follow these manuals and thus, fail to adhere to the scripted practices.

3: Vigilant Citizens Cluster

This cluster of small-scale WhatsApp groups (30-70 members) is characterised by vigilant behaviour in order to prevent break-ins. In contrast to the previous two clusters, the groups within the Vigilant citizens cluster attribute a narrow meaning to their practices. The group moderators do not allow practical or social matters to be discussed in the WNCP group. Moderator Louise (Village S) even decided to create an additional group to solve tensions similar to the aforementioned discussions about irrelevant content: “For all the other things that people deem important, we made a social app group.” Notably, the existence of a social group is an exception but emphasises the strict and focused character of the WNCP groups in the Vigilant citizens cluster: “The app is purely meant for fire, break-ins and real emergencies” (Louise, moderator Village S). The neighbours within the groups of this cluster display a vigilant mind-set and voluntarily engage in surveillance practices. Notably, they go further than the participants in other WNCP group clusters because they actively started to police their neighbourhood. The specific configuration of dimensions in this cluster can be characterised by narrow meaning and low competence levels because no police actors have been involved in initiating or moderating the groups. The vigilant surveillance practices in this cluster are a precarious form of participatory policing (Larsson, 2017).

Citizen initiated policing activities can affect passers-by who are (often unknowingly) subject of watchfulness practices. When walking through an unfamiliar neighbourhood, there is always the chance of being watched by residents. While spying eyes and suspicious glances might make one feel uncomfortable, WNCP surveillance practices augment this. Suspicions about passers-by are immediately materialised when a WNCP member sends a WhatsApp message to neighbours. This can lead to more neighbours peering through blinds or pulling back the curtains, which will invariably affect the experiences of passers-by. Furthermore, it is not just messaging but also the potential to be photographed. Cameras on smartphones can be used in unnoticeable ways and a picture of an allegedly suspicious person can be snapped in a split second. WNCP members often share images and some go great lengths to collect visual proof. Harold provides an example:

“My wife said: ‘what a strange man, he walked by and stopped to look at the houses’, so I jumped on my bike to see where he went. (...) And I held my smart phone camera in front of me, and made a picture of him. And these pictures I sent to the police. He didn’t notice at all that I was taking a picture of him.” (Harold, moderator Village H)

Photographing strangers and sharing pictures with police creates significant privacy concerns. These infringements are even bigger when smartphone pictures and surveillance camera footage are shared within WNCP groups. People who are photographed do not even have to raise suspicion. Not knowing someone can be reason enough to directly materialise even the smallest cause for doubt or distrust, i.e. to make a picture and share it in the WNCP group. However, not all moderators advocate this type of active behaviour. Some of them only allow pictures of people who are acting obviously suspicious or, as moderator Sven (Town B) explains: "Not until you clearly see that someone walks into gardens, looks into windows and is checking the gates." Likewise, an often-used rule is one that permits sharing of pictures within but prohibits the distribution of pictures outside of the WNCP group. However, the ease with which visual digital material can be distributed within and beyond WNCP groups presents privacy issues for all actors involved.

Whereas the moderators have communicated guidelines when they started the groups, these rules are often not strongly enforced. Events in Village H show that a low level of competence can cause risky situations. This group stands out because it is the most active group in the sample. During the interview and focus group, it became clear that a relatively large amount of warnings is exchanged in the group. At least once a week one of the neighbours notifies the group about an allegedly suspicious situation, break-in, or another event. Most groups are used far less often. Moreover, moderators and members of Village H frequently take immediate action when they receive a notification.

The abovementioned SAAR method is customarily ignored in highly vigilant neighbourhoods like Village H. Instead of informing the police before taking action, these citizens directly, and often impulsively, react to alerts. Member Vera (Village H) provides an example of a recent event when she distrusted a van that was parked in her street "…So my husband went to the car, and just said: 'Can I help you, is there something wrong with your car?' At that time another person came up, that jumped into the van, and they were gone." Vera reported the suspicious van in the WhatsApp group while her husband confronted the driver, but she did not contact the police until after the van left. In this and similar situations, the neighbours put themselves at risk. Interactions with suspicious persons might get out of hand or groups of members can get carried away in the heat of the moment. Vigilant actions of neighbours might be aimed at preventing break-ins, some neighbours simultaneously enjoy their participatory policing practices. Members Theo states: "It is just exciting (…) Life is one big risk." This excitement leads to behaviour that can put members and allegedly suspicious persons at risk.

Additionally, participatory policing and surveillance practices are likely to lead to widespread suspicion and ambivalence between neighbours (Reeves, 2012). In Village H, this became evident when moderator Betty saw a person running past her house at the beginning of the evening, a man wearing black clothes and a beanie. Betty did not trust the situation and got into her car: "So I drove around to see where that man went. And later, I posted it on the WNCP group, and then I received a message: 'Yes, that is the husband of one of the women in the WhatsApp group'". This example shows that vigilant surveillance practices can default to the lateral surveillance of neighbours not only in the form of monitoring but also by actively following them. Notably, whereas the activity and eagerness to take action are relatively high in Village H, this behaviour, the resulting suspicion, and risks are not limited to this neighbourhood. Multiple moderators shared stories of members that experience messages in the WhatsApp group as a direct call to action. This particularly concerning configuration of practices results in vigilant behaviour which blurs the boundaries between police and citizen responsibilities and puts neighbours and allegedly suspicious persons at risk.

4: Normalised Distrust Cluster

The Normalised distrust cluster includes four large groups with a narrow purpose focused on preventing break-ins, which stand out in their high level of competence visible in a level of professionalisation. In all four groups, community police officers were closely involved in the initiation, promotion, and construction of the groups and thus set the tone for these practices. Police officers assisted in the design of guidelines, and are connected to the moderators in a separate WhatsApp group in order to support them and to use the group's assistance during police actions. Furthermore, the moderators monitor multiple WhatsApp groups that function in compatible manner. Each group covers part of the area and is monitored by a local moderator (on street level) while the interview respondents, who can be seen as supra-moderators, hold an oversight position. Ron describes the design of his groups:

"We have nine WhatsApp groups which are named after their streets, and above these groups is another group and that is the Chief Control group. And in that group are, as I named them, the Ambassadors (…) and they are all part of that Chief Control group, but they are also the moderators of their own WhatsApp group. Let say that, in group [street name] there is a notification. I see that notification and I post it in the Chief Control group. All Ambassadors will see the notification and they take it up and forward it to their own WhatsApp group. This way, we can inform the whole village in three minutes." (Ron, Village N)

These professionalised groups are strictly moderated and comprise formalised practice bundles (Shove et al., 2012). The configuration of these combined efforts is documented in manuals and whereas they are constructed independently, these practice bundles coincide with municipality and police practices. The self-imposed power that supra-moderators provide themselves with creates a distinct hierarchy in neighbourhood networks. The functioning of multiple WNCP groups in a neighbourhood is determined and steered by one or two individuals who have created this position for themselves. The large-scale nature of these groups (300-3000 members) increases the impact of neighbours' everyday surveillance activities and presents similar issues as in the other clusters but on a larger scale. The rules and conventions of WNCP groups are still in development while they are simultaneously normalised.

When participating in these large-scale WNCP groups, neighbours become acutely aware of particular events which alters their experience of the neighbourhood. Activities in the street that previously would have gone unnoticed, are now materialised into WhatsApp group messages. Neighbours actively monitor their neighbourhood and share suspicions, they also work together in information searching practices (such as looking up a license plate registration online). WNCP participation sensitizes them to suspicious activities. A hitherto underexposed pressing concern of these practices is how determining what is seen as suspicious leads to the normalisation of categorisations and wrongful accusations. When neighbours engage in surveillance practices, they scan the street and assesses the situation at hand. This assessment is based on competence and meaning. The knowledge consulted when making a distinction between suspicious and non-suspicious persons and behaviour is guided by a categorisation of normal and deviant activities. Nevertheless, conventions about what counts as suspicious or deviant behaviour are ambiguous, there are no clear categories of normal/suspicious/deviant behaviour and persons.

In the interviews it became evident that the categories used by neighbours are often based on stereotypes and prejudice. A worrying number of incidents reported in all four clusters revolved around wrongful accusations based on intolerant categorisations. An example of a situation like this became visible in Suburb H (in the Scripted moderation cluster), where a neighbour consulted the group to voice concerns about a couple of men sitting in a car with a Polish license plate. Henry (member) recalls: "And then immediately someone replied: 'Yes, they belong here, they live here'." Dave (member) explains: "They were only waiting for someone to get in the car." This anecdote and similar cases show that intolerant categorisations are normalised in many WNCP groups. In the large groups in the Normalised distrust cluster, the scale of these discriminatory practices increases, and adequate responses of moderators and involved police seem to remain absent. The fact that these groups widely promote their WNCP templates and methods without addressing these issues, leads to the conclusion that the normalisation of surveillance in WNCP practices includes a concerning normalisation of suspicion, distrust and intolerance towards strangers as well as neighbours (Reeves 2012; Larsson 2017).

DIRECTIONS FOR FUTURE RESEARCH

This chapter explores the topic of WNCP practices and aims to provide a concise yet comprehensive overview of the benefits and drawbacks these practices bring, with a particular focus on lateral surveillance. Power is a key concept that is only touched upon briefly in the results section yet deserves more attention in future research. In WNCP as well as in similar neighbourhood watchfulness practices, power inequalities arise between active and engaged participants and unknowingly involved citizens. Moreover, an imbalance in information can arise, whereby it too often remains unclear which actors have access to which information – for instance in practices where police officers are involved. The question arises how this produces an arrangement of relationships within neighbourhood contexts, with commercial, law enforcement and citizen actors with unequal power structures and non-transparent information sharing.

Moreover, this study focuses on WNCP practices in the Netherlands, and while the sample is diverse, the practices are immersed in and influenced by the local context and culture. For future research, it would be good to see how the interplay between meaning, competence and material dimensions differs or is similar in WhatsApp watchfulness practices in other countries. For example, neighbourhood block watch groups in the UK, South Africa, and Australia also use WhatsApp for crime prevention[2]. Even though these initiatives are not as widespread as they are in the Netherlands, they might be dealing with similar issues as outlined in this paper. Future research should look into how cultural backgrounds influence neighbourhood watchfulness practices and how different contexts create different meaning and competence dimensions in these practices.

Moreover, especially in the US, commercial platforms have introduced similar networks which citizens can use to collaboratively safeguard their neighbourhoods. Nextdoor, a social networking platform for neighbours, includes a section on safety wherein law enforcement and municipalities can take part (see www.nextdoor.com). And recently Amazon introduced Neighbours (https://shop.ring.com/pages/neighbors), an app wherein citizens can exchange photos and videos of their smart doorbells in order to notify their neighbours of suspicious activities on their streets. Similar to Nextdoor, law enforcement can take part in these networks. The integration of law enforcement not only raises concerns about the privacy of alleged 'suspects' whose images are shared on the app without their knowledge, the commercial basis of these platforms amplifies these concerns. Because the commercial nature of these apps creates another dimension to watchfulness practice, a dimension that brings up urgent questions about data collection, and ownership of the images and videos shared on these platforms.

CONCLUSION

WhatsApp proved to be a technical layer in the amplification of surveillance practices guided by materialised suspicions. The implementation of WhatsApp to watchfulness practices changed neighbourhood dynamics and personal experiences. The existence of a WNCP group alters how neighbours experience their street because concerns and possible threats become more immanent. This urges many neighbours to take up their responsibility and contribute to surveillance practices in order to safeguard their neighbourhood. These practices are co-constructed as they are the result of collaborative efforts of neighbours, sometimes in conjunction with police or municipality actors. The immense popularity of this recently emerged phenomenon is of ambivalent nature and has precarious consequences. This study showed that while citizen participation in crime prevention leads to an increase in social support, feelings of safety, and the active prevention of break-ins, it is also accompanied by risky acts of participatory policing and vigilant behaviour. Furthermore, the monitoring of passers-by defaults to lateral surveillance practices, which impact the individual experience of citizens in their own homes and streets. The inscribed user of WhatsApp is often not compatible with the inscribed WNCP member that moderators have in mind. Caveats between scripts and daily practices lead to friction among neighbours.

The focus on WNCP practices enables an examination of lateral surveillance as a combination of digital and physical practices, a valuable contribution to existing literature focusing on digital interpersonal surveillance practices (Andrejevic, 2005, 2007; Lee et al., 2017; Trottier, 2012). Particular "suspicion-driven rituals of lateral surveillance' (Reeves, 2012, p. 238) are initiated by citizens whereby risky behaviour and intolerant attitudes are normalised and integrated in community life. The standardisation of these practices needs to be critically assessed, not only by social scientists but also by institutional actors involved such as municipal policy makers and police officers. Namely, these vigilant communities in the making raise significant issues for communities, police and municipalities and the transition of practice bundles across neighbourhoods raises questions about adjustments and standardisation processes.

Whereas this study is limited to an in-depth snap shot of particular Dutch practices with a limited sample, issues were identified that mirror the risky consequences of participatory policing practice in institutionally initiated practices in countries such as the US. The use of a practice theory lens proved to be particularly helpful in distilling the general issues of citizen participation in ICT-supported surveillance practices. The WNCP group clusters identified in this study were crucial in uncovering a variety of pressing issues arising from inherently diverse WNCP practices. This chapter shows that the emergence of WhatsApp amplified all three dimensions of a variety of neighbourhood watchfulness practices still in the process of stabilisation and normalisation.

REFERENCES

Akkermans, M., & Vollaard, B. (2015). *Effect van het WhatsApp-project in Tilburg op het aantal woninginbraken – een evaluatie*. Tilburg University. https://hetccv.nl/fileadmin/Bestanden/Onderwerpen/Woninginbraak/Documenten/Effect_van_het_WhatsApp-project_in_Tilburg_op_het_aantal_woninginbraken/tilburg-whatsapp_191015.pdf

Akrich, M. (1992). The De-Scription of Technical Objects. In W. E. Bijker & J. Law (Eds.), *Shaping Technology / Building Society: Studies in sociotechnical change* (pp. 259–264). MIT Press. https://mitpress.mit.edu/books/shaping-technology-building-society

Andrejevic, M. (2002). The Work of Watching One Another: Lateral Surveillance, Risk, and Governance. *Surveillance & Society*, *2*(4). Advance online publication. doi:10.24908s.v2i4.3359

Andrejevic, M. (2005). The Work of Watching One Another: Lateral Surveillance, Risk, and Governance. *Surveillance & Society*, *2*(4), 479–497.

Andrejevic, M. (2007). *Ispy: Surveillance and Power in the Interactive Era*. University Press of Kansas. https://www.bookdepository.com/ISpy-Mark-Andrejevic/9780700616862

Barnes, B. (2001). Practice as collective action. In T. R. Schatzki, K. Knorr Cetina, & E. von Savigny (Eds.), *The Practice Turn in Contemporary Theory* (pp. 25–36). Routledge.

Bellair, P. E. (2000). Informal Surveillance and Street Crime: A Complex Relationship*. *Criminology*, *38*(1), 137–170. doi:10.1111/j.1745-9125.2000.tb00886.x

Bervoets, E. (2014). *Een oogje in het zeil... Buurtpreventie in Ede: Onderzoek en een gefundeerd advies*. Bureau Bervoets. https://www.bureaubervoets.nl//files/pagecontent/OnderzoekEde%20Buurtpreventie-DEFDEF.pdf

Bervoets, E., Van Ham, T., & Ferwerda, H. (2016). *Samen signaleren Burgerparticipatie bij sociale veiligheid*. Platform 31. http://www.beke.nl/doc/2016/PL31-samen%20signaleren.pdf

Charmaz, K. (2014). Constructing Grounded Theory. *Sage (Atlanta, Ga.)*.

Church, K., & de Oliveira, R. (2013). What's Up with Whatsapp?: Comparing Mobile Instant Messaging Behaviors with Traditional SMS. *Proceedings of the 15th International Conference on Human-Computer Interaction with Mobile Devices and Services*, 352–361. 10.1145/2493190.2493225

Ciborra, C. U. (1992). From thinking to tinkering: The grassroots of strategic information systems. *The Information Society*, *8*(4), 297–309. doi:10.1080/01972243.1992.9960124

Corbin, J., & Strauss, A. (2007). *Basics of Qualitative Research: Techniques and Procedures for Developing Grounded Theory* (3rd ed.). SAGE Publications, Inc.

de Vries, A. (2016, February 1). BuurtWhatsApp: Goed beheer is complex. *Social Media DNA*. https://socialmediadna.nl/buurtwhatsapp-goed-beheer-is-complex/

Hampton, K. N. (2007). Neighborhoods in the Network Society the e-Neighbors study. *Information Communication and Society*, *10*(5), 714–748. doi:10.1080/13691180701658061

Huisregels. (2015, July 9). *WhatsApp Buurtpreventie*. https://wabp.nl/huisregels/

Karapanos, E., Teixeira, P., & Gouveia, R. (2016). Need fulfillment and experiences on social media: A case on Facebook and WhatsApp. *Computers in Human Behavior*, *55*, 888–897. doi:10.1016/j.chb.2015.10.015

Larsson, S. (2017). A First Line of Defence? Vigilant Surveillance, Participatory Policing, and the Reporting of 'Suspicious' Activity. *Surveillance & Society, 15*(1), 94–107.

Latour, B. (1992). Where Are the Missing Masses? The Sociology of a Few Mundane Artifacts. In W. E. Bijker & J. Law (Eds.), *Shaping Technology / Building Society: Studies in sociotechnical change* (pp. 225–258). MIT Press. https://mitpress.mit.edu/books/shaping-technology-building-society

Lee, E. W. J., Ho, S. S., & Lwin, M. O. (2017). Explicating problematic social network sites use: A review of concepts, theoretical frameworks, and future directions for communication theorizing. *New Media & Society*, *19*(2), 308–326. doi:10.1177/1461444816671891

Lincoln, Y. S., & Guba, E. G. (1985). *Naturalistic Inquiry* (1st ed.). SAGE Publications. doi:10.1016/0147-1767(85)90062-8

Lub, V. (2018). Watch Group 2: Countering Burglars. In M. Gill (Ed.), *Neighbourhood Watch in a Digital Age* (pp. 57–72). Palgrave Macmillan., doi:10.1007/978-3-319-67747-7_5

Lub, V., & De Leeuw, T. (2017). Perceptions of Neighbourhood Safety and Policy Response: A Qualitative Approach. *European Journal on Criminal Policy and Research*, *23*(3), 425–440. doi:10.100710610-016-9331-0

Lub, V., & De Leeuw, T. (2019). *Politie en actief burgerschap: Een veilig verbond? Een onderzoek naar samenwerking, controle en (neven)effecten* (No. 108; Politiewetenschap). Politie & Wetenschap. https://www.politieenwetenschap.nl/publicatie/politiewetenschap/2019/politie-en-actief-burgerschap-een-veilig-verbond-322/

Lyon, D. (2007). Surveillance studies: An overview. *Polity*.

Mann, S. (2016). *Surveillance (Oversight), Sousveillance (Undersight), and Metaveillance (Seeing Sight Itself)*. https://www.cv-foundation.org/openaccess/content_cvpr_2016_workshops/w29/html/Mann_Surveillance_Oversight_Sousveillance_CVPR_2016_paper.html

Mehlbaum, S. L., & van Steden, R. (2018). *Doe-het-zelfsurveillance. Een onderzoek naar de werking en effecten van Whatsapp-buurtgroepen* (No. 95; Politiekunde). Politie & Wetenschap. https://research.vu.nl/ws/portalfiles/portal/73988502/Mehlbaum_Van_Steden_2018.pdf

Mols, A., & Pridmore, J. (2019). When Citizens Are "Actually Doing Police Work": The Blurring of Boundaries in WhatsApp Neighbourhood Crime Prevention Groups in The Netherlands. *Surveillance & Society*, *17*(3/4), 272–287. doi:10.24908s.v17i3/4.8664

Niculescu Dinca, V. (2016). *Policing Matter(s)*. Universitaire pers Maastricht. https://cris.maastrichtuniversity.nl/portal/en/publications/policing-matters(b911f31c-f8e8-44e9-999c-5c8edcafd7b7).html

Pridmore, J., Mols, A., Wang, Y., & Holleman, F. (2018). Keeping an eye on the neighbours: Police, citizens, and communication within mobile neighbourhood crime prevention groups. *The Police Journal: Theory. Practice and Principles*, *92*(2), 97–120. doi:10.1177/0032258X18768397

Rastogi, N., & Hendler, J. (2017). WhatsApp Security and Role of Metadata in Preserving Privacy. In A. R. Bryant, J. R. Lopez, & R. F. Mills (Eds.), *Proceedings of the 12th International Conference on Cyber Warfare and Security* (pp. 269–274). Academic Conferences and Publishing International Limited. https://books.google.nl/books?hl=en&lr=&id=iXWQDgAAQBAJ&oi=fnd&pg=PA269&dq=whatsapp+end-to-end+encryption+privacy+users&ots=HVbblQPNqW&sig=_brz7_Y2SHFL6v1TH40Pqx6Pycs#v=onepage&q=whatsapp%20end-to-end%20encryption%20privacy%20users&f=false

Reckwitz, A. (2002). Toward a Theory of Social Practices. *European Journal of Social Theory*, *5*(2), 243–263. doi:10.1177/13684310222225432

Reeves, J. (2012). If You See Something, Say Something: Lateral Surveillance and the Uses of Responsibility. Surveillance & Society, 10(3/4), 235–248.

Schatzki, T. R. (2002). *The site of the social: A philosophical account of the constitution of social life and change*. The Pennsylvania State University Press. https://scholar.google.com/scholar?cluster=583341585722467485&hl=en&oi=scholarr

Shove, E., Pantzar, M., & Watson, M. (2012). The Dynamics of Social Practice: Everyday Life and how it Changes. *Sage (Atlanta, Ga.)*. Advance online publication. doi:10.4135/9781446250655

Smeets, M. E., Schram, K., Elzinga, A., & Zoutendijk, J. (2019). *Alerte burgers, meer veiligheid?: De werking van digitale buurtpreventie in Rotterdam*. https://surfsharekit.nl/publiek/inholland/87bc7b97-632b-43d0-906e-e68575308f56

Trottier, D. (2012). Interpersonal Surveillance on Social Media. *Canadian Journal of Communication*, *37*(2). Advance online publication. doi:10.22230/cjc.2012v37n2a2536

van der Land, M. (2017). Meer of minder eigenrichting door buurtwachten? In P. Ponsaers, E. Devroe, K. van der Vijver, L. G. Moor, J. Janssen, & B. van Stokkom (Eds.), *Eigenrichting* (pp. 63–72). Maklu.

Van der Land, M., van Stokkom, B., & Boutellier, H. (2014). *Burgers in veiligheid: Een inventarisatie van burgerparticipatie op het domein van de sociale veiligheid*. Vrije Universiteit. https://www.publicspaceinfo.nl/media/bibliotheek/None/LANDSTOKKO%202014%200001.pdf

Waterloo, S. F., Baumgartner, S. E., Peter, J., & Valkenburg, P. M. (2017). Norms of online expressions of emotion: Comparing Facebook, Twitter, Instagram, and WhatsApp. *New Media & Society*, *1461444817707349*. Advance online publication. doi:10.1177/1461444817707349 PMID:30581358

KEY TERMS AND DEFINITIONS

Lateral Surveillance: This is a form of interpersonal surveillance in which people actively monitor their peers, co-workers, family members or neighbours without a clear power separation between themselves and those that they are monitoring.

Neighbourhood Watch: This activity revolves around small groups of participating citizens devoting their time to neighbourhood safety through the ongoing observation of suspicious activity and by being physically present in neighbourhoods.

Participatory Policing: This practice indicates the points at which citizens (attempt to) assist law enforcement by specifically engaging in monitoring, information sharing, reporting, and preventative practices of and about criminal activities).

Practice Theory: This theoretical approach focuses on how society, culture, and social life are dependent upon and develop through practices, which are themselves dynamic bundles of activities that include material elements, competences, and meaning.

Surveillance: Surveillance pertains to the collection and processing of personal data for the purpose of influencing or managing other persons or systems.

ENDNOTES

1 The full CBS Degree of Urbanisation scale, based on the surrounding address density: 1. Extremely urbanised: 2,500 or more addresses//km^2; 2. Strongly urbanised: 1,500 to 2,000 addresses/km^2; 3. Moderately urbanised: 1,000 to 1,500 addresses/km^2; 4. Hardly urbanised: 500 to 1,000 addresses/km^2; 5. Not urbanised: fewer than 500 addresses//km^2. Source: https://www.cbs.nl/en-gb/our-services/methods/definitions?tab=d#id=degree-of-urbanisation.

2 Examples of WNCP practices across the world: - UK: https://www.guardian-series.co.uk/news/16108413.neighbourhood-watch-use-whatsapp-to-stop-crime/ - South Africa: https://gbnw.co.za/whatsapp-sign-up/ - Australia: https://www.abc.net.au/news/2019-02-02/neighbourhood-watch-these-days-are-all-about-whatsapp/10763624

Chapter 8
Reading and Collaboration:
Developing Digital Reading Practices With Computer-Assisted Text Analysis Tools

Andrew Klobucar
New Jersey Institute of Technology, USA

Megan O'Neill
New Jersey Institute of Technology, USA

ABSTRACT

The introduction of digital media into university writing courses, while leading to innovative ideas on multimedia as a rhetorical enhancement means, has also resulted in profound changes in writing pedagogy at almost all levels of its theory and practice. Because traditional approaches to examining and discussing assigned texts in the classroom were developed to help students analyze different genres of print-based texts, many university educators find these methods prohibitively deficient when applied to digital reading environments. Even strategies in reading and text annotation need to be reconsidered methodologically in order to manage effectively the ongoing shift from print to digital or electronic media formats within first year composition. The current study proposes one of the first and most extensive attempts to analyze fully how students engage with digital modes of reading to demonstrate if and how students may benefit from reading digital texts using computer-assisted text analysis (CATA) software.

INTRODUCTION

As writing instructors working with students in this digital age, we find ourselves facing a number of critical imperatives regarding reading as an assigned activity. Prior assumptions concerning print-based methods of comprehension and communication may not correspond as effectively once one shifts from page to screen. Accordingly, traditional modes of examining and assessing assigned texts in the classroom, developed to help students analyze different genres of printed texts, may be prohibitively deficient when applied to digital reading environments. As many initial studies into this possible discrepancy show, students engaging with electronic texts are more likely to avoid active notetaking, highlighting key passages

DOI: 10.4018/978-1-7998-5849-2.ch008

or comparing multiple works (Berry, 2012; Gold, 2012). It has further been demonstrated that higher levels of comprehension, including remembering crucial premises and text-specific-terminologies, are adversely affected (Wolfe, 2018). In addition, many of these tendencies seem to derive from issues in digital text interface design and its use as a mode of presentation. As screen-based media (usually PDF format), digital texts cannot easily be marked up or changed while being critically engaged. Hence, at one level, active notetaking, whether in class or on an individual basis seems somewhat indirectly prohibited. The very act of taking notes while working with screen media often requires the simultaneous use of both print and digital media formats while texts are actually in use. Faced with two different formats, a student may just as likely avoid reviewing one over the other.

As a result of these significant changes in how texts are actually distributed and consumed by learners in and outside the classroom, a range of new pedagogical challenges to re- develop critical reading strategies for online literary analysis has emerged. Thus, the major premises informing this project begin by acknowledging first that to "read" or interact with text in a digital format is to engage a profoundly different kind of tool for a new mode of knowledge construction.

At a more nuanced level, it seems that reading digitally, compared to any print-based scholarship practices, entails a much more sophisticated categorization and critical review of the rhetorical aims of texts, not to mention a distinct re-construction (again rhetorically, and epistemologically) of what a text actually is. In fact, much of our research respects Bruno Latour's (1994) materialist theories of technical mediation, allowing us to see digital texts as more than a mere adaptation of print, but rather as a wholly transformative set of tools that changes our relationship to language in terms of "our quality as subjects, our competences, our personalities" (p. 31). In Latour's view—and held by many contemporary philosophers and cultural theorists following him—texts and writing in general, being technical, mediate our environment, not so much as a semiotic system, as a set of social utilities, conveying in the process a distinct linguistic materiality or object-ness. In his words, language is best considered first and foremost in term of its technological structures:

[m]y word processor, your copy of Common Knowledge [the journal publishing his essay], Oxford University Press, the International Postal Union, all of them organize, shape, and limit our interactions. To forget their existence—their peculiar manner of being absent and present—would be a great error" (Latour, 1994, p. 50).

This utility-oriented perspective continues to inform this very study. Quite specifically, as our work with first year learners shows, when one reads texts on the screen as part of actual functioning digital networks, they tend broadly to understand them to have less authority as individually authored voices and more as participants in a broader community that includes him or herself as readers. Argument and proposition as rhetorical features accordingly change, requiring more pedagogical options and possibly the subsequent refinement of instructional methods.

Addressing the issue of and its growing significance in the humanities, this project initiated both a qualitative and quantitative study of select NJIT first year writing classes to determine possible deficiencies in reading-related exercises assigned in electronic format according to specific learning objectives. This research project will pursue two primary lines of study in order to determine a more a substantive and analytically accurate understanding of how first year university students may be responding to information and argument presented in electronic formats. First, basic levels of student reading comprehension and critical engagement with electronically assigned and distributed texts at NJIT will be determined through in-class examination and individual surveys. Second, the same texts will subsequently be assigned and electronically distributed through computer assisted text analysis (CATA) software developed specifically to improve readerly engagement with online and electronic texts. Working with a selection of NJIT undergraduates grouped according to individual course sections, this study aimed to analyze how specific CATA features like document mark-up and inter-reader commentary may impact and subsequently help instructors design new critical reading strategies for their classes. The study effectively captured the reading comments, various in-class comprehension analyses, and a series of related survey responses of students in first-year writing during a period of instructional transition from print-based writing to digitally based writing.

Reviewing these comments with these responses demonstrates that both the format and software delivering the texts play vital roles in influencing student comprehension levels.

BACKGROUND AND LITERATURE REVIEW

Electronic writing brings forth an array of interesting challenges when scrutinizing methods of argument and narrative structure in first year composition. On the one hand, because the current platform-based "web" operates simultaneously as a communication tool and a multimedia production space, it seems reasonable to consider online publication and distribution almost generically as ongoing, active social relationships. The web simply cannot be viewed as a singular, individual practice as writing as a print-based activity was; whether via cloud-based technologies, or the use of forums, electronic media places the writer in some form of dialogue the minute she logs on. At the same time, most web-based writing tools, even when supplemented with social media applications like discussion forums and chat boxes, appear incapable of fully integrating a more formally interactive or collaborative approach to writing and reading into their respective production spaces. Group exercises designed for commentary and analysis often remain tightly managed by individuals with administrative levels of access posting single responses as pre-written retorts. Hence, while web-based electronic writing continues to promote theories of multiple authorship, along with numerous modes of online public interaction, the resulting collections nevertheless take the form of quick summary comments that fail to generate proper dialogues with a sense of intellectual community. The CATA software used in our studies, including online tools like NowComment, CATMA (Computer Aided Textual Markup & Analysis), and Voyant each introduced a variety of different collaborative approaches to critical reading, text mark-up, and annotation. By combining annotation and mark-up devices with improved collaborative messaging tools, a functional set of synchronous and asynchronous exercises in critical analysis seem possible. Key to this study will be to demonstrate how such tools and their interfaces can be subsequently re- developed to facilitate improved modes of interactive engagement with screen-based user interface design and formal writing tasks. The academic field best suited to support and further advance electronic writing, both within and outside the classroom, would appear to be Digital Humanities (DH), an expansive, relatively new area of study specializing in introducing interactive media tools to literature and writing throughout the university. In fact, as many critics point out, electronic writing's particularly open-ended perspective on the use of multimedia in the literary arts easily complements the Digital Humanities emphasis on new and emerging media technologies. For Matthew Kirschenbaum (2012) the field commonly supports, regardless of specialization, an accepted understanding of media as a "free-floating signifier." Lauren Klein and Matthew Gold's (2016) recent anthology of DH methods and theories compares the field's ever-expanding and open-ended relationship to different media formats to Art History's inclusive, object-oriented study of contemporary sculpture in the postwar academy almost 40 years ago. Such revision, they remind us, rarely emerges without some level of critical controversy. At the peak of sculpture's inclusivism in both its practice and study, theorist Roselyn Krauss in her important 1979 critical essay "Sculpture in the Expanded Field" famously lambasted what she saw as the medium's increased ambiguity in form drawn from multiple modalities and contexts: "[n]othing," she observed, "could possibly give to such a motley of effort the right to lay claim to whatever one might mean by the category of sculpture. Unless, that is, the category can be made to become almost infinitely malleable" (p. 30). Klein and Gold are certainly more open to what they see as a similar malleability in digital media technology, in effect re-situating Krauss's original title of her essay. As an inherently "expanded field," especially in relation to modes of writing, the Digital Humanities offers many "alternative structuring devices" that are both critically and culturally productive. In their words:

the expanded field is constructed by the relationships among key concepts, rather than by a single umbrella term. And it is by exploring these relations—their tensions as well as their alignments—that the specific contributions of the range of forms and practices encompassed by the field can be brought to light."

Linking many of the essays collected in the Klein and Gold anthology is their shared emphasis on prominent, ongoing interactive relationships between multiple media forms and formats over pre-set genres and formats. A more thorough, and in some ways, precise investigation of digital media technologies and their current "expansive" effect on academic writing would thus feature a relational focus on form and practice. Historically speaking, this mode of media interactivity might be usefully traced to electronic writing's emergence as a hypertext in the early 1960s, and, of course, easily situated within the more recent practices of blogging and social media as they've developed over the last two decades. Consistent throughout this lineage is a very distinct reconsideration of writing itself as a media technology in terms of its capacity to connect, attach and build networks between different sources of content in a variety of formats.

Despite these changing understandings of writing as a screen-based, networking tool, literary and composition studies in the academy continue to de-emphasize modes of interactivity within electronic media, while maintaining prior discipline-based focuses on specific genres and formats like poetry, prose, script, etc. In fact, there are still few critical discourses that help us work with writing's increasingly complicated platforms and media interfaces. This essay will, in part, look specifically at this increasingly significant deficit in composition, as well as introductions to literature. Efforts to provide a critical discourse, it can be argued, might usefully begin by looking directly at how students are actually developing reading habits and effective writing practices when using electronic texts both within and outside the classroom.

Throughout this deep-rooted history of electronic writing and reading modalities, and consistent with the sheer variety of media formats they have sponsored, there remains an enduring question regarding the influence of new writing tools on literacy skills. In 2018, we initiated a study on how the dissemination of texts through text analysis tools (CATA) might affect reading comprehension. Literacy at the very least seems substantially transformed via composition methods based on computation as delivered through electronic writing tools. The very fact that print-based genres and formats continue to mandate at a cultural level how one interacts with digital writing testifies to the screen's ongoing ambiguity as a new type of textual artifact. Scrolling online from site to site, readers continue to read web "pages," understanding the term simply as a metaphor for text display; one stores documents in "folders," building "archives" supposedly as substantial as the libraries once treated by scholars and casual readers alike as second homes. As Lisa Gitelman (2014) pointedly notes:

[w]ritten genres in general are familiarly treated as if they were equal to or coextensive with the sorts of textual artifacts that habitually embody them. . . .Say the word 'novel,' for instance, and your auditors will likely imagine a printed book, even if novels also exist serialized in nineteenth century periodicals, published in triple-decker (multivolume) formats and loaded onto—and reimagined by the designee and users of—Kindles, Nook, and iPads." (p. 3)

Given contemporary culture's consistently nebulous relationship to digital writing delivered on-screen, it seems hardly a question of whether literacy has changed, so much as how or in what way. Connected to cloud-based servers, distributed over high speed WiFi, the digital screen streams visual and audio media in countless new formats more effectively and broadly than any prior medium dating back to the early 20th century; but how exactly this technological system serves new textual and lexical practices still remains open to further examination.

The transition from print-based to electronic media in the classroom has hardly been straightforward, and the challenges that continue to interfere with its full incorporation into writing and literature programs have been critically reviewed and analyzed on a number of fronts (Barry, 2012; Gold, 2012). From our own studies we can add to current scholarship the consistent observation that, while learners seem fundamentally open to screen-based, digital media as academic writing tools, they remain less disposed to using this technology for long- term reading exercises. Much of the difficulty observed regarding how students interact critically with electronic text in the classroom can be directly traced to the Humanities' consistent restraint as discipline to integrate multiple electronic, online formats into the liberal arts. While this hindrance would have had little effect on established theories of literacy both in terms of pedagogy and assessment up until the end of the 20th century, this is not the case now.

Part of the aim of this paper is to review past modalities in electronic writing in order to develop improved methods for incorporating them into current Humanities writing and literature programs. In doing so, we may able to address what has been formally designated by many pedagogical theorists as a growing "knowledge gap" in the academy between learners based upon the level of access they have to digital media tools. Such a gap, we argue, significantly determines how writing and reading continue to evolve as technique-oriented, tool-driven practices in our cultural landscape, while providing influential factors in how they are researched and taught within the academy. Our own study seems to confirm the wide, seemingly random choices in approaches and methods learners continue to apply to writing, reading, and the study of rhetoric when using digital formats. Equally familiar with analogue platforms like PDFs as well as a variety of interactive screen modalities, learners remain open to working through electronic documents in multiple media formats, though the pedagogical approaches guiding most assignments remain more traditionally print-based. Rarely do teachers when planning instruction take into consideration how different media tools might be actively used to help guide learners to choose certain modes of reading and response over others. One of the primary focuses of this very study seeks to compare at a fundamental level the effect of analogue, print-derived applications like PDFs on reading practices to digital tools that encourage a more interactive level of engagement with the text in question. PDFs, given their technical as well as symbolic use of the screen to replicate the cultural function and format of printed documents, provide a completely different concept of information distribution in relation to digital tools as texts. A summary describing these two modes of text access in the context of our research follows here.

Following the advancement of contemporary Computer Assisted Text Analysis (CATA) tools in overall usability, so too must evolve the methods used to distribute, interpret and instruct new modes of electronic writing. In terms of presentation and technical function, online writing can be systematically distinguished from print media in its inherent emphasis of a more analytical, quantitative, data-oriented relationship to text and writing in general. In fact, one might even note here that at a fundamental level, digital writing can be considered nothing short of a mode of computation. Even the earliest word processors distinguished themselves as more than a new mode of electric typewriter by offering new lexicographic capabilities in the form of search and replace engines as part of their basic functions. Copy and paste tools, also a standard addition in the history of word processing, has encouraged us to understand electronic writing at a cultural level as more than a relationship but an entire system for managing content and organizing references. Such capacities continue to be improved and made more robust, making contemporary word processing an increasingly sophisticated mode of text analysis. Amanth Borsuk (2018) aligns important cultural developments in linguistic or language-oriented media with the material transformation of the book itself as a technology for documenting and disseminating ideas of "law, commerce, religion, and cultural history." Whether the book in its form is constructed as a manuscript codex, or electronically as a digital tool coded to process and analyze linguistic content, its core cultural and epistemological purposes remain, Borsuk argues, essentially consistent. Borsuk's history provides a wealth of details regarding ongoing experiments to improve the book's epistemological capability, especially in relation to its recent adoption of computational linguistic functions. She includes, for example, references to the Dynabook, which in 1972 prototyped what would later be called a "laptop" or travel-based computer. Few predicted at that time that "linguistic media" would eventually become screen- based in terms of writing and reading practices, but the Dynabook, in both name and function, seems especially far-sighted.

These examples help contextualize some of the more recent advances in CATA tools, while demonstrating their prominent role in the book's unique historical evolution as a textual interface for combining word processing with information organization. In this way, CATA in digital writing usefully augments specific lexicographic and grammatical modalities that have been in development for many centuries. Recent critical studies of digital textual interfaces and their use in First Year Writing classes focus similarly on the significant transformation of common lexicographic devices when shifting from print to screen-based modalities. Kalir and Garcia's (2019) history of annotation in writing, simply titled *Annotation*, skillfully connects contemporary social media tools with more traditional literary procedures like "glosses, rubrics, scholia, and labels" under the category of "curated conversation," where any dialogue has long been "authored publicly, discussed and debated openly, and recursively rewritten by many." Khalir (2019) has also published case studies on web annotation technologies like the open source OWA tool Hypothesis for maintaining collaborative modes of learning among online communities. Likewise, other scholars (Mirra, 2018; Farber, 2019) have strove to define annotation as a means of maintaining public scholarship through constant, open dialogue between writers and readers. Both argue for annotation as a way, not just of engaging in public scholarship, but also for promoting equity and justice. Such studies seem particularly important, when much current cultural criticism on digital media networks tends sound a more cautionary note regarding the social and political capacity, or even significance, of social media as progressive instrument for building community online. Astra Taylor (2014), for example, effectively dismantles much of the promise originally aligned with the internet as an inherently "democratizing," perhaps even socially liberating, platform in her critical investigation *The*

People's Platform. The internet, she claims, has not been able to withstand many of the more oppressive social inequalities one continues to witness in every modern nation, regardless of its state of economic development. In many cases, as noted in numerous recent critiques of social media and digital networks, digital media technologies actually augment them (Tufekci, Z., 2017; Watts, R. & Flanagan, C. 2007). At the same time, current pedagogical explorations in collaborative learning within the academy feature a consistent critical interest in determining how best to integrate advanced technologies of social interaction and knowledge exchange in the classroom (Beach, 2012; Gutiérrez & Jurow, 2016).

The growing academic interest throughout the United States in collaborative learning techniques remains closely aligned with the media philosophies of theorists like Henry Jenkins (2016; Jenkins, Ito, & Boyd, 2015), beginning with what Jenkins, himself, has alternatively termed "convergence" or "participatory" forms of culture. Enyedy and Stevens (2014) follow this approach, confirming that collaboration has been central to the learning sciences from its foundation. Modern learning, they argue, quite literally begins at the point of information, exchange; however, the key to understanding collaboration in any critical manner is to accept its persistent variability in both form and function. To this end, Enyedy and Stevens outline a wide spectrum of methodologies inclusive of four basic approaches to participatory engagement in the classroom: one can see collaboration as a "window-on-learning" in terms of a related, but independent activity, collaboration "for-distal-outcomes," collaboration "for-proximal-outcomes," and finally collaboration as a form of learning in and of itself. Our methodological approach effectively follows this final approach to collaboration, where learning processes cannot be easily disassociated from the interactive devices employed in any classroom activities.

RESEARCH QUESTIONS AND METHODOLOGY

In nearly all contemporary studies of collaborative learning methodologies, textual interface technologies, whether derived from print or digital media, remain an essential factor. How a text organizes and displays its information seems just as significant to the learning process as the information, itself. The book, in form, as much as format, appears to have cultivated, if not initiated, capacity of texts to function as computational devices in their own right, formulating points, data, premises, etc. for permanent storage. Many of the current platforms mentioned here simply bring these technical features to forefront of the document through increasingly sophisticated interfaces. This paper examines a multi-staged effort over the course of one academic term to use CATA software in the instructional design and delivery of collaborative reading and writing exercises for first year composition. The use of common text analysis tools like annotation and highlighting, combined with different social media interfaces will subsequently be examined to determine how different electronic, screen-based reading modalities may improve critical reading skills at this level. After reviewing and assessing a set number of standard, in-class examinations designed to measure student reading comprehension levels for two separate electronic texts, several collaborative models of critical analysis in digital media will be proposed. The examinations themselves will be supplemented with individual student surveys on the use of CATA software as effective, perhaps even preferred reading devices, especially when compared to simply distributing PDFs as reading assignments across laptops and mobile devices. Consisting of a two-course sequence, catalogued as HUM 101 and HUM 102, first year writing is taught at NJIT by a core cohort of full-time instructors who both teach and assess students according to a pre-set single curriculum. When adjuncts are employed, they are mentored by the director of composition. All instructors involved in portfolio assessment have in-depth experience in both trait and holistic assessment and may be considered a community.

Our core sample consisted of nearly 200 (163) first year students newly enrolled in NJIT's writing programacross10 sections of HUM 101 and 102. The texts were selected primarily on the basis of readability as determined by the Flesch-Kincaid model with each indicating a general suitability for 10th to 12th grade readers (See Key Terms and Definitions). We worked with two different STEM-themed expository essays as delivered during the Spring 2018 term through a CATA tool that specializes in threaded comments and annotation. Data analysis was then used to collate and interpret student responses to either of the distributed essays in the form of additional comments and annotation.

As noted earlier, the tool we used was NowComment: a free and publicly available online text distribution and archiving tool that allows readers to comment and share their perspectives asynchronously on different articles and essays. The technical level and multiple formats of text analysis currently available through other programmable writing tools like CATMA and Voyant is superior to NowComment. Voyant is also a free tool which enables the reader to perform text analytics involving patterns, word frequency, and associations. CATMA is similarly robust, combining the functions of Voyant and NowComment, but has a particularly steep learning curve which made it impractical for our purposes. To compensate for NowComment's inability to provide better analytics information, we made use of a second software program called NVivo. Together, the two tools provided us with the capacity to record group discussions in the form of marginal comments and annotations and then analyze results.

NowComment announces itself on its own site as an online "group collaboration app," facilitating electronic discussions of uploaded texts through a fairly simple annotation interface. "Readers," while working through a single document can choose to highlight and subsequently comment on specific paragraphs or sentences. The comments are then arranged in a two-panel format to the right of the original document; selecting any one contribution simultaneously highlights it together with the corresponding part of the document (sentence or paragraph) it references. The tool's primary strength derives from its capacity to thread different comments from multiple users "in context" with the original reference. All corresponding replies to the opening comment appear immediately below it in the order submitted to produce an ongoing, often vibrant dialogue among multiple readers. These readers have been previously granted shared access to the document by the member who first uploaded it. For our project, this task was initiated by composition instructors, who invited individual classes to join as a single group. Hence, students in a particular class had the option of joining different conversations on the passage of their choice, and, as a result, fairly organic dialogues around certain paragraph-based premises, and even single statements could ensue.

The software's interface is specifically designed to organize threaded discussions around corresponding sections or parts of a text. While comparable to other more popular group editing tools like Google Docs, NowComment sports additional resources for uploading multimedia like video and images. A particularly useful service is a catalogue of different highlight colors corresponding to a spectrum of semantic values that can be applied to any part of the text.

Everything from individual words to whole paragraphs can be color coded according to five specific "heat maps," ranging in label from "Important" through "Unclear, Agree, Disagree" and simply "Like". Such tags can be used to prompt new annotations and comments quickly as a single group continues its ongoing, collective reading of a single document.

Two texts were uploaded into NowComment, and participating course instructors were asked to assign one of them as a reading task for a single class of students (typically a population of 20). Students were asked to read the text while responding to it via any NowComment features they might find useful. Although the annotation and commenting features were introduced to the classes, no specific requirement to apply them for evaluation followed. Rather, students were given the opportunity to explore annotation and discussion of the text using NowComment, before assessing this activity by way of face to face discussions of the essay in class.

The two texts used were "Welcome to Cancerland," published originally in Harper's in 2001 by the journalist and essayist Barbara Ehrenreich and "The Dark Bounty of Texas Oil," a more recent essay written by Lawrence Wright which was published in the New Yorker in January 2018. NowComment allows its members to upload most electronically formatted texts with much consistency, though the original layouts, font choices and even image placements can be challenging to reproduce. Readers are more typically given access to simplified single column texts featuring numbered paragraphs and minor shifts in font size. If the original text was especially complicated in its print design, unpredictable variations in font size and line spacing tended to follow its conversion to NowComment. How these variations may affect reading practices have yet to be studied, but extra care was taken to review and manually correct all major discontinuities in the text's final layout within the software itself.

The same texts were also distributed to the participating classes electronically as PDFs, a format that preserves original publication formats more accurately than a document in Word or other word processing program. Electronic commenting and highlighting are still possible when reading PDFs onscreen, but the software does not clearly feature these applications for ready use when distributing a text, nor are they uniformly available. Students were asked to perform critical readings of the PDF versions of the essays in a manner similar to the NowComment uploads, while, again, most parameters for this exercise were left open to the readers.

Highlighting was encouraged, as was group discussion. If a class read one of the two essays electronically as a PDF, the other was subsequently distributed to the same students via NowComment. Accordingly, each section participated in two distinct reading practices, allowing us to compare the use of CATA software in an assigned critical analysis with similar tasks applied to a PDF text.

A key objective of this study is to provide baseline observations regarding how learners may critically and rhetorically respond to texts delivered via different platforms. The distinction in formats is especially useful for this level of study since it emphasizes how both the technical and cultural understanding of the screen as a strict replication of print may, in fact, impair its use in critical reading exercises. Certainly, the PDF, despite the ongoing development of e-readers for tablet and mobile phone use, remains the most common analogue platform for document distribution within the academy. Specific statistics on PDF use in university courses are difficult to verify; yet it's clear this format can still be considered today synonymous with the concept of the "paperless" office. Because instructors are frequently reticent to assign expensive print- bound textbooks as required course material, electronic PDFs of selected articles and shorter book-length publications have become the default medium for document distribution. In fact, its current status as a kind of page-for-page substitute for print distribution within university courses and programs has already prompted a number of critical studies on its increasing effect on student learning (Baron, 2015; Howard, 2019; Wolf, 2018). Likely one of the best-known studies demonstrating key differences in reading ease and comprehension between print and PDF versions remains Naomi Baron's (2015) popular study "Words Onscreen: The Fate of Reading in a Digital World." In her work, Baron makes the still controversial claim that a whopping 92 per cent of students preferred to read from printed hard copy rather than screens. Baron observed two primary reasons for this consistent dislike of screen-based formats, namely: the problem of distraction, and a tendency towards physical discomfort, including eye strain and headaches. She also provides important details on the technical structure of the screen itself and how it negatively influences the reading practice: students, she insists:

like to know how far they've gone in the book. You can read at the bottom of the screen what percent you've finished, but it's a totally different feel to know you've read an inch worth and you have another inch and a half to go. Or students will tell you about their visual memory of where something was on the page; that makes no sense on a screen. (Robb, 2015)

Our survey for this study showed a similar personal preference for print over digital documents across campus, while also demonstrating a clear inclination to download and print out PDFs when assigned.

By using PDF as a primary electronic reading format in our own study, we are able to compare and assess specific CATA features like document mark-up and inter-reader commentary to less interactive, print-derived media displays in order to determine subsequently how electronic writing and reading technologies might be developed further to improve and possibly even enhance online learning strategies. The collection of data now available supports a thorough, summative report on electronic reading practices by looking at individual learner comments and dialogues submitted as marginal interjections in the text itself, along with follow- up, in-class comprehension exercises. Personal opinions on the technologies and formats used in the exercises will also be discussed based on a series of related survey responses the learners also submitted after working with both texts. By comparing individual responses derived from PDF texts to those distributed via NowComment software, we should be able to further calibrate and distinguish how well learners were able to respond to each of the texts' primary arguments based upon the format used.

Upon collecting what NowComment recorded as individual learner annotations and marginal comments, we can subsequently analyze all data both quantitatively and qualitatively with the additional CATA software NVivo. NVivo features an interface that allows users to code and organize data via word trees, word frequency queries, and word clouds. Generally speaking, for the purposes of our study, it provided a user-friendly, interactive tool to help us analyze students' annotations on the NowComment platform. The type of information NVivo provides seems especially useful for studies consistent with methodologies associated with the Digital Humanities, while providing equally relevant, qualitative comments on collaboration, highlighting, and annotation as key components of all electronic reading practices. One goal for our NowComment-based, NVivo analysis of the students' language use in annotations is to understand the ways in which students think about their own critical work as readers—both individually and collectively. Significant insights into how student comprehension may develop over the course of a single reading assignment seem to emerge by studying patterns in how students communicate with and guide each other as they collaboratively improve. Drawing on this focus, we examine the language used in the critical responses presently inputted as annotations to the original text. The term frequency tool in NVivo, while automatically excluding articles and other non-referential, lexical items such as numerals, salutations, and acronyms, allows us to align common textual references with their specific location in the document itself. For the purpose of the following section of this report and the results it discusses, such data will be referred to collectively as "corpus terms".

RESULTS

Critical readings of written works, regardless of genre, rarely include word frequency as a point of analysis. Key words in an essay, of course, can prove to be instrumental to understanding primary themes and dominant rhetorical approaches regarding the topic at hand, but a summary of different rates of lexical occurrence seems comparably less insightful. Writing analytics and the Digital Humanities in general, however, make strong critical use of word frequency in a single work, analyzing corpus terms qualitatively as important lexical indications of a particular writer's or writing network's verbal patterns in thinking. With these methodologies, analysts typically aim to determine levels of critical sophistication and accuracy in a written work by noting specific shifts from high frequency vocabularies, and thematic, topic- sensitive word-use to a reliance on more original, possibly even eccentric, lower frequency terms. Such distinctions, especially when employed over a number of texts by a single learner in a writing or literature course, can also tell us how effective reading strategies are in terms of sustaining significant expansions in that learner's working vocabulary, while providing pedagogical insight into the role lexicons typically play in developing formal modes of writing. Specific to our study, such approaches tell us on one level that if the words learners used when responding to either of the distributed essays simply repeat the lexicons of the primary texts, any corresponding learner vocabularies tend to mimic or mirror the terms in play, while possibly avoiding opportunities to add new terms of their own. This pattern in word use seems to reflect a rhetorically passive mode of engagement with each text's primary arguments, including the individual propositions informing them along with their support. At the same time, one might also observe in certain cases the effective linguistic adoption by one or more learners of a topic- specific lexicon, as well as various fundamental terms of rhetorical analysis. When a lexicon can be identified and located as such, it seems equally useful to review which of its words attracted the most attention of any one group of readers as they worked through their assigned documents. In addition, this information subsequently allows one to determine precisely where in an argument a single term appears to inspire the most critical interest. See Figure 1.

Armed with these methodologies, we typically began each NVivo-sponsored analysis with a ranked a list of entries identified by character length (i.e., number of letters), and according to "Weighted Percentage," or the frequency of the word relative to all other total words counted. These words and their corresponding frequencies are then further divided into one of two semantic categories: either premise-driven or context-driven. The categories are necessarily hierarchical with premise-driven terms being essential to a text's overall propositions and argumentative structure, and context-driven ones providing their semantic support.

As shown in Figure 1, the top words for all marginal comments made by learners across each essay appear to be "cancer," "oil," "breast," "Texas," "like," and "using." The first four terms seem immediately aligned with each of the essay's respective titles and primary objects of study. Ehrenreich's social critique "Welcome to Cancerland," written in part as a personal narrative, seems to lead learners quite effectively to a single center of focus, namely the disease of breast cancer. Comments thus tend to stay general in terms of subject matter. Learners typically emphasized the issue at hand when making their own additions. The same type of emphasis appears in comments on the Wright essay, "The Dark Bounty of Texas Oil," where the terms "Oil" and "Texas" dominate the various points added to the ensuing annotated discussion. For both essays, it seems, critical readings stay anchored to the key objects of study: again, breast cancer and Texas oil. Few synonyms or related terms were employed. The actual content of the annotations supports our quick review of the NVivo results. Overall throughout "Welcome to

Figure 1.

Word	Length	Count	Weighted Percentage
cancer	6	369	2.44%
oil	3	211	1.40%
breast	6	190	1.26%
texas	5	111	0.74%
like	4	102	0.68%
people	6	99	0.66%
fracking	8	97	0.64%
think	5	79	0.52%
one	3	72	0.48%
article	7	70	0.46%
also	4	61	0.40%
way	3	60	0.40%
many	4	58	0.38%
ehrenreich	10	51	0.34%
just	4	50	0.33%
economy	7	49	0.32%
point	5	49	0.32%
agree	5	48	0.32%
women	5	48	0.32%
edited	6	47	0.31%
shows	5	45	0.30%
disease	7	43	0.28%
even	4	41	0.27%
industry	8	41	0.27%
process	7	40	0.27%
reader	6	40	0.27%
something	9	40	0.27%
uses	4	40	0.27%

Cancerland," the learners seemed annotate more personally based on their own emotions, while offering speculations on how the author herself feels; the Wright article sponsored in contrast more academic-oriented marginal comments on the author's use of rhetoric and the argument at hand.

Looking further at the word frequency list, it appears that the term "cancer" emerges specifically 374 times in the annotations, giving it a weighted percentage of 2.48 with respect to the entire text. Closer inspection shows several measurable common features, which in turn supports earlier points regarding the focus of most annotated comments, while helping instructors to analyze and assess levels of reading comprehension in the class, along with elements of rhetorical style. Not only, for example, is the term "cancer" the most frequently used term among learners, but it recurs almost twice as often as the second most common word, "Oil," which surfaces 211 times. Although each article was read by an equal number of learners, the assigned essay, "Welcome to Cancerland," shows first and foremost more student engagement overall, while the level of engagement can be considered general in scope and focus. It seems that a larger number of learners found a point of access into the Ehrenreich essay compared to Wright's "The Dark Bounty of Texas Oil," and simply preferred to annotate actively when given the opportunity. The issue at hand being breast cancer thus inspired a level of considered interest beyond that of Texas oil. As the actual comments show, however, the level of analysis rarely moved off the primary topic of study. The author's name also appears frequently with a count of 51 times along with terms like "treatment," "women," "patients," and "disease." A similarly quick overview leads us to consider that learners tended to emphasize in their comments an accurate, but general focus on Ehrenreich's primary theme concerning the treatment of women as patients prone to a particular disease. Few annotations strayed from this point of concern.

Additional analysis needs to be initiated at this time looking more deeply into how many comments each individual student is making on both texts in order to determine what prompted one set of particular concerns over another. Collective focuses on a certain topic can indicate, as suggested previously, general levels of academic literacy in play, as well as certain types of audience or readership. One can see, for example, that, given the two topics of discussion, an active readership for "Welcome to Cancerland," seems fundamentally more STEM-oriented with a distinct social science-based appreciation for medical history. A similar approach rooted in the social sciences dominates readings of "The Dark Bounty of Texas Oil," albeit with a pronounced academic interest in economy, business, and finance. The different rates of response among these two readerships thus suggests that a larger STEM-oriented readership remained active throughout the assignment, comparing again the number of commentary terms that are context- specific to "Welcome to Cancerland" to those that refer to "The Dark Bounty of Texas Oil." Further analysis might show how even less premise-driven terms like "people," "think," etc. relate more consistently to the Ehrenreich essay over "The Dark Bounty of Texas Oil," while indicating new alliances between terms reproduced through different collaborative readings. It might also be worth noting here that the terms "breast" and "cancer" in and of themselves tend to provoke more of an

Figure 2.

Word	Length	Count	Weighted Percentage⌄
seems	5	38	0.25%
trying	6	38	0.25%
wright	6	38	0.25%
life	4	37	0.25%
good	4	36	0.24%
much	4	36	0.24%
well	4	36	0.24%
money	5	35	0.23%
support	7	35	0.23%
help	4	34	0.23%
interesting	11	32	0.21%
make	4	32	0.21%
now	3	32	0.21%
see	3	32	0.21%
throughout	10	31	0.21%
treatment	9	31	0.21%
get	3	30	0.20%
may	3	30	0.20%
patients	8	30	0.20%
state	5	29	0.19%
time	4	29	0.19%
feel	4	28	0.19%
first	5	28	0.19%
argument	8	27	0.18%
awareness	9	27	0.18%
fact	4	27	0.18%
made	4	27	0.18%
show	4	27	0.18%

emotional response than "Texas" (unless if one is a Texan), and are thus inherently more interesting to the participating learner population, especially given the likelihood that readers in the northeast United States may know or have known a person who experienced breast cancer rather than anyone with a connection to Texas oil. See Figure 2.

In general, it is not surprising that the Ehrenreich essay proved to be more accessible to learners participating in our study (many being pre-med) compared to Wright's piece, since the institution itself, a STEM university, offers no related programs in either petroleum engineering or anything, in fact, anything related to gas and oil industry management. Again, as much of the content from specific comments, along with further choices in word use indicate, many students easily related to the emphasis Ehrenreich placed on not being treated as a person by the same doctors who pursued a relentless attack on the cancer that was killing her. Other popular context-driven terms include "economy" with a weighted percentage of 0.35%, "industry," weighing 0.32%, and "wells," weighing 0.31%. Many students also commented on the currently popular press featured word "fracking," which, despite not appearing as a point of focus in the article, provided reference that students living outside the southwestern states could more easily apply to the topic of oil.

NVivo provides a number of different visual formats to display hierarchical patterns between a document's top word frequencies and their various affiliated terms. The number of specific term associations evident in "The Dark Bounty of Texas Oil," as Figure 3 shows, are double those found in "Welcome to Cancerland;" that said, word frequencies in general refer to "Welcome to Cancerland" twice as often as to the other essay. This discrepancy suggests that learners were generally less akin to comment on "The Dark Bounty of Texas Oil" than we see in relation to "Welcome to Cancerland," and worked to a greater percentage on the same key phrases. See Figure 3. It's clear from this study that the different patterns of engagement speak directly to different levels or even modes of reading comprehension.

Figure 3.

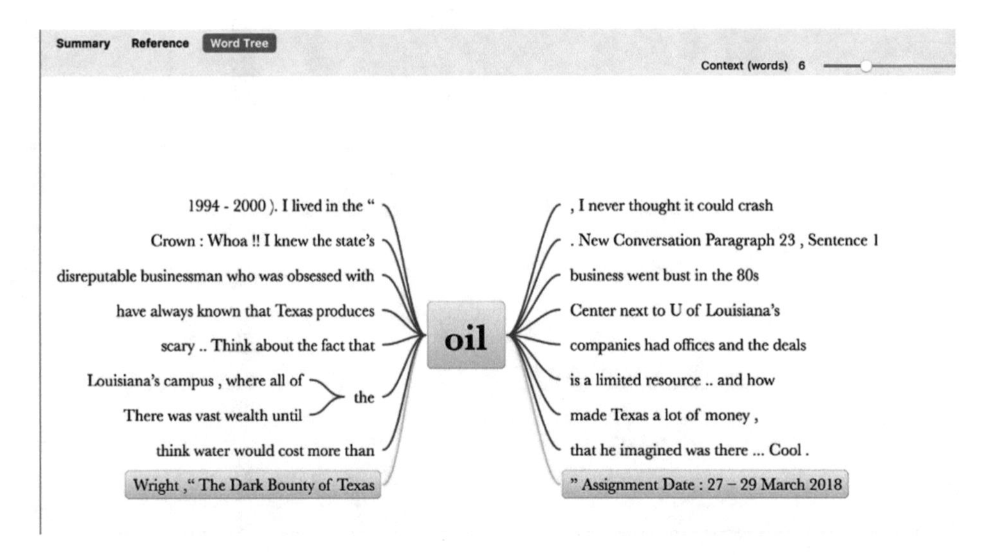

A few additional insights into how the students interpreted not just various topic areas relevant to the two essays, but the assignment in general can be garnered from word frequency levels that appear to reference specific writing tasks at hand. For example, the very term "argument" occurs throughout their comments with a frequency of 27 times, carrying a weighted percentage of 0.18; "think" occurs even more often at 44 times, and weighing 0.66%. This weight and corresponding frequency can tell us that the learners, when commenting and annotating the text, tended to focus on terms that reflect fairly common rhetorical tasks in essay writing. Terms like "fact" and "show" appear in similar numbers. Much more substantive language and substantiated form and structure of writing is used in the marginalia for Ehrenreich's piece, where the student did not use any titles for themselves, and only their full names. Correct usage of grammar, capitalization, and punctuation, as well as more sophisticated form—the usage of paragraphs, sentence structure, and lexicon. A clearer format for demonstrating word frequency in relation to text location as quantifiable data might be a "Word Cloud," where these quantities can be effectively visualized. See Figure 4.

Figure 4.

Considered together, especially after matching their verbal frequency to specific verbal contexts, these terms also demonstrate that our study's learners both collectively and individually remain dependent as readers on the traditional rhetorical strategies still dominant in first-year writing programs across the country. In fact, several of the more frequent words exemplify the common introduction of the essay in both form and genre as a linguistic tool of social critique.

Verbal contexts supporting the prominent use of the words "argument," and "think," for example, tend to house—however, indiscriminately—an equally large number of self-reflexive references to the reading process itself, as shown in words like "support," "agree," "process," "premises," and, of course, the very term "comment". When these terms are used to "support" various context-driven points as formal "premises," they also begin to designate verbal functions quite specific to rhetorical modes of analysis, and dialectical critique. Other rhetorical devices for evoking imagery and personification become immediately evident in very predictable patterns within the same contexts. Together all of these terms conform to traditional principles of rhetoric as they've been determined and studied throughout the Humanities for many centuries, classifying persuasion as a function of either "logos," "pathos," or "ethos." Nvivo's Word Tree tool supports this observation by collocating the terms used in both annotation corpuses to provide a kind of network analysis of all the responses.

Figure 5 visually organizes phrases as common links connecting different the learner annotations on the Wright essay.

Figure 5.

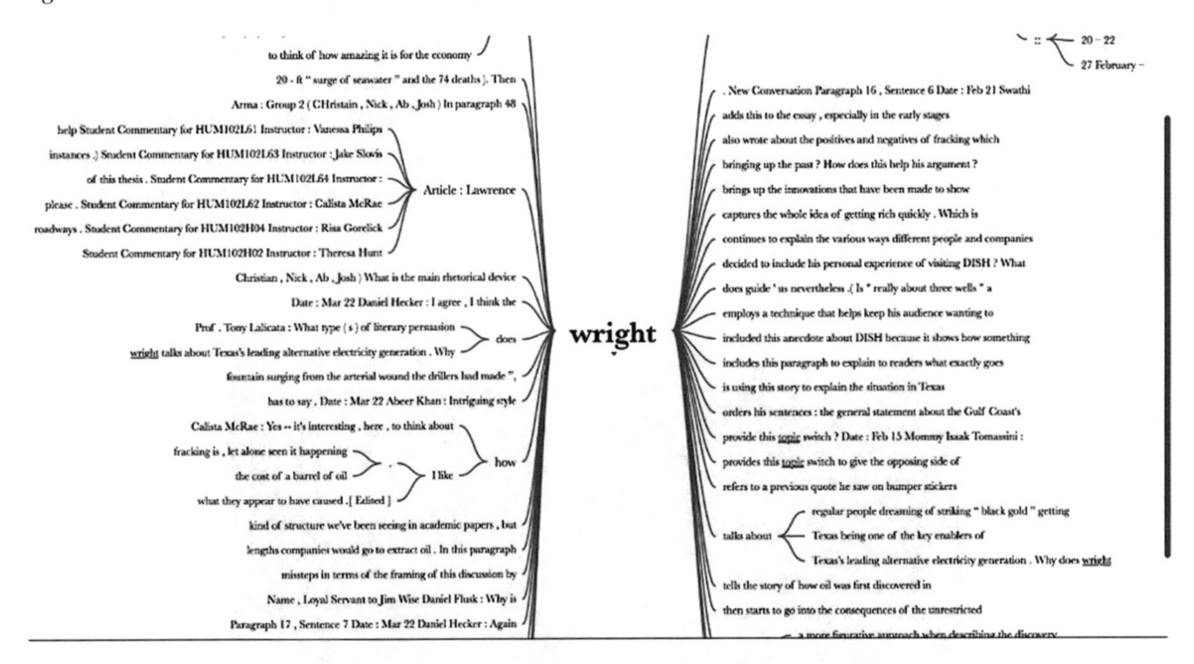

Note in the example featured here how the key term "oil" remains anchored via the preposition "of" to the phrases "trends in production," "as the pricing/price," and "major producers" to suggest a fairly dynamic, interactive "network" of shared comments about the term's (and referent's) prominent economic function in both Texas and the U.S. in general. The word tree represents a clear sign of learners organizing discussions around one of Wright's primary focuses. No doubt the word "oil" was quick to inspire learners to produce new academic summaries of Wright's own focus on the aggressive relationships Texans have historically maintained with this particular natural resource, treating it first and foremost as economic stock ripe for exploitation. Closer examination of these discussions as interactive, theme-based groups or networks may discover that only a handful of classes actually participated, effectively skewing the results; but the current prominence of these branches still subsequently suggests that no other discussions cohered as effectively. At this time, an extended effort to examine and compare how different sections work collaboratively on each article is in process.

Looking at a similar set of annotated discussions in dialogue form quickly confirms the formation of a kind response-based network, while showing what happens when learners avoid following prescribed analyses of the work in question. Alternatively, students begin instead to build together a type of counter-narrative based upon a much more socially playful, if attitudinal, set of voices and group perspectives with each other. In this example, the participant identified as "Instructor" begins the discussion with a simple question on authorial objectivity: "At this point, can you tell if the author is biased? How?" The responses of at least 14 students fill in accordingly:

Student 1: He's biased. Reasons: Because its biased Instructor: Bad idea. That's circular logic.
Student 1: y tho now im a sad boi:(Student 2: All hail Elon [Edited] Student 3: no
Student 4: Not enough nuances Student 5: No the author is not bias Student 6: sure. I mean I guess Student 7: no idears. I dont know? Student 8: comment here
Student 9: Because of his name. He must be absolutely crazy Student 3: He is weighing the pros and cons
Student 10: yse
Student 11: marshmallow; marshmallow
Student 12: what is bias???
Student 13: The stars at night They're dull and dim whenever they have to shine over dumb old stupid Texas.
Student 14: Bias. The author is biased towards his own opinion (ref.)

The "Instructor" makes an early attempt in the discussion to maintain an academic focus in all forthcoming comments, steering marginalia back to the question at hand, while rhetorically critiquing the first response as an example of "circular logic." For the learner being addressed most directly in this critique, the instructor's response instead, however, inspires a complete shift into irreverence: "y tho now im a sad boi:(." Further responses follow "Student 1's" lead, and an almost defensive (even borderline rude) discussion ensues, complete with the use of slang and emoticons. If there's to be a competition for attention among learners using these tools, it will be based on wit, derision, and mockery over intellectual prowess. And yet, such attitudes can also be assessed as an almost characteristic attempt by learners of this generation of learners to bond on their own terms, demonstrating their common cultural preference for advice from their peers over supposed authoritative figures like instructors and mentors. A similar discussion pattern occurs with a different instructor at another location in the same essay:

Student 1: Imagery Hey
Student 2: Why do you think the author uses such detail? Student 3: bhk?
Student 4: ???
Student 5: Stuff. This and that. Stuff and things.
Student 6: DBS My boy Frieza got rekt by Toppo
Student 7: noice noice
Student 8: Limit as x approaches 0

Even excluding the actual titles (designations ranging from "Mr.," "Sir,", and "Dr." to more exaggerated choices like "King," "President," "Lord," etc.) and the subsequent names the learners chose to identify themselves by, it's difficult to miss the playful, almost mischievous attitude dominating most of these discussions. In addition, it seems this penchant for roguish humor among learners was far more prevalent in marginal responses to the Wright essay compared to those complementing Ehrenreich's writing. The marginalia added to "Welcome to Cancerland" contains very little slang and few impish remarks, though its exchanges maintain a similarly explorative, open interest in discussing wide ranges of themes, not all of them directly related to the arguments at hand. Finally, noting the thorough, if broad, coverage of each essay by the commentary, moving swiftly from topic to topic, independent of most instructor-led focuses, one might acknowledge a unique autonomy among learners working with each other to complete all reading assignments. Learners seemed particularly more inclined to build their own respective approaches to the work, linking key arguments in the texts with their personal experience.

CONCLUSION AND FURTHER RESEARCH

Earlier established studies in the field indicate significant differences in reading practices between the use of electronic, screen-based texts and that of print documents. This study follows prior beta testings within NJIT's First Year Writing program that revealed additional distinctions in reading between various modes of electronic texts. An earlier, smaller study completed in 2017 suggested that learners reading electronic texts through different online, whether tablet or laptop-based, text analysis software experienced improved reading comprehension levels and better efforts at critical engagement than when using PDF formats. This more recent investigation, however, presents much less of a distinction between reading practices derived from the same two format types. Given the same articles formatted as either a PDF document or as content distributed through text analysis software like NowComment, learners exhibited very similar reading strategies and tendencies overall, even with access to improved levels and modes of reader-interaction.

The introduction of Computer Assisted Text Analysis as both a classroom tool, and a later mode of assessment immediately demonstrates many current pedagogical limitations in terms of examining and reviewing reading practices for first year writing. For this study, very few specific guidelines and/or learning strategies were distributed with either of the assigned texts. Each course section was given the same amount of time (48 hours) to consider each article individually to the best of their ability before being asked to review several comprehension questions in class for discussion. The lack of guidelines provided for some variation in how different instructors individually assigned the reading exercise, especially regarding annotation. Some instructors asked learners to make a minimum of three comments on each article. Others simply informed their respective classes that the articles were chosen and being distributed as part of a larger academic study (though the study's aims and overall purpose were not revealed); and any subsequent annotations or comments were to be considered particularly useful, if not valuable as data within said study. Whatever the context of the assignment, students were actively encouraged to annotate the text without being specifically guided or in any way advised on how the annotation should be recorded. As a result, the comments subsequently produced represent more or less spontaneous, individual reading strategies, as mediated by the respective format of the article itself, which is to say as a PDF or via the NowComment software.

Instructors were asked to review in-class comprehension levels for each of the articles in both formats with a series of questions designed to assess basic learner understandings of the essay's main arguments and rhetorical emphases. While we have on record within the NowComment software all annotated comments posted by individual learners according to course section, no formal review of learner engagement with PDFs has at this time been undertaken, save through general surveys on PDF notetaking.

Two separate sets of survey questions designed to collect individual perspectives and viewpoints on each of the articles as distinct reading experiences were distributed en masse to the sections towards the end of the spring term after all courses had completed these assignments. The surveys provide important information on how effective each of the two electronic formats (PDF and NowComment) was as a means for providing learners with a clear and readable copy of the assigned article; at the same time, very few of the questions were designed to assess levels of reading comprehension, whether in terms of either article's primary argument or even their respective structures as prose writing. It's expected that future studies, with better guidance from the instructors distributing both the software and the texts to be read, might be able to show more complex modes of interactivity and text analysis in both composition and literary criticism. Importantly, this review shows that text analysis features can improve reading comprehension, specifically by allowing students to interact more critically and collaboratively with the document at hand.

Learners took up the surveys in-class as printed assignments, giving us a nearly 100 percent response rate: 159 students out of 163. The actual survey questions are reproduced according to their distributed format in Appendix A of this article. What follows here is quick summary of several key indicators in our study regarding how functional or useful individual learners regarded the two respective electronic formats used for each assigned article. First it can be noted that, overall, respondents to this survey seem to follow prior recorded trends in reading format choices, preferring print documents over electronic, screen-based ones by at least 60 percent; interestingly the ratio of preference is substantially narrower than the near total disdain for electronic formats one finds in Baron's survey some years ago. However, just over half of the respondents or 56 percent claimed to enjoy reading as a general activity to varying degrees of affirmation (agree, strongly agree, or very strongly agree); hence a substantial proportion of learners in this survey clearly do not consider themselves readers regardless of the format in which the text is distributed. Close to 90 percent of the learners surveyed confirmed that most, if not all texts read over the course of a single day are digital and thus consumed via some kind of electronic device. Asked whether these electronic devices were preferred to print documents, 66 percent disagreed, claiming that print was easier to read than screen-based writing. This ratio seems to correspond with the broader preference for print in general over electronic documents, though, once again, the overall ratio remains fairly narrow. Between 30 and 40 percent of the learners surveyed found electronic formats to be perfectly effective, if not even easier for reading.

When asked more specific questions about reading and working with the two different electronic formats used for each of the distributed articles (PDF and NowComment), very similar ratios occur regarding how many learners actively finished each text and engaged in annotation and general notetaking. Just over 41 percent of the learners confirmed that they read the PDF while taking notes on their own and highlighting important passages. 42 percent of the learners reading the same articles via NowComment could make the same assertion. Various prior surveys, including those informing Baron's study, have indicated that electronic texts, regardless of format or screen device being used, tended to be left incomplete, and read with less attentive focus or scrutiny than expected. Baron, among other researchers in this area, does not hesitate to point out that the general practice of reading by screen, especially when online, proceeds much more intermittently than with print documents, as readers take to multitasking with a plethora of interfaces at a single time, while skimming quickly through each and any text at hand. As electronic interfaces, both the PDF and NowComment effectively underscore the exact same reading habits, where the majority of learners failed to complete the full article. Just under a quarter of the learners surveyed successfully read the text in its entirety when distributed as a PDF; even fewer, or less than 23 percent, were able to finish the text when displayed online within the NowComment interface. Given a PDF, a full third of those surveyed could complete at least 75 percent of the document, while a slightly higher ratio or 35 percent could complete the same amount when distributed via NowComment. Comparable readership percentages occur between the two formats for each of the remaining document proportions: whether the document was distributed as a PDF or as a NowComment text, just over 16 percent completed half the text, 10 percent completed close to a quarter of the text and only 5 percent admitted to never starting the assignment. The rough third who were able to complete at least 75 percent of the document represents thus the largest single group of readers engaging with the text to some degree or another. It should be mentioned at this point that neither text format displays exact percentage points marking how much of a document may have been read on screen (as one commonly finds in Kindle texts), making it difficult for readers in our study to calibrate accurately at any one time the precise amount of text consumed. Responses to the above survey questions remain at best in these instances personal estimates by the readers involved.

The past five years have brought forth a number of important advances in text analysis and markup software with many of the most significant ones involving interface design, analytics and social media networking. The text, itself, from both a cultural perspective and as a digital object clearly represents a startlingly different type of construct than what is typically identified as a print work. For this study, each of the distributed articles was purposely presented as a non-digital, print-based object converted for screen. In other words, learners were consistently aware that the task at hand was to read and annotate a set of print articles, even though they were being delivered electronically for onscreen interaction. Any learning aims associated with the assigned PDFs were thus similarly based in traditional, print-oriented formats for reading and writing in the academy. At no time during these assignments were learners asked to review or even consider the documents in terms of certain electronic features, including, for example, in-text links, or specific highlighting, selection, and annotation tools. Hence, it seems reasonable to consider at this point that if an assigned text is introduced and distributed to readers under guidelines reserved for print documents, little critical attention will be paid to any electronic features that may be affecting its presentation. The fact that learners in this study tended to review either of the texts more or less identically regardless of the electronic format in which it was distributed likely testifies to the dominant role print played in helping readers identify and interact with the specific cultural object under review. In this context, common readerly relationships to essays and arguments as print works quite effectively, it seems, overrode most of the technical aspects differentiating the PDF version of the article from its NowComment design. As the surveys indicated, if there was any critical reflection among learners regarding the format of the essay, it typically served to emphasize a sense of shared inferiority in both the PDF and the NowComment interface compared to possible print versions of each work.

Looking again to future studies of electronic reading practices in the classroom, it seems that a clearer, more interface specific set of reading tasks might help learners better differentiate and respond to certain screen-based features when working with such texts. Even though, as stated earlier, the software developer, himself, made an effort to work with participating instructors before assigning the texts on different approaches to forming discussion groups within NowComment, no actual reading assignments were developed for in-class use. At the same time, it is immediately apparent that many students benefitted from the annotation and discussion tools offered by NowComment, taking up the opportunity to highlight specific paragraphs and sentences for asynchronous group dialogues. The idea of constant social interaction, perhaps even organized group discussion while reading an assigned text can, and in some ways did, lead to an array of new pedagogical directions in these writing courses. How well and fully these students used such resources as they presented themselves remains at this point incompletely analyzed.

The single, most evident quality upon reviewing all learner comments posted through NowComment remains the typically playful, non-hierarchical framework for discussion that seems to evolve in just about every section. Few individual learners, it appears, took it upon themselves to initiate a point of control or direction within the emerging discussions to emphasize particular issues they felt worthy of analysis. In this way, the general interface in the process of being explored and actively used rarely, if ever, inspired an organized, critical investigation of topics and ideas on its own. This point, in itself, is interesting in that the producers responsible for NowComment's unique interface openly prioritized developing a more or less self-guiding, autonomous set of tools for reading and discussion. Working with these devices on their own, however, learners typically refrained from actively building or devising specific debates about the texts in any capacity. As noted previously, the learner surveys seem to confirm this pattern, as few learners cited experiencing many advantages when using screen-based tools like NowComment to read and analyze electronic texts.

Important critical work within the relatively nascent field of Digital Humanities continues to introduce a broad range of interactive devices for the digital screen, conceiving it in the process as nothing less than a new mode of textual representation and thus a very different style of reading practice. The NowComment interface for reading and writing, as noted earlier, offers a limited set of annotation and highlighting options compared to a tool like NVivo, for example; yet the self-apparent design of its interface allows for the software's swift integration within university writing and literature curricula. That said, this initial study still raises a number of questions regarding the importance of instructor-led sessions for introducing the electronic screen formally and directly to learners when distributing texts in this manner. All annotation tools, including highlighting and forum creation, combined with online communication media greatly expand how readers can actively work with each other, building ongoing discussions and debates around just about any type of shared document. As this study shows, learners are clearly capable of self-organizing into highly enthusiastic, autonomous discussion groups, often independent of the instructor's own attempts to intellectually guide and lead them. And yet in order to prevent or at least somehow limit these activities from devolving swiftly into personal, often irreverent social media feeds, it seems necessary to explore and possibly develop improved formats for electronic critical dialogues. The sheer variety—not to mention consistently lively character—of numerous group discussion formats displayed across different course sections confirms that the use of CATA tools can help us read and engage more effectively with electronic texts. Subsequent work in this area will aim accordingly to prepare better methods of critical analysis that take into consideration how these devices actually present and, in part, determine how the text is to be processed by different readers: in other words, what techniques and interface options readers need to take account of when interacting with the text as a type of digital object. To these ends, the project's next phase of development will introduce more quantified approaches to text analysis, along with complementary data visualization applications, as demonstrated by tools like Voyant and NVivo. Guided by a more consistently Digital Humanities informed methodology, the reading practice itself emerges anew, providing in the course of this development a very different kind of critical relationship to texts in general as active sites of collective analysis and technological interplay.

REFERENCES

Baron, N. (2015). *Words onscreen: The fate of reading in a digital world*. Oxford University Press.

Beach, R. (2012). Constructing digital learning commons in the literacy classroom. *Journal of Adolescent & Adult Literacy*, *55*(5), 448–451. doi:10.1002/JAAL.00054

Berry, D. (2012). *Introduction: understanding the digital humanities. In Understanding Digital Humanities*. Palgrave Macmillan. doi:10.1057/9780230371934

Borsuk, A. (2018). *The book*. MIT Press. doi:10.7551/mitpress/10631.001.0001

Ehrenreich, B. (2001, November). Welcome to cancerland. *Harper's Magazine*. https://harpers.org/archive/2001/11/welcome-to-cancerland/

Enyedy, N., & Stevens, R. (2014). Analyzing collaboration. In R. E. K. Sawyer (Ed.), *The Cambridge handbook of the learning sciences* (2nd ed., pp. 191–212). Cambridge University Press., doi:10.1017/CBO9781139519526.013

Farber, M. (2019). Social annotation in the digital age. *Edutopia,* 22. https://www.edutopia.org/article/social-annotation-digital-age

Gitelman, L. (2014). *Paper knowledge: Towards a media history of documents*. Duke University Press.

Gold, M. K. (2012). *The digital humanities moment. Debates in the digital humanities*. University of Minnesota Press.

Gutiérrez, K., & Jurow, A. S. (2016). Social design experiments: Toward equity by design. *Journal of the Learning Sciences*, *25*(4), 565–598. doi:10.1080/10508406.2016.1204548

Howard, N. (2009). *The book: The life story of a technology*. Johns Hopkins University Press.

Jenkins, H. (2006). *Convergence culture: Where old and new media collide*. New York University Press.

Jenkins, H., Ito, M., & Boyd, D. (2015). *Participatory culture in a networked era*. Polity Press.

Kalir, J. (2019). Open web annotation as collaborative learning. *First Monday*, *24*(6). Advance online publication. doi:10.5210/fm.v24i6.9318

Kalir, J., & Garcia, A. (2019). *Annotation*. MIT Press. https://bookbook.pubpub.org/annotation

Kirschenbaum, M. (2012). What is digital humanities and what is it doing in English departments? In M. Gold (Ed.), *Debates in the digital humanities*. University of Minnesota Press., doi:10.5749/minnesota/9780816677948.003.0001

Klein, L. R., & Gold, M. (2016). Digital humanities: The expanded field. In M. Gold & L. R. Klein (Eds.), *Debates in the digital humanities 2016*. University of Minnesota Press., doi:10.5749/j.ctt1cn6thb.3

Krauss, R. (1979). Sculpture in the expanded field. *October, 8*, 30–44.

Latour, B. (1994). On technical mediation. *Common Knowledge*, *3*(21), 29–64.

Mirra, N. (2018). Pursuing a commitment to public scholarship through the practice of annotation. *The Assembly: A Journal for Public Scholarship on Education, 1*(1). https://scholar.colorado.edu/concern/articles/xw42n8680

Robb, A. (2015, January 14). 92 percent of college students prefer reading print books to e-readers. *New Republic*. https://newrepublic.com/article/120765/naomi-

Taylor, A. (2014). *The people's platform: Taking back power and culture in the digital age*. Metropolitan Books.

Tufekci, Z. (2017). *Twitter and tear gas: The power and fragility of networked protest*. Yale University Press.

Watts, R., & Flanagan, C. (2007). Pushing the envelope on civic engagement: A developmental and liberation psychology perspective. *Journal of Community Psychology*, *35*(6), 779–792. doi:10.1002/jcop.20178

Wolf, M. (2018). *Reader, come home: The reading brain in a digital world*. Harper Collins.

Wright, L. (2018, January 1). The dark bounty of Texas oil. *The New Yorker*. https://www.newyorker.com/magazine/2018/01/01/the-dark-bounty-of-texas-oil

ADDITIONAL READING

Adler, M. (1940). How to mark a book. *Saturday Review*, (July), 11–12.

Anderson, S. (2011, March 4). What I really want is someone rolling around in the text. *The New York Times*. https://www.nytimes.com/2011/03/06/magazine/06Riff-t.html

Ávila, J., & Pandya, J. (Eds.). (2013). *Critical digital literacies as social practice: Intersections and challenges*. Peter Lang.

Barney, S. A. (Ed.). (1991). Annotation and Its Texts. Publications of the University of California Humanities Research Institute. Oxford University Press.

Berners-Lee, T. (2010). Long live the web. *Scientific American*, *303*(6), 80–85. doi:10.1038cientificamerican1210-80 PMID:21141362

Bush, V. (1945). As we may think. *The Atlantic*. https://www.theatlantic.com/magazine/archive/1945/07/as-we-may-think/303881/

Carnegie Corporation of New York & Center for Information & Research on Civic Learning and Engagement. (2003). *Guardian of Democracy: The civic mission of schools*. https://www.carnegie.org/publications/guardian-of-democracy-the-civic-mission-of-schools/

Chen, B. (2019). Designing for networked collaborative discourse: An UnLMS approach. *TechTrends*, *63*(2), 194–201. doi:10.100711528-018-0284-7

Cope, B., Kalantzis, M., & Abrams, S. (2017). Multiliteracies: Meaning-making and learning in the era of digital text. In F. Serafini & E. Gee (Eds.), *Remixing multiliteracies: Theory and practice from New London to New Times* (pp. 35–49). Teachers College Press.

DeRosa, R., & Jhangiani, R. (2017). Open pedagogy. In E. Mays (Ed.), *A guide to making open textbooks with students* (pp. 6–21). Rebus Community Press.

Fairclough, N. (2013). *Critical discourse analysis: The critical study of language*. Routledge. doi:10.4324/9781315834368

Gee, J. (2014). *An introduction to discourse analysis: Theory and method* (4th ed.). Routledge. doi:10.4324/9781315819679

Grafton, A. (1997). *The footnote: A curious history (revised)*. Harvard University Press.

Gutiérrez, K., & Vossoughi, S. (2010). Lifting off the ground to return anew: Mediated praxis, transformative learning, and social design experiments. *Journal of Teacher Education*, *61*(1-2), 100–117. doi:10.1177/0022487109347877

Jackson, H. J. (2001). *Marginalia: Readers writing in books*. Yale University Press.

Kalir, J. (2018). Equity-oriented design in open education. *International Journal of Information and Learning Technology*, *35*(5), 357–367. doi:10.1108/IJILT-06-2018-0070

Kalir, J., & Dillon, J. (2020). Educators discussing ethics, equity, and literacy through collaborative annotation. In K. H. Turner (Ed.), *The ethics of digital literacy: Developing knowledge and skills across grade levels*. Rowman & Littlefield.

Kalir, J., & Perez, F. (2019). The marginal syllabus: Educator learning and web annotation across sociopolitical texts and contexts. In A. Reid (Ed.), *Marginalia in modern learning contexts* (pp. 17–58). IGI Global. doi:10.4018/978-1-5225-7183-4.ch002

Kozan, K., & Richardson, J. C. (2014). Interrelationships between and among social, teaching, and cognitive presence. *The Internet and Higher Education*, *21*, 68–73. doi:10.1016/j.iheduc.2013.10.007

Ladson-Billings, G. (2006). From the achievement gap to the education debt: Understanding achievement in U.S. schools. *Educational Researcher*, *35*(7), 3–12. doi:10.3102/0013189X035007003

Lankshear, C., & Knobel, M. (2011). *New literacies: Everyday practices and classroom learning*. Open University Press.

Lizárraga, J., & Gutiérrez, K. (2018). Centering Nepantla literacies from the borderlands: Leveraging “in-betweenness” toward learning in the everyday. *Theory into Practice*, *57*(1), 38–47. doi:10.1080/00405841.2017.1392164

Luke, A. (2012). Critical literacy: Foundational notes. *Theory into Practice*, *51*(1), 4–11. doi:10.1080/00405841.2012.636324

Marshall, C. (1997). Annotation: From paper books to the digital library. In *Proceedings of the second ACM international conference on digital libraries* (pp. 131-140). ACM Press. doi:10.1.1.29.2013&rep=rep1&type=pdf 10.1145/263690.263806

Mattern, S. (2017). *Code and clay, data and dirt: Five thousand years of urban media*. University of Minnesota Press.

Morgan, B. (2010). New literacies in the classroom: Digital capital, student identity, and third space. *International Journal of Technology, Knowledge and Society*, *6*(2), 221–239. doi:10.18848/1832-3669/CGP/v06i02/56094

Morrell, E. (2013). *Critical media pedagogies: Teaching for achievement in city schools*. Teachers College Press.

Orgel, S. (2015). *The Reader in the Book: A Study of Spaces and Traces*. Oxford University Press.

Piper, A. (2012). *Book was there: Reading in electronic times*. University of Chicago Press. doi:10.7208/chicago/9780226922898.001.0001

Pomerantz, J. (2015). *Metadata*. MIT Press. doi:10.7551/mitpress/10237.001.0001

Reagle, J. M. (2016). *Reading the comments: Likers, haters, and manipulators at the bottom of the web*. MIT Press.

Reid, A. J. (Ed.). (2018). *Marginalia in modern learning contexts*. Information Science Reference.

Schacht, P. (2015). Annotation. In *MLA commons digital pedagogy in the humanities: Concepts, models and experiments*. https://digitalpedagogy.mla.hcommons.org/keywords/annotation/

Serafini, F., & Gee, E. (Eds.). (2017). *Remixing multiliteracies: Theory and practice from New London to new times*. Teachers College Press.

Share, J. (2015). *Media literacy is elementary: Teaching youth to critically read and create media* (2nd ed.). Peter Lang.

Taylor, K., & Hall, R. (2013). Counter-mapping the neighborhood on bicycles: Mobilizing youth to reimagine the city. *Technology. Knowledge and Learning*, *18*(1-2), 65–93. doi:10.100710758-013-9201-5

Teeters, L., Jurow, A. S., & Shea, M. (2016). The challenge and promise of community co-design. In V. Svihla & R. Reeve (Eds.), *Design as scholarship: Case studies from the learning sciences* (pp. 41–54). Routledge. doi:10.4324/9781315709550-4

Udell, J. (2017, January 17). A hypothesis-powered toolkit for fact checkers. *Hypothes.is*. https://web.hypothes.is/blog/a-hypothesis-powered-toolkit-for-fact-checkers/

Whaley, D. (2017, February 24). Annotation is now a web standard. *Hyposthes.is*. https://web.hypothes.is/blog/annotation-is-now-a-web-standard/

KEY TERMS AND DEFINITIONS

Annotation: The term annotation refers to any extra information included within a document that denotes or references a particular point being made by the author(s) in an argument. Typically, annotations are placed in the margins of a printed book page, and can be added by either readers or the original authors. For this reason, they often mark a point of necessary discussion between readers and writers on a specific topic. Digital media formats continue to develop better methods of including annotations online or on mobile devices.

Computer-Assisted Text Analysis (CATA): This term is used throughout our paper to denote a variation in the programs and methods currently used to process computer-assisted qualitative data analysis (CAQDA). As a methodology, CAQDA generally depends upon specific search tools, mapping or networking tools, query tools, and annotation tools, among other software to assist with common data analysis in the fields of psychology, social sciences, and even much market-oriented research. The programs central to these tasks are able to perform often sophisticated data comparisons and other breakdowns from within the research document on the point of compilation. Advances in linguistic programming and digital lexicography have made it possible to extend these capacities to conventional word processing software, allowing writers to analyze diction, semantics, and other linguistic features in a document as the text is being written. Common CATA software includes programs like NowComment (used in our research), CATMA (Computer Aided Textual Markup & Analysis), and Voyant, Nvivo.

Digital Humanities: DH is a growing field of research able to combine methodologies and theories of Humanities scholarship with computing and/or digital media technologies in general. The field commonly emphasizes ways and methods of applying digital resources throughout the liberal arts in an effort to analyze new modes of literacy and literary practices. Pedagogies in the area are distinct for highlighting techniques that are collaborative, transdisciplinary, and critically engaged with new, screen-based media formats.

Flesch-Kincaid: The Flesch-Kincaid model uses a numerical value between 0 and 100 to measure factors of reading ease among students at different grade levels. A text with a Flesch-Kincaid rating of 90, for example, is far more difficult to process and interpret by readers than one with a value of 10.

Knowledge Gap: Explorations of a possible divide within society concerning how effectively different classes of citizens may access knowledge remain based in the hypothesis that knowledge, itself, like wealth is differentially distributed throughout a single social system. Academic concerns regarding such distributive trends hold that mass media networks typically issue information at varying rates depending on the socio-economic status of different populations.

Lexicography: Refers to the study and practice of compiling formal lexicons like dictionaries and encyclopedias. Research in this area of scholarship analyzes and helps standardize semantic, syntagmatic, and paradigmatic relationships found in language and language use.

STEM: STEM is an acronym denoting Science, Technology, Engineer, and Mathematics as heavily emphasized areas of study in any school institution.

Chapter 9
Varieties of Sharing:
Action Frameworks, Structures, and Working Conditions in a New Field

Christian Papsdorf
Chemnitz University of Technology, Germany

Markus Hertwig
Chemnitz University of Technology, Germany

ABSTRACT

This chapter focuses on one element in the digitalisation of work: the forms and conditions of working in the so-called 'sharing economy' (SE). Based on an analysis of 67 SE platforms, it distinguishes three segments, each of which constitutes a distinctive institutional sphere within the sharing economy: these are an 'exchange and gift economy', a 'niche and sideline economy', and the 'platform economy'. In a further step, the study then identified and compared five dimensions of work within these segments: the type of activity, the form of compensation or recompense involved (monetary or non-monetary), skills and competencies required, the role of technology, and control mechanisms. Each segment is associated with a particular pattern of these dimensions. The chapter then discusses the shift in the traditionally understood determinants of work now observable in the sharing economy. While some of these determinants are being added to by new factors, others are being displaced by internet communities and the socio-technical structures and strategies of the platform providers.

1. INTRODUCTION

The recent past has witnessed an accumulation of economic phenomena that, at first sight, seem to represent a movement beyond the logic and business models seen as defining late-capitalism: individuals now give away goods they no longer need, lend tools to strangers, rent out private rooms, or offer lifts to strangers. These, together with many other activities, have been collectively brought together under the concept of the 'sharing economy' (SE), a catch-all term covering internet-based platforms and market places that connect suppliers and customers offering and buying a dizzyingly wide range of products

DOI: 10.4018/978-1-7998-5849-2.ch009

and services through processes that operate outside the conventions of traditional business models. For both the media and academic research, the SE is frequently associated with customer gains in terms of efficiency and quality, the emergence of innovative services and products, and economy in the use of scarce resources. There is also virtual unanimity that the SE marks a radical departure from the 'old economy' and is potentially superior to it in many respects.

At the same time, there is growing scepticism about whether the sharing economy is what it purports to be, given that many apparent web-based 'new' providers or apps often seem to be no more than a front for old-fashioned, and unapologetic, profit-driven businesses. An increasing number of studies (Hall and Krüger, 2015) have also highlighted the risks of precariousness to which those engaged in 'digital work' are exposed, including in what would appear to be the free and innovative, but also in some respects barely regulated, SE (see too Benner, 2015).

Some contributions to the debate on the SE (see Section 2) suggest that this represents an ambivalent and heterogeneous field in terms of the diversity of services on offer, the conditions under which the individuals involved work, and the institutional context and regulation of the social practice of sharing in this form. As yet, however, there have been few academic studies that have engaged with the evident diversity and heterogeneity of the platforms, actors and activities that make up the SE. This applies, in particular, to the content, conditions and consequences of SE work.

This shapes the context for the present study, which has two main aims; firstly, to investigate the different forms of the sharing economy, drawing on two dimensions to categorise sharing economy platforms and identify distinctive SE segments; and secondly, in an exploratory approach, to gather information on the conditions of work of providers in the sharing economy using five dimensions: the types of activity involved, skills and competencies, compensation (monetary and non-monetary), the role of technology, and control mechanisms. These two aims build on each other inasmuch as the study will compare and contrast work in each segment. The study argues that one of the key structuring factors for work in the SE is how it is institutionally embedded in each of the SE segments, including the strategies pursued by platform providers, the socio-technical structures that transport these platforms, and the cultural expectations of the internet communities involved.

The section that follows begins by reviewing the current state of research. Section 3 sets out the theoretical framework for the study. Section 4 outlines the methodology, with the two following Sections (5 and 6) presenting the findings in relation to the two research questions. Section 7 contains the discussion.

2. THE 'SHARING ECONOMY:' CONCEPT AND STATE OF RESEARCH

As Martin (2016: 151) has shown, the concept of the SE embraces and is used to cover a wide range of phenomena, making it difficult to arrive at a single and consistent definition. In practice, it continues to represent a somewhat ambiguous notion embracing a long list of synonyms, not all of which are equivalent in meaning (Martin 2016; Belk 2014; Dillahunt and Malone 2015).[1] Nonetheless, it is possible to discern some consensus between the various attempts at a definition. On this, the SE can be seen as a set of business models, platforms and exchange relationships (Alen and Berg 2014, following Daunoriene et al. 2015), in which resources, services or access to material goods (on a temporary basis) are gifted, exchanged, lent or purchased (Dillahunt und Malone 2015; Richardson 2015) between private individuals (peer-to-peer) in return for a fee or other forms of compensation (Belk 2014) via web-based social media platforms (Hamari et al. 2015).

This definition leads on to five characteristics that both frame the SE and distinguish it from the 'old economy'. Firstly, services are provided to a significant degree on a peer-to-peer basis, that is by non-professional internet users. Secondly, supply and demand, and to some extent the provision of the services themselves, are coordinated via digital media. Thirdly, the SE is evidently distinctive in that it is characterised by an alternative form of value-chain based on the assignment of rights to usage rather than ownership (although some selling platforms might also be included in the SE). Fourthly, the economic subjects engaged in the SE exhibit a wide spectrum of motivations. This includes not only a desire to earn money: the practices of exchange, gifting and lending also point to altruistic or post-material motives. This is also connected with the fifth characteristic: that the SE is constituted as an alternative and opposition – both conceived in socially progressive terms – to the capitalist logic of profit and exploitation. On this, reasonable doubts can be raised as to whether this is the true guiding principle behind the SE or rather whether it serves as a means to legitimate and distinguish this form of economic activity (DiMaggio and Powell 1983). Critical perspectives, for example, refer to 'pseudo-sharing' (Belk 2014: 1597) or 'sharewashing' (Light and Miskelly 2015: 49). In fact, the SE can be seen as a hybrid phenomenon that includes both sharing as well as selling, together with a range of gradations between these two poles (Scaraboto 2015; Belk 2014), and as both a part of and an alternative to capitalist forms of economic activity (Richardson 2015: 121).

The spread and growing intensity of the use of SE online platforms has also generated greater differentiation in research into this field, with five discernible strands: (1) research on sustainability aspects; (2) political scope for regulation; (3) user motivation; (4) developing an appropriate conceptual apparatus; and (5) effects on employment and the broader economy. This chapter is particularly concerned with the latter two of these.

The notion of shared resources is intrinsically inseparable from sustainability considerations, as a number of authors have emphasised. For example, sharing can make use of excess capacity (Benkler 2004), cut greenhouse gas emissions (Firnkorn 2011), and reduce urban car traffic (Glotz-Richter 2012). More recent research has suggested that ecological arguments now have a lower profile and are seen as being of only minor importance (Barnes und Mattsson 2016; Shaheen 2016; Hamari et al. 2015). Martin (2016: 149), for example, has asked whether the SE represents 'a pathway to sustainability or a nightmarish form of neoliberal capitalism'. One consequence of this shift has been increased concern with regulatory aspects of the SE, with the aim of offering greater security to those engaged in the market for services (Hartl et al. 2015; Weber 2014), securing an optimal combination of private and public providers (Cohen and Kietzman 2014), or elaborating minimum standards for providers' pay, skills and legal liability (Ranchordás 2015). The reasons why and under what circumstances certain individuals make use of the SE to obtain goods or services vary considerably depending on the product. For example, 'land sharing' initiatives are typically motivated by health concerns (McArthur 2015), ride-sharing by everyday practical considerations (Gargiulo 2015), and space sharing by financial factors along with other motives (Möhlmann 2015).

Other studies have aimed at developing typologies of the diversity within the SE. Based on a sample of 18 sharing initiatives, for example, Cohen and Munoz (2015) have proposed five types, using two dimensions (consumption/production and self-interest/common good): these are sustainable consumption platforms, sustainable production platforms, localised sustainable consumption, localised sustainable production, and complete hybridisation. By contrast, Richardson (2015) identified three modes through which existing capitalist practices can be suspended but which allow activities of an economic nature to take place. This occurs through the formation of communities, access to individual resources (sharing), and collaboration between users. Scaraboto (2015) has proposed an alternative categorisation that sees the SE as a hybrid between market-based and non-economic forms of exchange. This can embrace a variety of forms for exchanging resources and values.

Although SE phenomena might often bear strong economic connotations, only a few studies have been carried out that directly address the economic effects and characteristics of work performed in the SE. Dillahunt and Malone (2015), for example, discussed the potential of the SE for people with social disadvantages and concluded that the SE could lower many of the barriers to employment that they encountered. Fang (2015) and Geron (2012) also noted the positive effects of AirBnb for tourism. Hall and Krueger (2015) researched work in a narrower sense in their investigation into the characteristics, motives, earnings and working hours of Uber drivers.

Both of these research strands have exposed a research gap that this chapter addresses. For example, rather limited samples or case studies are used to generate core distinguishing features - such as sustainability, structural characteristics, or forms of exchange. Such an approach is not sufficient to capture the diversity of SE platforms. And only the study by Hall and Krueger (2015) expressly deals with work in the SE. Working conditions associated with other platforms have only been researched for phenomena understood as 'crowdworking' (Leimeister et al. 2016) and this only partially overlaps with the SE.

3. THE TRANSFORMATION OF WORK AND THE SHARING ECONOMY

3.1 The Transformation of Work and Employment

The transformation of work and employment as one outcome of a range of new socio-economic megatrends has been a core thesis of the sociology of work since at least the 1990s (Trinczek 2011). These include the growing globalisation and financialisation of exchange relations, sectoral change (tertiarisation, the knowledge society), shifts in the structure of employment, and raised employee aspirations for more 'satisfying' forms of activity. These megatrends also include deregulation, a development that has profoundly changed the structure of incentives facing both employees and those seeking work as well as employers, leading to an expansion in atypical forms of employment, again from the 1990s. The more recently discussed phenomena associated with the digitalisation of work, the focus of this study, can also be placed in this canon, with evident links to trends that have already been noted, such as the transnational character of digital labour markets for crowdwork.

This would suggest that activities in the SE diverge markedly from the classic idea of gainful employment, with the most notable difference being that these do not necessarily involve *gainful* employment, given that they are often not, or not mainly, motivated by money. At the same time, as a purposive social practice that entails expending time and effort these activities also exhibit typical features of work (Voß 2010). Nevertheless, even where 'sharing' activities have a commercial purpose (and can be seen as 'gainful'), they are not necessarily intended to meet the basic needs of those engaged in them. Many forms of SE work might consequently be categorised as 'atypical employment' (Keller and Seifert 2013; Hohendanner and Walwei 2013).

3.2 Constitutive Dimensions of Work

Although the sociology of work can draw on a wide range of middle-range theories to explain many individual aspects of work (Maurer 2004; Pfeiffer 2010), recent research in this field lacks both a consistent or shared concept of 'work' as such and its 'determinants' – that is, those dimensions or influences that might systematically explain the observable 'variance' in the phenomenon of work.[2]

Industrial sociology has customarily identified several dimensions that are held to be constitutive of 'work' in its widest sense: that is, they describe what 'work' is and are typically subject to change as a result of technological development (Pfeiffer 2010). They generally focus on work activities and work contents, which in turn are associated with the skills or competencies needed to (successfully) perform them. Specific aspects of work, often subsumed under descriptions of the terms and conditions of employment, can also be subject to these pressures for change. These include forms of remuneration or income, and how work is directed and controlled. This list can be extended with numerous further examples.[3] There are number of reasons why the study confines itselve here to five dimensions (activity, skill, role of technology, compensation, and direction/control). Firstly, they are typically subject to change in the context of rationalisation strategies – in line with classic studies of the transformation of industrial work (e.g. Pfeiffer 2010). Secondly, they are at the core of the current debate about how digitalisation affects work (Hirsch-Kreinsen 2015). And thirdly, there is a research imperative to focus on dimensions that can be validly captured using the given methodology.[4]

The issue of the influence of technology or new technology-based rationalisation strategies has a long tradition in the discipline (Pfeiffer 2010). Over the course of the debates in this field, the once dominant view in classic studies of work and technology ('rationalisation debate', 'automation debate': see Minssen 2006), according to which there is a 'determinism' between work and technology with new technologies invariably forcing work to change in a particular direction, has yielded to a more contingent and triangular constellation between work, organisation and technology in which, in the final analysis, it is actors that determine how new technologies are configured at the workplace (Kern and Schumann 1984; Pries et al. 1990).

Previous studies suggested that technical change typically affects certain dimensions of work. For example, change in the contents of work activities might be associated with new skill requirements. As a rule, technical systems imply new forms of direction and control. While direct control includes personal subordination to a line manager, bureaucratic control is based on codified organisational rules. In contrast, ideological-cultural control is effected via institutionalised norms and expectations and, in some instances, how such norms are internalised by actors (see too Marrs 2010).

The SE is characterised by a unique set of institutional conditions under which labour-power is expended, as digitally-mediated work is performed neither solely as gainful employment nor within the institutional arrangements of normal employment relationships. Many sharing activities are one element of work that is either not directed at obtaining payment or is limited to a sideline within a wider freelance existence. In this sense, work in the SE lies outside the traditional hierarchical forms of workplace integration and subordination that characterise managerial prerogatives in capitalist firms. Yet at the same time, these activities are digitally mediated and structured by large firms operating on a profit-seeking basis (Kirchner and Beyer 2016). 'Sharing' activities are, as consequence, always the product of new corporate strategies and are subject to the influence of the platform providers who use technical and bureaucratic rules to steer and control users' social practices (work) and with this also influence a range of key aspects of work, such as the scope to earn money and what hours individuals work (as in the case of Uber).

The institutional conditions under which labour-power is expended also have an ideological-cultural component as SE platforms are associated with a specific pattern of legitimation that guides how services should be provided, both expressly creating a connection to an overarching notion of 'sharing' and underpinning their offer by appealing to the specific values of the community involved.

4. METHODOLOGY AND SAMPLE

Categorising the diversity of SE phenomena and characterising them in terms of the conceptual register of the sociology of work called for a very heterogeneous sample of sharing platforms. One immediate problem in this respect was that no usable listing of SE platforms has been compiled, either by official bodies or trade association. Obtaining a sample therefore had to begin by consulting the lists of sharing initiatives and platforms put together by journalists or companies interested in documenting SE initiatives and showcasing them to potential users. Of these, the list produced by justpark.com[5] is both the most up-to-date and suitable in terms of capturing SE initiatives that meet the definition adopted here. At the time of the research, it contained 817 initiatives and platforms.

The population addressed by the study comprises economically independent organisational units, such as various forms of corporate organisation, and platforms that have not been established on a commercial basis and are not profit-seeking. Only organisations and platforms based online were included. A stratified sample was then extracted from this overall population by economic branch (Statistisches Bundesamt 2008; see Table 1) in order to enable as broad a range as possible to be investigated in terms of activities and service provision processes. Cases were randomised within each category, with cases added to the sample until a point of theoretical saturation had been reached in terms of the two research dimensions: groups of actors and whether organisations were profit or non-profit. This point was reached at 67 cases.

The investigation of SE activities and working conditions was carried out via a study of the platforms' websites[6] together with a content analysis of selected supplier profiles (Mayring 2008). Data was analysed using MaxQDA. Using structured content analysis, the material was assigned to pre-determined categories. In this respect, the primary aim was not to conduct an inductive-style search to 'discover' new factors influencing or affecting work in the SE, but rather to analyse the forms encountered along a number of pre-determined dimensions. These included how platforms operated, the products and services on offer, skill requirements for service provision, the status of service providers (freelance, dependent employees), and pricing. This also included analysing the extent to which services were provided either by private individuals or professionals engaged in a commercial operation, and whether services were offered with the aim of obtaining financial reward or for other reasons (the pleasure of the experience, altruism etc.). The software enabled us to categorise the cases into distinct segments and thus characterise the nature of work in each segment. The software was also used to check intercoder reliability and update the categories as the analysis proceeded. However, website analysis was not able to provide data on issues such working hours, work intensity, or the maintenance of boundaries between work and private life.

Although the 817 cases (and the 67 analysed in detail) cannot be deemed representative, they do offer a first impression of the spread of SE initiatives are across branches. Most are concentrated in five of these: commerce, vehicle repair and maintenance, transport and storage, hospitality, financial and insurance services, and miscellaneous commercial services. Many SE platforms offer the sale or rental

Table 1. Structure of the survey population

Branch*	Typical Activities	No. in Population	No. in Sample
Agriculture, forestry and fishing	Growing vegetables; production of eggs for human consumption	2	2
Manufacturing	Manufacture of hand-made or made-to-measure clothing and consumer goods	16	7
Energy	Use of solar panels	1	0
Wholesale and retail trade; repair of motor vehicles	Sale of second-hand clothes and other objects; import of consumer goods	92	5
Transport and distribution	Transport of people and goods; provision of parking spaces	120	8
Accommodation and food service activities	Accommodation rental; food and beverage provision	91	5
Information and communication	Access to telecommunication and IT services	10	3
Financial and insurance services	Provision of credit, financial services	114	4
Real estate activities	Renting private accommodation	20	6
Professional, scientific and technical activities	Occupational and personal advice	54	6
Other business services	Lending vehicles and other equipment; skilled repair and maintenance services	201	5
Health and social care	Massage; personal care	2	1
Arts, entertainment and recreation	Events management; provision of artists and artistic services	14	4
Other service activities	Household services; non-institutional educational services	56	6
Non-classified activities, due to absence of commercial purpose	Exchange and donation of clothes and other goods.	24	5

Source: authors.

* The branch classification is taken from the German Federal Statistical Office classification of economic activities, referred to as WZ-2008, which is aligned with NACE Rev. 2.

of clothing, other goods, vehicles and accommodation, simple services, and the provision of credit. By contrast, other important sectors, such as health, are only marginally represented. The final row of Table 1 is also of interest. This lists 24 platforms and organisations that could not be subsumed under the usual economic classification as they neither sold nor rented anything and nor did they offer a service in return for monetary payment. These initiatives are confined to the exchange or gifting of goods and, because they are non-profitmaking, cannot be assigned to any of the usual economic categories. However, if these are classified in terms of their 'core business' – their principal activity, analogously to commercial enterprises – they do have an identifiable 'operational objective' in terms of products or services, yielding an interesting finding: all can be assigned to commerce, although as a rule this entails the free-of-charge transfer or exchange of goods.

In a first step, the platforms were analysed in terms of the two core dimensions: profit/non-profit and service provider status, allowing us to identify three segments into which platforms can be assigned. In a second step, platforms were analysed in terms of the typical features that characterised work in each segment, drawing on an analytical schema based on the five key dimensions of work noted above: that is, activities, skills, compensation, the role of technology, and direction/control.

5. THREE SEGMENTS OF THE SHARING ECONOMY

Figure 1 sets out the segments of the SE, based on the two dimensions identified as being of particular interest, given the current state of research and the definition adopted here. The horizontal axis denotes the profit/non-profit dimension, and therefore reflects the antitheses between both the common good and self-interest as well as between market-based and non-market exchange. Initiatives that are less profit-led and more community-oriented are also those in which actors provide access, perform services or gift goods without looking for any form of recompense. There are also exchange relationships that involve reciprocity, but not any intent to secure a profit. The other end of the spectrum consists of initiatives in which the service providers (users, peers) aim to make a gain, either in the form of money or as non-monetary benefits, such as benefits-in-kind. Between these two poles, there are various initiatives that are both profit-led and community-oriented or which combine both modes of operation (sale/rent vs gift/exchange).

The vertical axis set out the various types of service provider, focussing, in line with the state of current research, on the tensions between peer-to-peer logic and professionalisation as reflected in the various

Figure 1.
Note: The size of the circles corresponds to the number of platforms
Source: authors.

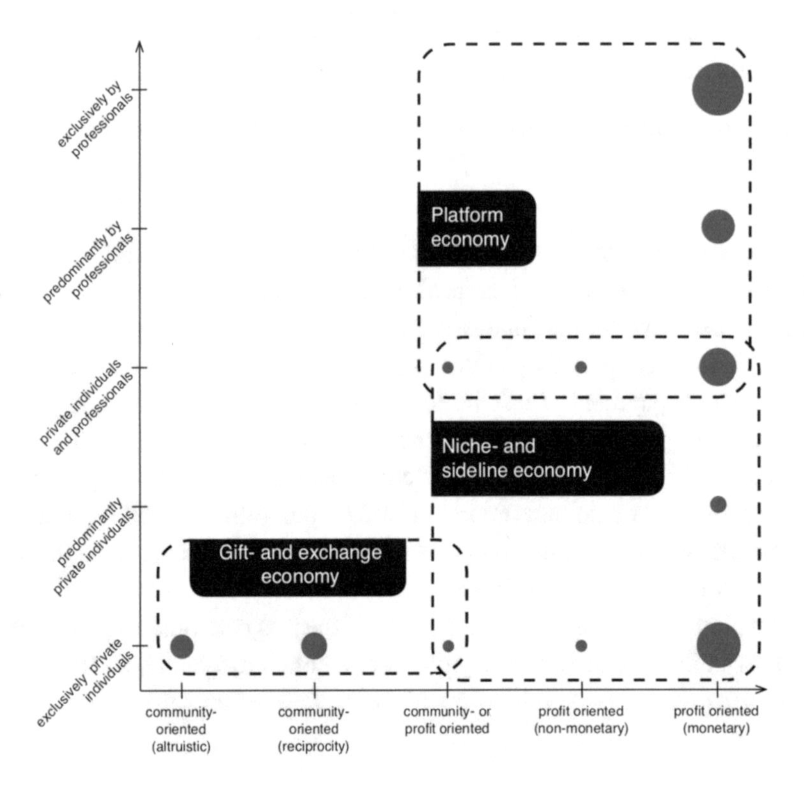

groups of actors involved. Some service providers are either exclusively or predominantly private ('lay') individuals: this term refers to internet users who either do not have specific qualifications, who lack extensive experience, or who do not operate in a formal professional setting. The fact that such individuals are classified as 'lay' or 'amateur' in terms of formal occupational status does not necessarily imply that they cannot provide the services they claim to or do so inadequately, as the example of user-generated Web 2.0 content has shown (Papsdorf 2009). By contrast, the category of 'professional' comprises those service providers who are formally economically active in their field (either freelance or employed) and possess the corresponding skills and qualifications. The intermediate category is applied to initiatives for which the services are provided to a substantial extent by both 'lay' individuals and professionals.

Based on our research, the study divides the SE into three segments: a gift and exchange economy, a niche and sideline economy, and the platform economy. The core element of the first is the practice of sharing; the second is defined by its complementary relationship to the classic labour market; and the third is characterised by the use of new channels to deliver familiar services and products.

1. The gift and exchange economy best embodies the original and authentic notion of sharing. Resources, such as books, tools, food or land, are made over to other users, or offered temporarily, on ecological, altruistic, or other non-economic grounds. This can take the form of lending free-of-charge, donation, or swapping, with no need for there to be any equivalent in terms of value or costs incurred. One notable exemplar is neighborgoods.net, which operates under the slogan: 'Save money and resources by sharing stuff with your friends'. This would also include 'Sharing Groups' that resemble offline networks such as associations, neighbourhood groups and organisations in which participating individuals can request to borrow or offer to lend tools or other things free of charge. Users post a profile that contains personal information, evaluations of other users, the number of transactions they have engaged in, and the monetary value that other users have saved by borrowing. These relationships based on providing mutual favours are innovative in that, while they exist between individuals who do not know each other, they are also direct and personal. As a rule, such favours are exchanged within social networks such as families, circles of friends or between neighbours. By contrast, in the SE favours are offered to complete strangers. At present, peer-to-peer sharing networks are limited to providing access to certain resources and do not include activities that might, for example, be performed in the context of various types of personal services, often performed on a voluntary basis, such as caring, teaching or social work.
2. Services in the niche and sideline economy are also predominantly provided by users – that is 'lay' people and amateurs – but in this instance with the aim of obtaining a monetary reward. This can take the form of lending or selling physical goods or offering basic trade skills or household services, such as plumbing or electrical repairs, for payment. A good example is manyship.com. This is an online network in which users willing to transport packages post their travel dates and carrying capacity, while other users put up details of what goods they would like to have transported and where. For air transport across national borders, delivery using such a platform can be quicker and, in particular cheaper, than the postal service or a commercial freight forwarder.

What is novel about this segment is that resources previously unavailable to the economic system, such as those in private ownership and various forms of potential labour input, become commodified. This often involves economic niches that have the potential to function as sidelines, mobilising new categories of atypical employee who offer activities outside their usual occupation. Activities previously confined to the personal or private sphere and performed without reward, such as doing shopping or offering lifts, can also now be made available on markets.

3. The platform economy[7] consists of initiatives in which services are provided by individuals who are formally economically active: that is the employed and self-employed. Service providers lend resources and perform simple, but possibly also complex and demanding, services in return for monetary reward, typically involving resources and services already available on the market. They are assignable to the SE, either by providers themselves or others, because internet platforms are used to enable these services or products to reach a wide circle of potential customers by transcending the barriers of time, space and organisational constraint. One example is fiverr.com, which allows freelancers and firms, including the latters' employees, to offer services such as the design of logos, search machine optimisation, translation, animation, programming or the preparation of advertising materials. Service providers post a very precise definition of their services and can be booked immediately without the need for further formal arrangement; in some cases, providers might note the thousands of orders already delivered on that platform as a reference.

In some cases, providers in the platform economy are engaged in 'sharewashing', where standard services are offered using sharing rhetoric. This segment might also be dubbed the digitalised or 'platformed' 'old economy'. Such arrangements also can allow resources to be used more efficiently, and this serves as one of the core legitimating arguments for the SE. In contrast to the first two segments, this third segment is not about fostering novel forms of economic activity that are enabled by the medium of the internet; rather, in the SE, platforms are used as electronic market places – as means to coordinate supply and demand – to enable familiar forms of economic activity to take advantage of new technical and media opportunities. As Figure 1 shows, many cases can be assigned to both the second and third segments as both encompass professional services and those provided by private ('lay') individuals. No professional providers in the sample offered services or resources free-of-charge: as a consequence, the upper left quadrant in Figure 1 is empty. Nonetheless, although such arrangements are not infrequent, with examples including food donations by supermarkets or pro bono services by lawyers, these have evidently not yet been offered online.

6. WORK IN THE THREE SEGMENTS OF THE SHARING ECONOMY

Based on the segments outlined above, this section now explores whether there are differences between them in terms of activities or work performed, focusing on the five dimensions noted in Section 3: work contents, skills, technology, recompense/compensation, and the direction and control of work performance.[8] The findings are illustrated with examples from the research.

(1) Activities

There are some notable differences in terms of typical activities between the first segment – the gift and exchange economy – and the other two segments. While the latter two are, by definition, characterised by activities than can be understood as 'gainful employment' (Voß 2010), providing services for monetary reward does not feature in the first segment. At the same time, activities performed in the gift and exchange economy do exhibit features seen as characteristic of work. This is not, however, work 'for others', as this segment mainly entails making resources available, but rather preparatory and subsequent activities. This includes administrative tasks, such as posting and maintaining a provider profile, responding to queries, and clearing up and cleaning afterwards. For example, at neighborgoods.net, the real service includes the lending of and, if necessary instruction in, tools and equipment that are provided to the requester on a temporary basis and free-of-charge.

NeighborGoods.net is the leading social platform for peer-to-peer borrowing and lending. Need a ladder? Borrow it from your neighbor. Have a bike collecting dust in your closet? Lend it out and make a new friend. By sharing with your neighbors, you can save money while reducing waste and strengthen your local community in the process (SEP17: 37).

In the two profit-oriented segments, the most typical activities fit with those of classic gainful employment, embracing a wide range of activities. One notable feature of the niche and sideline economy is that this is usually conducted by private individuals. It also often involves activities or resources that have not generally been available on the market. For example, a platform called vrumi.com allows homeowners 'to make extra money from renting out their unused space [...]'.

There are thousands of global businesses, consultants, writers, life coaches, therapists, small businesses owners and other professionals all looking for a flexible space in homes just like yours, for meeting, thinking, creating, writing and more' (SEP11: 111-112).

The platform economy also embraces classic dependent employment as well as freelancers. The difference to traditional offline business is frequently simply the channel of communication. 'Urban Massage is the new way for people to book and experience a professional massage at home, work or hotel' (SEP38: 18). Some platforms also offer a wide range of traditional offline services, not confined to any particular sector or branch. For example, fiverr.com expressly states the range of potential services that can be offered on the platform: 'Be creative! You can offer any service you wish as long as it's legal and complies with our terms. There are over 100 categories you can browse to get ideas' (SEP29: 142).

(2) Skill Requirements

As with the activity profiles, typical skill requirements vary as between the segments. In the gift and exchange economy, competency requirements tend to be low. Those specified by the platforms are confined to capacities such as a 'friendly and helpful' manner and basic preconditions for work, such as a driving licence (for ride-sharing). Work in the niche and sideline economy is also typically performed by private individuals, on occasions with some minimal checks on suitability.

These are the individuals who cook for their families and friends everyday already and are just cooking extra to share with you and your family. All Mouthful Home cooks go through a step-by-step vetting process to ensure only quality and safety-conscious chefs are admitted (SEP5: 98).

Selling homemade food on the platform requires formal registration and typically authorisation attesting that hygiene standards have been complied with. In general, however, there is no restriction on *who* is allowed to provide services.

And by 'everyone', we mean everyone — regardless of age, gender, religion, political beliefs, social standing, educational background, work experience, and other factors (SEP1: 27).

In line with the diversity of activities performed in the platform economy, there is a correspondingly wide spectrum of skill requirements. This is made clear, for example, at ontask.in,[9] platform that connects requesters and providers across a range of services.

Our service providers come with varied backgrounds and skills; from young professionals to stay-at-home parents, from retirees to students looking to earn some extra cash (SEP29: 71).

By contrast, other platforms work exclusively with highly-skilled individuals, with the highest level of skill requirements of all the segments.

Our friendly, scientifically trained staff – many with master's degrees and PhDs – can help you understand provider and requester needs (SEP31: 16).

(3) Role of Technology

As far as the significance of technical infrastructure for the labour process is concerned, there are some slight differences between the three segments. Although all three are technically based and would not exist without internet platforms, technology appears to have hardly any direct and marked effects on the labour process in the sample. There are some initial signs, however, that technology is beginning to exercise a structuring and regulative role. By contrast, the direct effect of algorithms on work processes is very familiar for some platforms, such as Uber, where they can determine which customer requests are forwarded to which service provider (Rosenblat and Stark 2016).

In both the gift and exchange economy as well as the niche and sideline economy, platforms highlight the central benefit of their organisation and its technology as one of 'simplification'.

Sharetown was created to simplify the life of admins. It empowers users with useful tools and functionalities to make groups run smoothly (SEP9: 11).

This platform, which operates in the niche and sideline economy, emphasises how technology enables organisational activities to be carried out more simply and efficiently. Work in the more narrow sense, such as household services, is performed offline and without direct help from the platform. One further benefit noted by platforms is security.

Patriot List successfully solves a major problem faced by other peer-to-peer market platforms by removing anonymity through its validation/verification process (SEP1: 3).

A similar function can be discerned in the gift and exchange economy:

NeighborGoods helps facilitate transactions with a reservation calendar, automated reminders, wishlist alerts, and private messaging between neighbors. NeighborGoods keeps track of all your stuff (SEP17: 70).

Platforms in the niche and sideline economy simplify the relationship between service providers and requesters in terms of establishing trust, which research has indicated is especially challenging on the Internet (Lim et al. 2006). By contrast, the platform economy focuses instead on the notion of 'connection'. 'As a marketplace venue, EcoHabitude's role is to connect shoppers (buyers) with independent shops (sellers)' (SEP21: 197). One user noted that outsourcing customer acquisition to the platform is sufficient to warrant paying what can be quite substantial charges.

SmartShoot has been instrumental in allowing me to be successful, spending less time marketing my business and more time doing exactly what I love: capturing the world through the camera lens (SEP33: 66).

(4) Compensation

By definition, the issue of *monetary* compensation does not arise in the gift and exchange economy. Nonetheless, some platforms have developed 'currencies' for coordinating exchange processes. BookMooch, a platform that enables book swapping between peers and can be assigned to the gift and exchange economy, has installed a points system intended to ensure that exchanges are reciprocated given that a 'reward', other than access to books, is not possible.

Give & receive: Every time you give someone a book, you earn a point and can get any book you want from anyone else at BookMooch. ... In order to keep receiving books, you need to give away at least one book for every two you receive (SEP14: 11-13).

The niche and sideline economy is mainly dominated by low-paid activities, such as household services. In theory, providers could set their own charges; in practice, the platforms interpose themselves through recommendations and stipulations that prevent providers earning an income sufficient to live on.

Each cook has the authority to price their own meal. Depending on portion size, we recommend cooks to price meals between $8 – $15. ... Mouthful sets a limit at 10 for each meal (SEP5: 112-114).

In contrast, the platform economy tends to consist of professional services of various sorts, such as programming or consultancy, for which there is market demand at a corresponding price. There is a strong orientation towards the market with an emphasis on direct competition between providers. 'Post your photo or video job for free and receive competitive bids from professionals within minutes' (SEP33: 4, SmartShoot). Successful providers will have a good earnings potential and can make a living through activities offered via the platform. One photographer, for example, posted on their personal webpage on the platform.

The extra income and connections I made through SmartShoot enabled me to quit my full-time job and focus on photography as a full-time career (SEP33: 64).

There is a good deal of variation in this area, however (see, for example, Hall und Krueger 2015), with some providers managing only very modest earnings. Even the successful cannot be certain of long-term stable incomes, given their exposure to competition and the vagaries of the market.

(5) Direction and Control

There are very substantial differences between the three SE segments in terms of the interplay of different forms of direction and control and in particular the significance of technological control, bureaucratic control (through formalised mandatory rules) and ideological-control (via the social norms and expectations of an internet community) (Edwards 1979; Friedman 1987; see too Marrs 2010). In general, direction by means of bureaucratic rules plays a relatively minor role in the gift and exchange economy, compared with the other two segments, with control predominantly exercised through ideological and cultural means. By contrast, the other segments are characterised by a variety of forms of control, of which the most important is the bureaucratic. The fact that one form of control is dominant in a segment does not mean, however, that others are absent.

The importance of ideological and cultural forms of control in the gift and exchange economy is especially evident in the detailed descriptions of the motivations and activities that these platforms sometimes post on their websites. These both set out the underlying rationale for the service offered as well as specifying the behaviours expected of users, not usually as formal rules but rather more often as 'tips' or 'advice' for operating successfully on the platform.

[E]ach time you receive a book, you can leave feedback with the sender, just like how eBay does it. If you keep your feedback score up, people are most likely to help you out when you ask for a book (SEP14: 17, Bookmooch.com).

Unambiguous rules (bureaucratic control) for service delivery are only rarely specified, and mainly where legal or security aspects are concerned or for the technical requirements needed to be able to participate and obtain payment. One example is offering ride-sharing on the site tridid.ph, which requires participants to register their vehicles.

One significant instrument for control is the use of evaluations that service providers receive from service requesters – that is, other platform users and/or 'peers'. These function as a means of control by creating an incentive to comply with the rules or expectations of the platform community. In some instances, they are expressed as community norms: 'Still, we find it's helpful to outline some general expectations for the community. As a member of NeighborGoods, we expect you to share according to these guidelines. The 3 Golden Rules of Sharing' (SEP17: 18-22). Although platforms typically define possible – and expected – infringements, none of the websites in the sample in this segment indicated any mechanism for imposing sanctions.

Compared with this, the degree of bureaucratic control in the niche and sideline and platform economy segments is relatively high with a variety of formalised rules outlining the tasks and duties of service providers, sometimes in considerable detail. These platforms also have extensive terms and conditions of business that might, for example, specify 'proscribed behaviours' (SEP33; 211, SEP21: 197-225).

The platforms also set out sanctions, of varying severity, including exclusion for repeated infringements. These rules cover a range of issues, addressing the behaviour of providers and requesters and how customers should be treated, but also pricing and providers' skills. The platform ecohabitude, for example, stipulates the following on dealing with customers.

We expect you to provide timely, friendly customer service. Our expectations of customer service are that you: Respond to message inquiries from buyers or potential buyers within 72 hours but we'd suggest you aim for a 24-hour turnaround time on message inquiries or purchases. Acknowledge each order and thank your customers for choosing your store and product. Create policies for your store as a whole and/or on a per product basis that clearly communicate your policies on returns, refunds, exchanges, shipping, and turnaround time for production etc. (SEP21: 134-175).

Forms of control also include stipulations on the appropriate presentation of products for the communities involved. Bureaucratic rules are often combined with options for technical forms of direction and control in that the provisions are embedded in the technical infrastructure, giving service providers no choice but to comply. At ecohabitude, for example, providers must post certain product information before that product can be put online.

All items listed in your shop must be for sale and be tagged with at least 2 product footprint. You will not be able to upload products to your shop if they do not include at least 2 product footprints and a description for each tag. For each product footprint you choose, you must explain in your own words how the product meets the requirements of this product footprint (SEP21: 140; the authors emphasis).

As far as sanctions for infringements are concerned, the procedure might have several stages, as in the case of Patriot List, an exchange platform in the niche and sideline economy. 'Consequences: First offense: Warning, Patriot List will step in to mediate. Second offense: 90-day suspension. Third offense: Full account termination' (SEP1: 187). The fairly high degree of rule-setting in the niche and sideline economy, and even more so in the platform economy, is attributable to the fact that the platforms in these segments are profit-based, with profitability dependent on the economic success of the service providers, whether professional freelancers or private individuals. Moreover, poor performance or quality problems will damage platforms' images. For the platform operators, this poses the classic transformation problem of ensuring that work performed by service providers does actually comply with the platform's requirements, in terms of quality, price and customer service. The fact that these platforms represent profit-seeking organisations consequently obliges them to professionalise and standardise the conditions under which 'non-professional' providers operate. To do this, they draw on the full repertoire of direction and control: formal bureaucratic rules, on occasions embedded in the technology (such as permitted price ranges) and sanctions for infringements. Platforms will also use cultural and ideological forms of direction to address and resolve the transformation problem. Table 2 summarises the main differences between work in the three SE segments.

Table 2. Comparison of five dimensions of work in SE segments

Dimension	Gift and Exchange Economy	Niche and Sideline Economy	Digitalised Old Economy
Activity	Donate, exchange (free), lend	Lend (for reward), sell, provide services	Lend (for reward), sell, produce, provide services
Skills	Low	Low	Low to high
Role of technology	Simplification	Simplification	Mediation
Compensation	Unpaid	Negligible-low	Low to high
Direction and control			
- Direct	Low	Low	High
- Financial	Low	Medium to high	High
- Social norms	High	Medium to high	Low

Source: authors

7. CONCLUSION

The growing trend towards the digitalisation of social spheres has had a marked impact on the world of work, with the specific features of internet technology transforming both the contextual conditions for work (such as better access to employment opportunities) and also bringing about innovations (such as sharing platforms) that have created novel foundations for economic activity, and with this for work. The Sharing Economy (SE) represents a notable instance of this development in two respects: firstly, it has placed 'classic' business models on a new technological footing; and, secondly, it has led to the emergence of new platforms that, by facilitating the matching of providers and requesters, has enabled products and services to enter into marketised relationships. This has created new markets and facilitated the commodification of products and services previously provided solely on a private basis.

The SE is not a homogeneous segment of alternative economic activity, something occasionally implied in both the media and by academic research. Rather, it consists of a number of differing institutional spheres that can be distinguished in terms of the status of service providers, the logic with which they operate, and how their activities are directed.

On the issue of forms of work, the segments exhibit both common features and differences in terms of the five dimensions considered here, with work in each segment characterised by a distinctive configuration of these. This suggests the particular significance of how work activities are embedded in a specific institutional context – here each SE segment – that characterises the fundamental logics, structures of meaning, and norms of work and exchange. In this respect, each SE 'variant' contains different institutional regulations that shape the social practices of work and hence also affect the conditions under which service providers act. These define both the range of subjective meanings and objective logics of each segment. In both the platform economy and the niche and sideline economy, the guiding maxim for action is monetary reward. In the gift and exchange economy, by contrast, this is represented by the common good or the structures of meaning established by the community involved. These differences correspond with segment-specific modes of direction and control. Whereas work in the gift and exchange economy is steered by social norms, the cultural patterns of the community involved, and peer evaluation, the other segments are dominated by bureaucratic rules together with direction and control exercised by firms and organisations – the platform providers – which, on some issues, embed their rules into the technical infrastructure.

Considering the SE as a whole, the conclusion – and also a hypothesis for further research – is that a shift has taken place in the determinants of work. Influences seen as central in structuring work in decades of research in industrial sociology have been found to be dominant in just one segment of the SE: the digitalised part of the 'old economy' that is termed the 'platform economy'. In the other segments, factors such as the branch, establishment size or form and style of management play a generally subordinate role, not least because of the absence of a tangible workplace – a core determinant of the nature of work. These factors have been replaced, firstly, by community and, secondly, by the platform – each of which is located in a mutually dynamic relationship. And even if the platform is seen as a substitute or 'successor' to the classic workplace, the relationship between the platform and the service provider nonetheless differs from the classic employer-employee relationship in a number of respects.

This raises a number of points of relevance for future research. For example, empirical analysis of the activities of peers and other service providers in the SE would be needed to confirm and specify our findings that have been derived from an analysis of websites. Quantitative research could also offer a more representative picture of the various facets of work in the different SE segments. And finally, the transnational structure of many social media platforms suggests looking at the issues dealt with here from an internationally comparative perspective in order to identify the possible influences of national work cultures and institutional systems.

REFERENCES

Barnes, S. J., & Mattsson, J. (2016). Understanding current and future issues in collaborative consumption: A four-stage Delphi study. *Technological Forecasting and Social Change*, *104*, 200–211. doi:10.1016/j.techfore.2016.01.006

Belk, R. (2014). You are what you can access: Sharing and collaborative consumption online. *Journal of Business Research*, *67*(8), 1595–1600. doi:10.1016/j.jbusres.2013.10.001

Benkler, Y. (2004). 'Sharing Nicely': On Shareable Goods and the Emergence of Sharing as a Modality of Economic Production. *The Yale Law Journal*, *114*(2), 273–358. doi:10.2307/4135731

Benner, C. (2015). *Crowdwork - zurück in die Zukunft? Perspektiven digitaler Arbeit*. BUND Verlag.

Boltanski, L., & Chiapello, È. (2007). *The New Spirit of Capitalism*. Verso.

Cohen, B., & Kietzmann, J. (2014). Ride On! Mobility Business Models for the Sharing Economy. *Organization & Environment*, *27*(3), 279–296. doi:10.1177/1086026614546199

Cohen, B., & Munoz, P. (2015). Sharing cities and sustainable consumption and production: Towards and integrated framework. *Journal of Cleaner Production*, *134*, 87–97. doi:10.1016/j.jclepro.2015.07.133

Daunoriene, A., Draksaite, A., Snieska, V., & Valodkiene, G. (2015). Evaluating Sustainability of Sharing Economy Business Models. *Procedia: Social and Behavioral Sciences*, *213*, 836–841. doi:10.1016/j.sbspro.2015.11.486

Dillahunt, T. A., & Malone, A. R. (2015). The Promise of the Sharing Economy among Disadvantaged Communities. *CHI 2015 - Proceedings of the 33rd Annual CHI Conference on Human Factors in Computing Systems: Crossings,* 2285–2294.

DiMaggio, P. J., & Powell, W. W. (1983). The Iron Cage Revisited: Institutional isomorphism and collective rationality in organizational fields. *American Sociological Review, 48*(2), 147–160. doi:10.2307/2095101

Dolata, U. (2015). Volatile Monopole. Konzentration, Konkurrenz und Innovationsstrategien der Internetkonzerne. *Berliner Journal fur Soziologie, 24*(4), 505–529. doi:10.100711609-014-0261-8

Dörre, K. (2009). Die neue Landnahme. Dynamiken und Grenzen des Finanzmarktkapitalismus. In K. Dörre, S. Lessenich, & H. Rosa (Eds.), *Soziologie, Kapitalismus, Kritik. Eine Debatte* (pp. 21–86). Suhrkamp.

Edwards, R. C. (1979). *Contested Terrain.* Basic Books.

Fang, B., Ye, Q., & Law, R. (2015). Effect of sharing economy on tourism industry employment. *Annals of Tourism Research, 57,* 264–267. doi:10.1016/j.annals.2015.11.018

Firnkorn, J., & Müller, M. (2011). What will be the environmental effects of new free-floating carsharing systems? The case of car2go in Ulm. *Ecological Economics, 70*(8), 1519–1528. doi:10.1016/j.ecolecon.2011.03.014

Fligstein, N., & McAdam, D. (2012). *A Theory of Fields.* University Press. doi:10.1093/acprof:oso/9780199859948.001.0001

Friedman, A. (1987). Managementstrategien und Technologie. Auf dem Weg zu einer komplexenTheorie des Arbeitsprozesses. In E. Hildebrandt & R. Seltz (Hrsg.), Managementstrategien und Kontrolle. Eine Einführung in die Labour-Process-Debatte (pp. 99–131). Berlin: Edition sigma.

Gargiulo, E., Giannantonio, R., Guercio, E., Borean, C., & Zenezini, G. (2015). Dynamic ride sharing service: Are users ready to adopt it? *Procedia Manufacturing, 3,* 777–784. doi:10.1016/j.promfg.2015.07.329

Geron, T. (2012). Airbnb Had $56 Million Impact On San Francisco: Study. *Forbes.* https://www.forbes.com/forbes/welcome/?toURL=https://www.forbes.com/sites/tomiogeron/2012/11/09/study-airbnb-had-56-million-impact-on-san-francisco/&refURL=https://www.google.de/&referrer=https://www.google.de/

Glotz-Richter, M. (2012). Car Sharing – 'Car on call' for reclaiming street space. *Procedia: Social and Behavioral Sciences, 48,* 1454–1463. doi:10.1016/j.sbspro.2012.06.1121

Hall, J., & Krueger, A. (2015). An Analysis of the Labor Market for Uber's Driver-Partners in the United States. *IRS Working Papers, 587.*

Hamari, J., Sjöklint, M., & Ukkonen, A. (2015). The sharing economy: Why people participate in collaborative consumption. *American Society for Information Science and Technology Journal, 67,* 2047–2059.

Hartl, B., Hofmann, E., & Kirchler, E. (2015). Do we need rules for 'what's mine is yours'? Governance in collaborative consumption communities. *Journal of Business Research, 69*(8), 2756–2763. doi:10.1016/j.jbusres.2015.11.011

Hirsch-Kreinsen, H. (2015). Einleitung. Digitalisierung industrieller Arbeit. In H. Hirsch-Kreinsen, P. Ittermann, & J. Niehaus (Eds.), *Digitalisierung industrieller Arbeit* (pp. 9–30). Nomos. doi:10.5771/9783845263205-10

Hohendanner, C., & Walwei, U. (2013). Arbeitsmarkteffekte atypischer Beschäftigung. *WSI-Mitteilungen*, *66*(4), 239–246. doi:10.5771/0342-300X-2013-4-239

Keller, B., & Seifert, H. (2013). *Atypische Beschäftigung zwischen Prekarität und Normalität. Entwicklung, Strukturen und Bestimmungsgründe im Überblick*. Berlin: edition sigma.

Kern, H., & Schumann, M. (1984). *Das Ende der Arbeitsteilung? Rationalisierung in der industriellen Produktion: Bestandsaufnahme, Trendbestimmung*. C.H. Beck.

Kirchner, S., & Beyer, J. (2016). Die Plattformlogik als digitale Marktordnung. Wie die Digitalisierung Kopplungen von Unternehmen löst und Märkte transformiert. *Zeitschrift für Soziologie*, *45*(5), 324–339. doi:10.1515/zfsoz-2015-1019

Kleemann, F., & Voß, G. G. (2010). Arbeit und Subjekt. In F. Böhle, G. G. Voß, & G. Wachtler (Eds.), *Handbuch Arbeitssoziologie* (pp. 415–450). Springer. doi:10.1007/978-3-531-92247-8_14

Leimeister, J. M., Durward, D., & Zogaj, S. (2016). *Crowd Worker in Deutschland: Eine empirische Studie zum Arbeitsumfeld auf externen Crowdsourcing-Plattformen*. Hans-Böckler-Stiftung.

Light, A., & Miskelly, C. (2015). Sharing Economy vs. Sharing Cultures? Designing for social, economic and environmental good. *Interaction Design and Architecture(s). Journal*, *24*, 49–62.

Lim, K., Sia, C., Lee, M., & Benbasat, I. (2006). Do I Trust You Online, and If So, Will I Buy? An Empirical Study of Two Trust-Building Strategies. *Journal of Management Information Systems*, *23*(2), 233–266. doi:10.2753/MIS0742-1222230210

Marrs, K. (2010). Herrschaft und Kontrolle in der Arbeit. In F. Böhle, G. G. Voß, & G. Wachtler (Eds.), *Handbuch Arbeitssoziologie* (pp. 331–356). Springer. doi:10.1007/978-3-531-92247-8_11

Martin, C. J. (2016). The sharing economy: A pathway to sustainability or a nightmarish form of neoliberal capitalism? *Ecological Economics*, *27*, 149–159. doi:10.1016/j.ecolecon.2015.11.027

Maurer, A. (2004). Von der Krise, dem Elend und dem Ende der Arbeits- und Industriesoziologie. Einige Anmerkungen zu Erkenntnisprogrammen, Theorietraditionen und Bindestrich-Soziologien. *Soziologie*, *33*, 7–19.

Mayring, P. (2008). *Qualitative Inhaltsanalyse: Grundlagen und Techniken*. Beltz.

McArthur, E. (2015). Many-to-many exchange without money: Why people share their resources. *Consumption Markets & Culture*, *18*(3), 239–256. doi:10.1080/10253866.2014.987083

Minssen, H. (2006). Arbeits- und Industriesoziologie. New York: Campus.

Möhlmann, M. (2015). Collaborative consumption: Determinants of satisfaction and the likelihood of using a aharing economy option again. *Journal of Consumer Behaviour*, *14*(3), 193–207. doi:10.1002/cb.1512

Papsdorf, C. (2009). Wie Surfen zu Arbeit wird: Crowdsourcing im Web 2.0. New York: Campus.

Pfeiffer, S. (2010). Technisierung von Arbeit. In F. Böhle, G. G. Voß, & G. Wachtler (Eds.), *Handbuch Arbeitssoziologie* (pp. 231–261). Springer. doi:10.1007/978-3-531-92247-8_8

Pries, L., Schmidt, R., & Trinczek, R. (1990). *Entwicklungspfade von Industriearbeit. Chancen und Risiken betrieblicher Produktionsmodernisierung*. Westdeutscher Verlag. doi:10.1007/978-3-322-85577-0

Ranchordás, S. (2015). *Does Sharing Mean Caring? Regulating Innovation in the Sharing Economy*. Tilburg Law School Legal Studies Research Paper Series, 6.

Richardson, L. (2015). Performing the sharing economy. *Geoforum*, *67*, 121–127. doi:10.1016/j.geoforum.2015.11.004

Rolf, A., & Sagawe, A. (2015). *Des Googles Kern und andere Spinnennetze. Die Architektur der digitalen Gesellschaft*. UVK.

Rosenblat, A., & Stark, L. (2016). Algorithmic Labor and Information Asymmetries: A Case Study of Uber's Drivers. *International Journal of Communication*, *10*, 3758–3784.

Sauer, D., & Döhl, V. (1997). Die Auflösung des Unternehmens? Entwicklungstendenzen der Unternehmensreorganisation in den 90er Jahren. In Jahrbuch Sozialwissenschaftliche Technikberichterstattung 1996. Schwerpunkt: Reorganisation (pp. 19-76). Berlin: edition sigma.

Scaraboto, D. (2015). Selling, Sharing, and Everything In Between: The Hybrid Economies of Collaborative Network. *The Journal of Consumer Research*, *42*(1), 152–176. doi:10.1093/jcr/ucv004

Shaheen, S. A., Mallery, M. A., & Kingsley, K. J. (2012). Personal vehicle sharing services in North America. *Research in Transportation Business & Management*, *3*, 71–81. doi:10.1016/j.rtbm.2012.04.005

Statistisches Bundesamt. (2008). *Übersicht über die Gliederung der Klassifikation der Wirtschaftszweige*. Ausgabe 2008 (WZ2008).

Trinczek, R. (2011). Überlegungen zum Wandel von Arbeit. *WSI-Mitteilungen*, *11*(11), 606–614. doi:10.5771/0342-300X-2011-11-606

Voß, G. G. (2010). Arbeit als Grundlage menschlicher Existenz: Was ist Arbeit? Zum Problem eines allgemeinen Arbeitsbegriffs. In F. Böhle, G. G. Voß, & G. Wachtler (Eds.), *Handbuch Arbeitssoziologie* (pp. 23–80). Springer VS. doi:10.1007/978-3-531-92247-8_2

Weber, T. A. (2014). Intermediation in a Sharing Economy: Insurance, Moral Hazard, and Rent Extraction. *Journal of Management Information Systems*, *31*(3), 35–71. doi:10.1080/07421222.2014.995520

ADDITIONAL READING

Arcidiacono, D., Gandini, A., & Pais, I. (2018). Sharing what? The 'sharing economy' in the sociological debate. *The Sociological Review*, *66*, 275–288. doi:10.1177/0038026118758529

Arvidsson, A. (2018). Value and virtue in the sharing economy. *The Sociological Review*, *66*(2), 289–301. doi:10.1177/0038026118758531

Benjaafar, S., Kong, G., Li, X., & Courcoubetis, C. (2019). Peer-to-Peer Product Sharing: Implications for Ownership, Usage, and Social Welfare in the Sharing Economy. *Management Science*, *65*(2), 477–493. doi:10.1287/mnsc.2017.2970

Cheng, X., Fu, S., Sun, J., Bilgihan, A., & Okumus, F. (2019). An investigation on online reviews in sharing economy driven hospitality platforms: A viewpoint of trust. *Tourism Management*, *71*, 366–377. doi:10.1016/j.tourman.2018.10.020

Cohen, B., & Munoz, P. (2015). Sharing cities and sustainable consumption and production: Towards and integrated framework. *Journal of Cleaner Production*, *134*, 87–97. doi:10.1016/j.jclepro.2015.07.133

Hall, J., & Krueger, A. (2015). An Analysis of the Labor Market for Uber's Driver-Partners in the United States. IRS Working Papers, 587.

Hamari, J., Sjöklint, M., & Ukkonen, A. (2015). The sharing economy: Why people participate in collaborative consumption. American Society for Information Science and Technology. *Journal*, *67*, 2047–2059.

Hartl, B., Hofmann, E., & Kirchler, E. (2015). Do we need rules for 'what's mine is yours'? Governance in collaborative consumption communities. *Journal of Business Research*, *69*(8), 2756–2763. doi:10.1016/j.jbusres.2015.11.011

Jiang, B., & Tian, L. (2018). Collaborative Consumption: Strategic and Economic Implications of Product Sharing. *Management Science*, *64*(3), 1171–1188. doi:10.1287/mnsc.2016.2647

Lee, S., & Kim, D.-Y. (2018). The effect of hedonic and utilitarian values on satisfaction and loyalty of Airbnb users. *International Journal of Contemporary Hospitality Management*, *30*(3), 1332–1351. doi:10.1108/IJCHM-09-2016-0504

Lee, Z. W. Y., Chan, T. K. H., Balaji, M. S., & Chong, A. Y.-L. (2018). Why people participate in the sharing economy: An empirical investigation of Uber. *Internet Research*, *28*(3), 829–850. doi:10.1108/IntR-01-2017-0037

Lutz, C., & Newlands, G. (2018). Consumer segmentation within the sharing economy: The case of Airbnb. Journal of Business Research, 88, 187¬–196.

Martin, C. J. (2016). The sharing economy: A pathway to sustainability or a nightmarish form of neoliberal capitalism? *Ecological Economics*, *27*, 149–159. doi:10.1016/j.ecolecon.2015.11.027

McArthur, E. (2015). Many-to-many exchange without money: Why people share their resources. *Consumption Markets & Culture*, *18*(3), 239–256. doi:10.1080/10253866.2014.987083

Möhlmann, M. (2015). Collaborative consumption: Determinants of satisfaction and the likelihood of using a aharing economy option again. *Journal of Consumer Behaviour*, *14*(3), 193–207. doi:10.1002/cb.1512

Perren, R., & Kozinets, R. V. (2018). Lateral Exchange Markets: How Social Platforms Operate in a Networked Economy. *Journal of Marketing*, *82*(1), 20–36. doi:10.1509/jm.14.0250

Rosenblat, A., & Stark, L. (2016). Algorithmic Labor and Information Asymmetries: A Case Study of Uber's Drivers. *International Journal of Communication*, *10*, 3758–3784.

Scaraboto, D. (2015). Selling, Sharing, and Everything In Between: The Hybrid Economies of Collaborative Network. *The Journal of Consumer Research*, *42*(1), 152–176. doi:10.1093/jcr/ucv004

So, K. K. F., Oh, H., & Min, S. (2018). Motivations and constraints of Airbnb consumers: Findings from a mixed-methods approach. *Tourism Management*, *67*, 224–236. doi:10.1016/j.tourman.2018.01.009

Taeuscher, K., & Laudien, S. M. (2018). Understanding platform business models: A mixed methods study of marketplaces. *European Management Journal*, *36*(3), 319–329. doi:10.1016/j.emj.2017.06.005

Wood, A. J., Lehdonvirta, V., & Graham, M. (2018). Workers of the Internet unite? Online freelancer organisation among remote gig economy workers in six Asian and African countries. *New Technology, Work and Employment*, *33*(2), 95–112. doi:10.1111/ntwe.12112

KEY TERMS AND DEFINITIONS

Dimensions of Work: Work is an abstract concept which comprises dimensions that build typical patterns. Central dimensions analyzed here are the type of activity, the form of compensation or recompense involved (monetary or non-monetary), skills and competencies (qualification) required, the role of technology, and control.

Gift and Exchange Economy: This segment of the SE best embodies the original and authentic notion of sharing, where resources are made over to other users, or offered temporarily, on ecological, altruistic, or other non-economic grounds. This can take the form of lending free-of-charge, donation, or swapping, with no need for there to be any equivalent in terms of value or costs incurred.

Niche and Sideline Economy: Actors in this segment of the SE are predominantly 'lay' people and amateurs who seek a monetary reward. This can take the form of lending or selling physical goods or offering basic trade skills or household services, such as plumbing or electrical repairs, for payment.

Platform Economy or Digitized Old Economy: This segment of the SE consists of initiatives in which services are provided by individuals who are formally economically active (as employees or self-employed). Service providers lend resources and perform simple, but possibly also complex and demanding, services in return for monetary reward, typically involving resources and services already available on the market.

Sharing Economy: The sharing economy is regarded a distinct area of social life, which comprises social activities mediated by internet-based platforms and market places that connect suppliers and customers offering and buying a wide range of products and services through processes that operate outside the conventions of traditional business models.

Work: Work is defined here in a wide sense as a purposive social practice that entails expending time and effort, while the form of compensation involved can be monetary or non-monetary. Thus, work does not necessarily have a commercial purpose (as 'employment').

ENDNOTES

[1] Examples include 'collaborative consumption', 'the mesh', 'commercial sharing systems', 'peer-to-peer economy', 'gig economy' und 'collaborative economy'. 'SE' has now established itself as the overarching denotation (Hamari et al. 2015).

[2] Hirsch-Kreinsen (2015), who draws on socio-technical systems theory, is an exception to this but focuses principally on the workplace level and the recent application of so-called 'Industry 4.0' technologies.

[3] These include where and when work is to be performed, gender differences, work demands, occupational issues, political control, and workplace social integration.

[4] For example, methods such as qualitative interviews or participant-observation research would be needed to obtain reliable results on work demands, average working hours or forms of 'debordering'.

[5] JustPark itself in an SE platform that also maintains a comprehensive listing of SE platforms at www.justpark.com/creative/sharing-economy-index.

[6] These websites offer an informative description of the service, FAQs, terms and conditions of service, some best practice examples, together with what the platform provides and scope for making applications.

[7] Rolf and Sagawe (2015) use the term 'platform economy' in the context of a 'digital society'. The logic of this matches the approach adopted here: providers and requesters are coordinated via new platforms (often independent of time and place). Nevertheless, the study is not concerned with the diversity of platform-based business models (such as crowdfunding) and media (such as social media) but concentrates on those platforms that can be assimilated to the SE.

[8] Based on typical schemas for analysing work, what is immediately apparent is that some dimensions cannot sensibly be applied to SE activities. This is especially the case for the gift and exchange economy, in which providers often act on altruistic motives. Moreover, the range of activities is typically concentrated on less neatly delineable tasks, such as posting up offers or assigning goods that are lent or given to peers. As a consequence, it is not appropriate to investigate pay (but not other forms of recompense), job security, or scope for employment involvement or training.

[9] This website has ceased operations since the research was carried out.

Chapter 10
The Role of Big Data and Business Analytics in Decision Making

Pedro Caldeira Neves
Polytechnic of Coimbra, Portugal

Jorge Rodrigues Bernardino
https://orcid.org/0000-0001-9660-2011
Polytechnic of Coimbra, Portugal

ABSTRACT

The amount of data in our world has been exploding, and big data represents a fundamental shift in business decision-making. Analyzing such so-called big data is today a keystone of competition and the success of organizations depends on fast and well-founded decisions taken by relevant people in their specific area of responsibility. Business analytics (BA) represents a merger between data strategy and a collection of decision support technologies and mechanisms for enterprises aimed at enabling knowledge workers such as executives, managers, and analysts to make better and faster decisions. The authors review the concept of BA as an open innovation strategy and address the importance of BA in revolutionizing knowledge towards economics and business sustainability. Using big data with open source business analytics systems generates the greatest opportunities to increase competitiveness and differentiation in organizations. In this chapter, the authors describe and analyze business intelligence and analytics (BI&A) and four popular open source systems – BIRT, Jaspersoft, Pentaho, and SpagoBI.

INTRODUCTION

New advances of Information and Communication Technologies (ICT) continue to rapidly transform how business is done and change the role of information systems in business and our daily life. The amount of data in our world has been overwhelmingly exploding. Enterprises are flooded with ever-growing data of all types, easily amassing terabytes and even petabytes, of data. Analyzing such so-huge amounts of

DOI: 10.4018/978-1-7998-5849-2.ch010

data is a keystone of competition, synonym for productivity growth, and innovation. With the emergence of new data collection technologies and analytical tools, Big Data offers a way to make businesses more agile, and to answer questions that were previously considered beyond reach. Increasing competition, demand for profits, contracting economy, and savvy customers all require companies and organizations to make the best possible decisions. With the fast advancement of both business techniques and technologies in recent years, knowledge has become an important and strategic asset that determines the success or failure of an organization (Wit & Meyer, 2010). Studies show that a competitive advantage in the business environment depends on the accessibility to adequate and reliable information in shortest time possible – sometimes even in real-time – and the high selectivity in the creation and use of information. An effective instrument to create, aggregate and share knowledge in an organization has therefore become a key target for management.

The need to implement decision support systems in organizations is not only an unavoidable reality but also a prerequisite for todays' companies (Arsham, 2015). Currently, the majority of organizations have platform services, designed to record and store massive amounts of data resulting from the operational activity (Wang et al., 2019). This dataset is then transformed in information and all that information will lead to knowledge useful for the organizations.

The terms Business Intelligence (BI) and Business Analytics (BA) have been widely used in various contexts, but there seems to be no commonly accepted definition of what, at least BA is. Hindle et al., (2019) define BA in the same way that Davenport and Harris (2007) defines BI, which is "the extensive use of data, statistical and quantitative analysis, explanatory and predictive models, and fact-based management to drive decisions and actions".

On this study, we follow a terminology defended by a few researchers, who call this area BI&A which stands for Business Intelligence and Analytics and represents a merge of Business Intelligence techniques and systems with a Business Analytics strategy (Torres et al., 2018).

In addition, in a competitive environment, traditional decision-making approaches no longer meet the requirements of organizations for decision-making; organizations must make good use of information system tools such as BI&A systems to quickly acquire desirable information from huge volume of data to reduce the time and increase the efficiency of decision-making procedure. Different researchers have different definitions for business intelligence and analytics systems, for example Sharda et al., (2014) defined these as "an umbrella term that encompasses tools, architectures, databases, data warehouses, performance management, methodologies, and so forth, all of which are integrated into a unified software suite".

Business Intelligence and Analytics is one of the few forms of sustainable competitive advantage left (Burstein & Holsapple, 2008). For example, any two well-funded competitors in a market have near real access to capital, technology, market research, customer data, and distribution. People and the quality of the decisions that they make are the primary competitive differentiators in the Information Age (Lin et al., 2009). The implementation of an effective BI&A strategy is the key to sustaining long-term competitive advantage as so, is normally driven by the top management as a whole instead of just the information technology department as usual.

Several studies have shown how BI&A investments impact enterprise performance (Albright & Winston, 2016; Evans, 2019). In the earlier steps of BI&A implementation, the selection of the most convenient strategy is very important: organizations implementing a BI&A strategy and solutions need to consider several factors such as data management and quality policies, data retention, data frequency and metadata stored required. Afterwards, an architecture to cope with the data strategy agreed needs to be defined and detailed upon the functionalities that each software provide; how each system fit within the organization's data model; and how expensive is the total cost of ownership (total cost of acquiring and implementing a BI&A solution). This is critical for organizations which would like to enter into the new market and operate on a global scale. Thus, open source BI&A systems can trigger immense possibilities of accelerating knowledge acquisition, intensifying entrepreneurship development and improving business skills, therefore, leading to business sustainability. In this context, open source BI&A can be seen as another form of open innovation, which can be used by business communities. In this work, we address the importance of Business Intelligence and Analytics in revolutionizing knowledge towards economics and business sustainability and describe the top key systems to implement an effective open source BI&A in organizations: BIRT, Jaspersoft, Pentaho, and SpagoBI, improving our previous paper (Bernardino & Neves, 2019).

The rest of the chapter is organized as follows. First, we describe the problem of growing data volumes that organizations have to deal with. Second, we introduce the concept of BI&A and address its importance in revolutionizing knowledge to enhance organization's response in making better and more efficient business decisions, also increasing innovation. Some BI&A resources are also introduced and the advantages of using the open source model are discussed. After, we present the top BI&A software vendors to implement open BI&A within organizations. Some future research directions are also pointed out. Finally, the concluding remarks are presented in conclusions section.

BIG DATA AND DECISION MAKING

Business Intelligence and Analytics provides a set of methodologies, processes, architectures, and technologies, which transform raw data into valuable information to enable more effective strategic, tactical, operational insights and decision-making (Evelson & Norman, 2008). As we know, BI&A applications used to gather data from a data warehouse or a data mart (Bernardino & Madeira, 2001; Bernardino et al. 2002). However, the situation begun to change a few years back. In recent years, with the widespread use of use of wiki and the popularity of micro blog and other Web 2.0 applications for business, there has been an explosive increase in the amount of data in different types of enterprises, which even exceeds the rate of Moore's Law. For example, Walmart, one of the famous worldwide supermarkets, collects more than 2.5 petabytes (PB) of data every hour from its customer transactions, and it also has related 40 Billion photos held by Facebook alone in order to facilitate the marketing.

The data explosion is globally recognized as a key Information Technology (IT) concern. Structured information in databases is just the tip of the iceberg; some important milestones document the explosive growth of unstructured data as well: According to market intelligence company IDC, the 'Global Datasphere' in 2018 reached 18 zettabytes. This is the total of all data created, captured or replicated. The vast majority of the world's data has been created in the last few years and this astonishing growth of data shows no sign of slowing down. In fact, IDC predicts the world's data will grow to 175 zettabytes in 2025 (Gantz & Reinsel, 2018).

Complex business processes are increasingly expected to be executed through a variety of interconnected systems. Integrated sensors and probes not only enable continuous measurement of operational performance, but the interconnectedness of many systems allows rapid communication and persistence of those measures.

Today, more than 5 billion consumers interact with data every day – by 2025, that number will grow up to 6 billion. This exponential growth is due to the billions of IoT devices connected across the globe, which are expected to create over 90 zettabytes of data in 2025 (Gantz & Reinsel, 2018).

We can conclude from these few examples of rapid expansion of the amount of digital information that exciting new views can be opened up as never before through combined analysis of structured and unstructured data. New and improved means of data analysis allow organizations to identify new business trends, innovate, assess the spread of disease, or even fight crime, among many other opportunities. It appears that we are reaching the point where information is becoming the most significant focus of the business, where "statisticians mine the information output of the business for new ideas" (Bates et al., 2009). And these new ideas are not just lurking in structured databases. Rather the analysis must encompass unstructured data artefacts as well.

Nevertheless, as the volume of data grows, so does the complexity of finding those critical pieces of information necessary to make those business processes run at their optimized level. The issue is no longer the need to capture, store, and manage that data. Rather, the challenge is distilling out and delivering the relevant pieces of knowledge to the right people at the right to time to enhance the millions of opportunities for decision-making that occur on a daily basis.

Undoubtedly, Big Data brings big opportunities. The integration and analysis of Big Data can help enterprises glean deeper insights into the internal and external forces that affect their performance, anticipate development trends, and respond more quickly to changes. Until now, more and more companies have recognized that there is a lot of treasure contained in these huge datasets, indicating that Big Data adoption goes main stream in enterprises.

Big Data poses Big challenges, including scalability and storage bottleneck, noise accumulation, spurious correlation, and error measurement. These challenges and opportunities associated with Big Data necessitate rethinking many aspects of data management software.

In essence, this can be summarized as the desire to integrate BI&A in a pervasive manner into both the strategic and the operational processes across all functions and levels of the organization and whether this means notifying senior management of emerging revenue opportunities, providing real-time insight into corporate performance indicators, or hourly realignment of field repair team schedules to best address customer service outages. The ability to accumulate, transform, and analyze information to provide rapid, trustworthy analyses to the right people at the right time can enhance growth opportunities and competitiveness, leading to a sustained open business model of value chain. All this can be done selecting the most appropriate BI&A system to the organization.

BUSINESS INTELLIGENCE AND ANALYTICS

Business Intelligence was first introduced in 1989 by Howard Dresner who defined it as "a set of concepts and appropriate methods to support decision making, using the data provided by support systems to business process" (Power, 2007). However, some authors' claim Dresner re-appropriated the term to rebadged what was then called DSS-Decision Support Systems. They say H.P. Luhn actually invented the term Business Intelligence, not in 1989, but in 1958 in an IBM Journal article that pretty accurately predicts BI&A systems today. The original definition of Business Intelligence from (Luhn, 1958) is: "business is a collection of activities carried on for whatever purpose, be it science, technology, commerce, industry, law, government, defence, et cetera. The communication facility serving the conduct of a business (in the broad sense) may be referred to as an intelligence system". The notion of intelligence is also defined here, in a more general sense, as "the ability to apprehend the interrelationships of presented facts in such a way as to guide action towards a desired goal".

Perhaps the first probable reference to BI&A was made in Sun Tzu's "Art of War" (Tzu, 1963), who was born five centuries BC, where he claimed that to succeed in war, full knowledge on one's strengths and weaknesses as well as the strengths and weaknesses of the enemy must be known. Applying this to the modern business world, BI&A becomes the art of wading and sieving through tons of data, and presenting the overloaded data as information that provides significant business value in improving the effectiveness of managerial decision-making (Camm et al., 2020). As such, BI&A is carried out not just for gaining sustainable competitive advantages, but it also has a valuable core competence in most instances. A variety of businesses have used BI&A for activities such as customer support and service, customer profiling, market research and segmentation, product profitability, inventory and distribution analysis, etc.

The concept of "business intelligence and analytics" is mainly strategical and include tools and techniques supporting a collection of user communities across an organization, as a result of collecting and organizing numerous (and diverse) datasets to support both management and decision making at operational, tactical, and strategic levels. Through data collection, aggregation, analysis, and presentation, BI&A can be delivered to best serve a wide range of target users. Organizations that have matured their data warehousing programs allow those users to extract actionable knowledge from the corporate information asset and rapidly realize business value (Santos et al., 2011).

But while traditional data warehouse infrastructures support business analyst querying and canned reporting or senior management dashboards, a comprehensive program for information insight and intelligence can enhance decision-making process for all types of staff members in numerous strategic, tactical, and operational roles. Even better, integrating the relevant information within the immediate operational context becomes the differentiating factor. Offline customer analysis providing general sales strategies is one thing, but real-time business intelligence and analytics can provide specific alternatives to the sales person talking to a specific customer based upon customer's interaction history in ways that best serve the customer while simultaneously optimizing corporate's profitability. Maximizing overall benefit to all parties involved, ultimately improves sales, increases customer and employee satisfaction and improves response rate while reducing the cost of goods sold – a true win-win for everyone.

The wide ranges of analytical capabilities all help suggest answers to a series of increasingly valuable questions (Loshin, 2011):

- *What?* – Predefined reports will provide the answer to the operational managers, detailing what has happened within the organization and various ways of slicing and dicing the results of those queries to understand basic characteristics of business activity (e.g., counts, sums, frequencies, locations, etc.). Traditional BI&A reporting provides 20/20 hindsight – it tells what has happened, it may provide aggregate data about what has happened, and it may even direct individuals with specific actions in reaction to what has happened.
- *Why?* – More comprehensive *ad hoc* querying coupled with review of measurements and metrics within a time series enables more focused review. Drilling down through reported dimensions lets the business client get answers to more pointed questions, such as the sources of any reported issues, or comparing specific performance across relevant dimensions.

Figure 1. A range of techniques benefits a variety of consumers for analytics

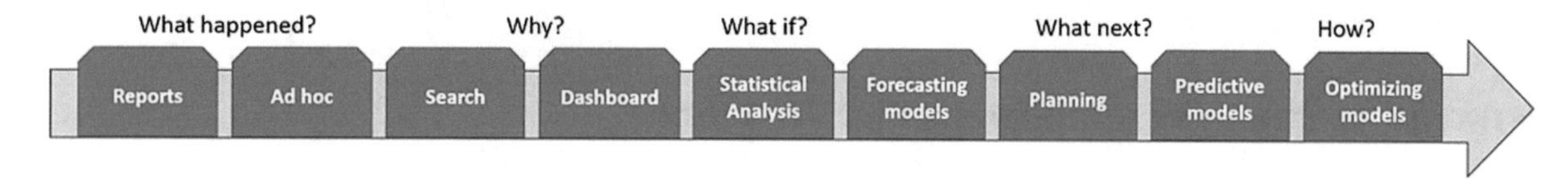

- *What if?* – More advanced statistical analysis, data prediction models, machine learning, and forecasting models allow business analysts to consider how different actions and decisions might have impacted the results, enabling new ideas for improving the business.
- *What next?* – By evaluating the different options within forecasting, planning, and predictive models, senior strategists can weigh the possibilities and make strategic decisions.
- *How?* – By considering approaches to organizational performance optimization, the senior managers can adapt business strategies that change the way the organization does business.

Information analysis makes it possible to answer these questions. Improved decision-making processes depend on supporting business intelligence and analytics capabilities that increase in complexity and value across a broad spectrum for delivering actionable knowledge (as shown in Figure 1). As the analytical functionality increases in sophistication, the business client can gain more insight into the mechanics of optimization. Statistical analysis will help in isolating the root causes of any reported issues as well as provide some forecasting capabilities should existing patterns and trends continue without adjustment. Predictive models that capture past patterns help in projecting "what-if" scenarios that guide tactics and strategy towards organizational high performance.

Intelligent analytics and business analytics are maturing into tools that can help optimize the business. That is true whether those tools are used to help (Loshin, 2011):

- C-level executives (CEO, CIO, CFO, …) review options to meet strategic objectives;
- Senior managers seeking to streamline their lines of business;
- Operational decision-making in ways never thought possible.

These analytics incorporate data lake, warehousing, data science, multidimensional analysis, streams, and mash-ups to provide a penetrating vision that can enable immediate reactions to emerging opportunities while simultaneously allowing one to evaluate the environment over time to discover ways to improve and expand the business.

Business Intelligence, Analytics and Decision Making

A BI&A systems allow an organization to gather, store, access and analyze corporate data to aid in decision-making. Generally, these systems will illustrate business analytics in the areas of customer profiling, customer support, market research, market segmentation, product profitability, statistical analysis, and inventory and distribution analysis to name a few.

Most companies collect a large amount of data from their business operations. To keep track of that information, a business would need to use a wide range of software programs, such as Excel, Microsoft Access and different database applications for various departments throughout their organization. Using multiple software programs makes it difficult to retrieve information in a timely manner and to perform analysis of the data.

Benefits of Business Intelligence and Analytics

Initially, BI&A reduces IT infrastructure costs by eliminating redundant data extraction processes and duplicate data housed in independent data marts across the enterprise. For example, 3M justified its multimillion-dollar data warehouse platform based on the savings from data mart consolidation (Watson et al. 2004).

BI&A also saves time for data suppliers and users because of more efficient data delivery. End users ask questions like "What has happened?" as they analyze the significance of historical data. This kind of analysis generates tangible benefits like headcount reduction that are easy to measure; however, these benefits typically have local impact (Lapa et al., 2014).

Over time, organizations evolve to questions like "Why has this happened?" and even "What will happen?" As business users mature to performing analysis and prediction, the level of benefits become more global in scope and difficult to quantify. For example, the most mature uses of BI&A might facilitate a strategic decision to enter a new market, change a company's orientation from product-centric to customer-centric, or help launch a new product line.

The spectrum of BI&A benefits can be summarized in the following (Torres et al., 2018):

- Cost savings consolidation;
- Time savings for data suppliers;
- Time savings for users;
- More and better information;
- Better decisions;
- Improvement of business processes;
- Support for the accomplishment of strategic business objectives.

However, success with BI&A isn't automatic and organizations are more likely to be successful when certain facilitating conditions exist. According to (Cheng et al, 2018) the top Roadblocks to BI success is the fact that the traditional BI&A strategy focused on technical profiles instead of encompassing all the organization. In the next section, we explain some of the BI&A available resources.

BI&A RESOURCES

Resources on BI&A are freely available on the Web for both practitioners and academics. Here, we describe four well-known resources: Coursera, Teradata University Network and TDWI-The Data Warehousing Institute.

Coursera

Coursera (https://www.coursera.org/) is the most known multi-language cross e-learning platform that provides both free and paid online courses taught by several universities. The platform provides several video contents as well as reading material, exercises and sandbox environments so that students can deep dive on the course.

Coursera also provide certified specialization courses and integrate with several other platforms such as LinkedIn, allowing its students not only to have access to a variety of materials from the courses, but naturally to make connections within the business sector, making easier for them to get a job and for head-hunters to recruit more assertively.

By 2019, Coursera now holds roughly 20 million students over a wide of 3200 courses within 1800 business areas and 310 specializations from management, to communication; from data management and big data to BI&A, data science and Artificial Intelligence. Coursera is undoubtedly an authority regarding sharing BI&A knowledge both through academia and business areas and accounts the most majority of online/e-learning courses.

TUN - Teradata University Network

Teradata University Network (www.academics.teradata.com) is a web-based portal for faculty and students in big data, data warehousing, business analytics/decision support, and database that is provided at no cost. This teaching portal is composed of practitioners and renowned academics recruited from around the world, united by their dedication to sharing innovative, proactive applications of authentic technology for data-driven decisions (Teradata, 2019).

The portal also support introduction to IT courses at the undergraduate and graduate levels. Teradata University Network (TUN) offers various free BA-related resources: course syllabi used by other faculty, book chapters, articles, research reports, cases, projects, assignments, PowerPoint presentations, software, large data sets, Web seminars, Web-based courses, discussion forums, podcasts, and certification materials.

Via the companion Teradata Student Network web site, professors design assignments using databases with millions of records — all without the issue of scale. TUN also provides students with resources that would normally be prohibitively expensive for a university to develop and maintain, thereby creating opportunities to learn realistic business analytics using Teradata data warehousing technology. The software such as the Teradata database is available through an application service provider arrangement so that schools do not have to install and maintain BA software.

The Teradata University Network currently has more than 5,000 registered faculty members, from more than 2,500 universities, in more than 100 countries, with thousands of student users, according to its web site.

TDWI - The Data Warehousing Institute

TDWI (The Data Warehousing Institute), a division of 1105 Media, is the premier provider of in-depth, high-quality education and research in the business analytics, data warehousing, and analytics industry according to the information in its web site (TDWI, 2019). TDWI provides BI&A training, research, and networking opportunities to its members around the world.

Starting in 1995 with a single conference, TDWI is now a comprehensive resource for industry information and professional development opportunities. TDWI offers vendor-neutral educational opportunities at quarterly conferences, on-site classes, and regional events and through its web site. Each year, TDWI sponsors a BI&A best practices competition that recognizes organizations that have achieved significant success in BI&A. TDWI offers five major World Conferences, topical seminars, onsite education, a worldwide membership program, business analytics certification, live Webinars, resourceful publications, industry news, an in-depth research program, and a comprehensive Web site: tdwi.org. The TDWI Business Analytics publications provide fresh ideas and perspectives to help organizations operate more intelligently. They also provide actionable insight on how to plan, build, and deploy business analytics and data warehousing solutions.

ADVANTAGES OF OPEN SOURCE MODEL

The economic shift occurred in the past decades, introduced with the rise of the Internet. The Internet provides connectivity, which in turn connects demand to supply; this removes the capital equation and shifts conditions in the market. People with computers all over the world provide the means of conduction, and with easy ways for these people to connect they can easily collaborate on projects with very little infrastructure. This, in turn, led to the popularity of Open Source Model.

A variety of authors (e.g., Pollock, 2009; Lakhani & von Hippel, 2003; Marinheiro & Bernardino, 2015; Bessen, 2006) have pointed out that an open-source approach may offer substantial efficiency advantages – for example by allowing users to participate directly in adding features and fixing bugs – and this is particularly true where the information good is complex and transaction costs are high.

Open Source Model has had many consequences for the enterprise software market, primarily by making it a commodity market and by driving down price. It also allows people who have a desire to do something to make money from doing it. Production inputs are widely distributed and the raw materials are there to produce things in new ways. Many argue that this shift has been about software licensing, but actually it is largely about the production and distribution of software; it is therefore a change that is more about economics than about ideology. That economic base – the combination of connectivity, of easier ways to market, produce and distribute software, and the fact that the Internet acts as a giant copying machine – changes the conditions under which software can be sold. That is why Open Source continues to disrupt the market and to force vendors to come up with new ways to cope with it. Enterprise software is not going away; rather, it has been forced to shift.

A 2008 survey by North Bridge Venture Partners (Skok, 2008) asked: "Which sector of the software industry is most vulnerable to disruption by Open Source?" At the top of the resulting list was web publishing and content management, a market that's currently highly fragmented and therefore perfect for introducing Open Source products. The second sector was social software, an entirely new category that has been difficult to define, and difficult currently for commercial vendors to make money from; again, Open Source Model is the ideal solution.

Business intelligence was the third sector listed in the North Bridge survey. This is understandable, since the majority of BI and big data systems are exactly that: tools. While end users consume the outputs, they're not themselves often put into the hands of end users; rather, IT sets up the meta-data, defines metrics, etc. Which means Open Source is targeted at the right audience; there is a platform in a typical data architecture that has multiple layers, any one of which can be Open Source or proprietary. Those modular boundaries make Open Source an ideal fit.

Why Open Source?

In addition to examining the sectors left vulnerable by Open Source, the North Bridge Venture survey also asked (Skok, 2008): "What makes Open Source attractive?" Price was the most popular answer. In addition to the capital cost of licensing, Open Source drives down acquisition cost, as companies no longer have to go through heavy proof of concept. It's also easier to trial things, and if license costs are less then maintenance costs are also less. Finally, unbundling maintenance support and service means that businesses can choose from a menu of services instead of being forced into the traditional approach of a 20 percent single lump fee.

The study also showed that companies found that Open Source offered freedom from vendor lock-in. In addition, it is flexible, with easy access to commodity code, in order to develop, embed and build content into websites.

Another 2008 study, by The 451 Group (2008), gave a list of what organizations saw as benefits to Open Source once it was already introduced. At that point, flexibility came out over cost. In this instance, though, the definition of flexibility was different; respondents were impressed by the fact that there is no vendor that dictates when they have to upgrade, how they have to upgrade, and why they have to upgrade. For a commercial vendor at end of life or desupporting versions of their software, a sales and marketing schedule will often require a release every Q1, whether they are ready or not. As a result, companies will be forced into an upgrade they don't want, that was released simply because marketing dictated it and not necessarily because it was ready. Open Source is typically not under those same types of pressures.

Buyers were also impressed by scaling costs; if an organization has 50 initial licenses and pushes it out to 150 more people, in BI&A lowered costs come in beyond the initial capital costs. As well, companies are not dependent on vendors for service or for additional pieces of technology.

However, misconceptions still exist around Open Source, based largely on the idea that these projects are being developed by teenagers in their parents' basement; a myth that is more than a decade outdated, as today Open Source projects are generally run by commercial Open Source providers with paid developers. Still, a lack of information can prove a problem for organizations introducing Open Source. This often plays out in the legal department, which reviews all contracts and may not be informed when it comes to Open Source software licenses. An Open Source license will be missing half the standard clauses the lawyers are used to; clauses of restriction that dictate how to deploy, what can be deployed, where it can be deployed, and how the software can be used. Most Open Source licenses do not have these same clauses and lawyers will want to know why they're missing. That can be another road bump in getting the software acquired.

Support can also be more confusing. Most major projects have commercial vendor support behind them. But unbundling is also available, which means buyers don't have to purchase support, but instead can buy the subscription model that covers only bug fixes, integrated patch testing and certification against databases. This is because enterprise support is often overrated; Moreover, not internal support solves more problems than the vendor does.

Chesbrough & Crowther (2006) also proclaim the benefits of openness and at the same time it is closely related to entrepreneurship by focusing on new knowledge creation, too. The authors support the idea of transferring the open source development philosophy to entrepreneurship, i.e. the concept of open innovation. Open innovation at its core is the increasing usage of external sources for creating and developing new ideas, which lead to innovation. In contrast to a closed innovation paradigm, companies try to include customers, users, universities and even competitors in different stages of their new product development processes (Chesbrough, 2003). As every open source software innovation process is based on the desire of integrating external knowledge, it is not astonishing that open source development is often referred to the open innovation paradigm (Dahlander & Magnusson, 2005; West & Gallagher, 2006; West & Lakhani, 2008) and extends open business strategies.

OPEN BUSINESS ANALYTICS SOLUTIONS

In this section, we present the BI&A systems from the vendors that follow the open source model. We describe the top open source BI&A tools: BIRT, Jaspersoft, Pentaho, and SpagoBI. After, we evaluate the tools using OSSpal methodology (Wasserman et al., 2017).

BIRT

Actuate Corporation, which was now sold to OpenText (OpenText, 2019) was founded in 1993 and had been selling commercial BA software since its formation. In 2004, Actuate became the first public BI&A software company to enter the open-source space by petitioning for membership into the Eclipse Foundation and proposing a new open XML report design specification and accompanying toolset.

The Business Intelligence and Analytics Tool (BIRT), founded and co-lead back then by Actuate, is based on the Eclipse framework. Under the BIRT project, Actuate made a small subset of its commercial enterprise reporting solution available under an open-source license.

BIRT offers a variety of products and services, supporting users through an annual subscription model for the open-source version. As such, the BIRT offering must be considered an attempt to attract highly technical do-it-yourself audiences and get a "foot in the door". Customers looking for the full functionality BI&A platform are required to upgrade to the commercial license.

BIRT's advanced and highly interactive reporting functionality, made highly scalable and function-rich, propelling BIRT to the leadership position. BIRT is also expanding its offering into end-to-end document management (or ILM — Information Life-cycle Management) capabilities. Producing reports often starts a report life cycle, where a report needs to be distributed, stored, secured, and archived.

Eclipse's BIRT project is a flexible, open source, and 100% pure Java reporting tool for building and publishing reports against data sources ranging from typical business relational databases, to XML data sources, to in-memory Java objects. BIRT is developed as a top-level project within the Eclipse Foundation and leverages the rich capabilities of the Eclipse platform and a very active open source community of users. Using BIRT, developers of all levels can incorporate powerful reporting into their Java, J2EE and Eclipse-based applications. Figure 2 shows an example of a report.

BIRT has two main components: a report designer based on Eclipse, and a runtime component that we can use to integrate BIRT reports into our applications. BIRT also has a charting engine that lets include charts into BIRT reports or add standalone charting capabilities to Java applications.

BIRT has also the Spreadsheet Designer (see Figure 3), an Excel-like authoring environment that gives report developers and Excel experts the power to automate the task of creating and updating Excel spreadsheets with the latest data.

BIRT Spreadsheet automates and centralizes spreadsheet production, maintenance, archiving, and security, eliminating version discrepancies and curbing the proliferation of multiple silos of Excel workbook data.

Figure 2. BIRT report design
(source: https://www.eclipse.org/birt/)

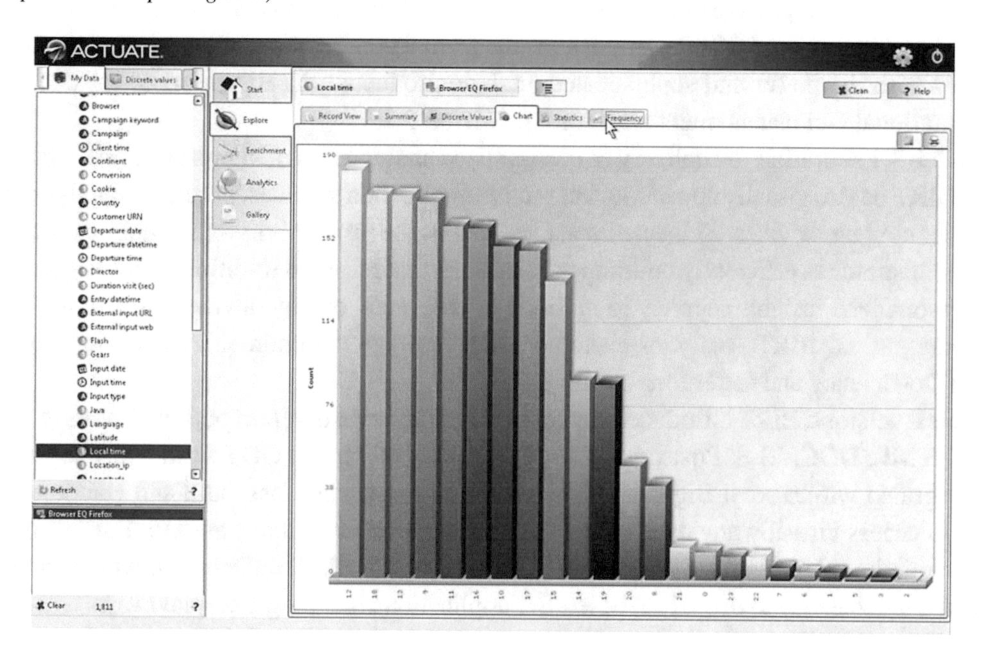

Figure 3. BIRT Spreadsheet Designer
(source: https://www.eclipse.org/birt/)

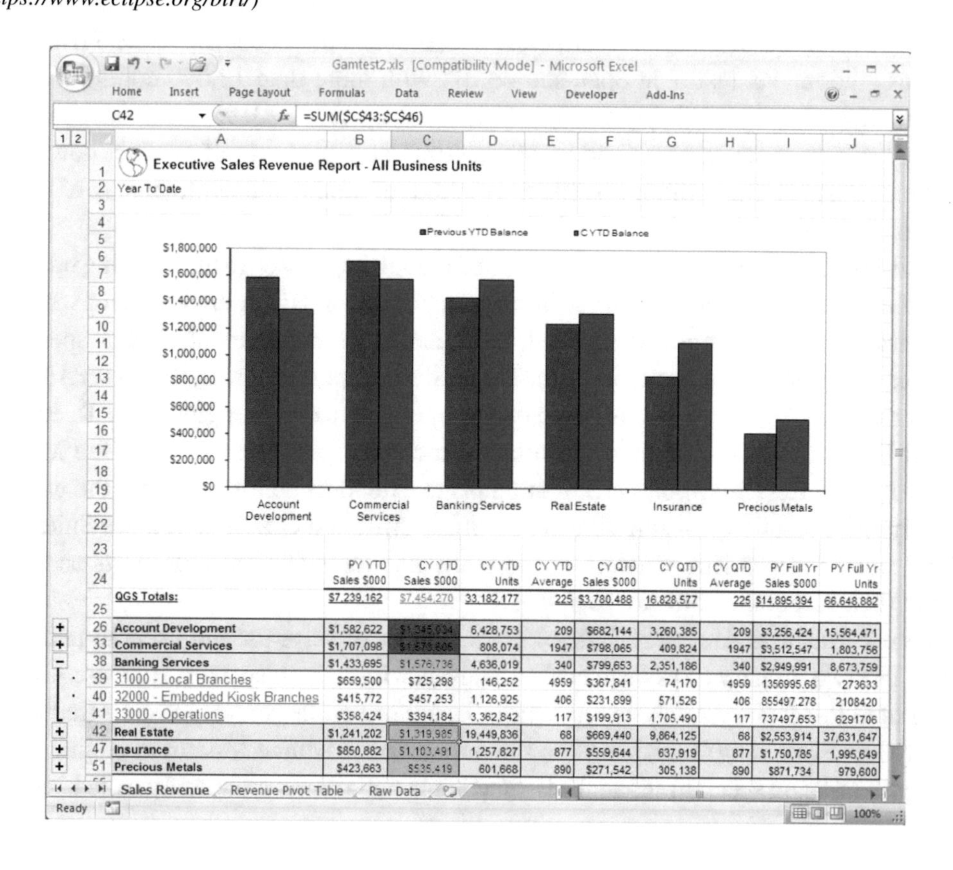

	PY YTD Sales $000	CY YTD Sales $000	CY YTD Units	CY YTD Average	CY QTD Sales $000	CY QTD Units	CY QTD Average	PY Full Yr Sales $000	PY Full Yr Units
QGS Totals:	$7,239,162	$7,454,270	33,182,177	225	$3,780,488	16,828,577	225	$14,895,394	66,648,882
Account Development	$1,582,622	$1,345,034	6,428,753	209	$682,144	3,260,385	209	$3,256,424	15,564,471
Commercial Services	$1,707,098	$1,578,806	808,074	1947	$798,065	409,824	1947	$3,512,547	1,803,756
Banking Services	$1,433,695	$1,576,736	4,636,019	340	$799,653	2,351,186	340	$2,949,991	8,673,759
31000 - Local Branches	$659,500	$725,298	146,252	4959	$367,841	74,170	4959	1356995.68	273633
32000 - Embedded Kiosk Branches	$415,772	$457,253	1,126,925	406	$231,899	571,526	406	855497.278	2108420
33000 - Operations	$358,424	$394,184	3,362,842	117	$199,913	1,705,490	117	737497.653	6291706
Real Estate	$1,241,202	$1,319,985	19,449,836	68	$669,440	9,864,125	68	$2,553,914	37,631,647
Insurance	$850,882	$1,103,491	1,257,827	877	$559,644	637,919	877	$1,750,785	1,995,649
Precious Metals	$423,663	$535,419	601,668	890	$271,542	305,138	890	$871,734	979,600

BIRT Analytics, a visual data mining and predictive analytics tool for identifying customer insights within the organization, provides business users with fast, free-form visual data mining and predictive analytics. The uniqueness of BIRT Analytics results from a combination of the ease of use of data discovery tools with the power and sophisticated analytic products typically reserved for data scientists, and the operational and management reach of BIRT iHub 3.1.

BIRT iHub 3.1 simplifies the delivery of personalized analytics and insights via a single platform that integrates BIRT based, visually appealing, interactive application services, predictive analytics services, and customer content services. This release also enhances the productivity of the application developer and the administrator in efficiently building applications that leverage traditional and Big Data sources to serve personalized insights securely to millions of end users, on any device at any time. With BIRT iHub 3.1 integration, BIRT Analytics customers can leverage common shared services, resulting in improved IT efficiency and faster time to value.

In the latest versions, BIRT introduced support for a broad range of report output formats (e.g.: HTML, paginated HTML, DOC, XLS, Postscript, PPT, PDF, ODP, ODS, and ODT formats). The software can now be integrated with several Big Data platforms such as Apache Cassandra and Hadoop, providing the ability to access virtually any data source that is structured or contains an API. Furthermore, BIRT can now be purchased as platform as a service (PaaS) through BIRT onDemand, which came provide an easy and available way to use BIRT-based BA applications in a plug-and-play way.

Jaspersoft

Jaspersoft is a well-established brand in the open source BA market. Founded in 2001, the company was bought in 2014 by TIBCO software. Nevertheless, Jaspersoft, now under the name TIBCO Jaspersoft claims itself as "the market leader in open-source BI" with more than 12,000 commercial customers worldwide and more than 11 million product downloads (Jaspersoft, 2019). Still, that claim is difficult to substantiate as many open-source vendors are quoting large numbers of product downloads, often hundreds of thousands or even millions, while the number of production deployments of their community editions is unclear.

TIBCO Jaspersoft provides a Web-based, open, and modular approach to the evolving business analytics needs of the enterprise, providing a first-in-class, multi-tenant BI&A environment while providing a common platform for on-premise, virtualized, SaaS, and cloud deployments. The JasperSoft product family includes JasperServer, JasperReports, the JasperStudio report designer, the JasperAnalysis OLAP analysis server and JasperETL, which is based on the open-source ETL engine from Talend.

JasperSoft has established a partner network that includes companies such as Sun Microsystems, MySQL, Novell, Red Hat Unisys. JasperSoft attempts to extend its community into the world of applications with specific solutions for salesforce.com, SugarCRM and Oracle eBusiness Suite (Jaspersoft, 2019). Many small independent software vendors are also including JasperReports as the reporting component in their respective software packages.

The Jaspersoft Business Analytics Suite offers a number of ways for end users to perform interactive analysis. The two primary tools dedicated to analysis within the Jaspersoft suite are JasperAnalysis and the integrated In-Memory Analysis tool within JasperServer. JasperAnalysis is a traditional OLAP (Online Analytical Processing) tool, based on the popular open source Mondrian engine. OLAP tools perform analytic tasks on specialized, analysis-tuned data collections commonly called "cubes."

Figure 4. A Jaspersoft dashboard sample
(source: www.jaspersoft.com)

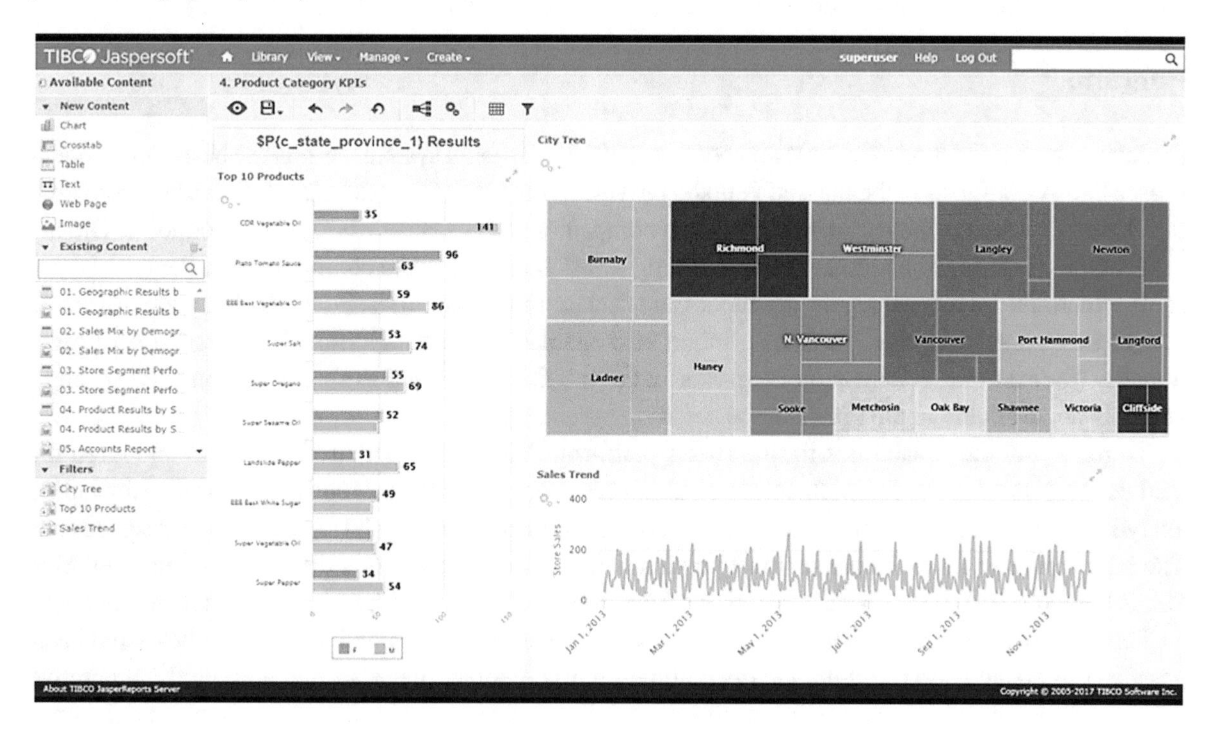

JasperServer provides an integrated suite of BI&A capabilities, including *ad-hoc* query and reporting, dashboarding and analysis. All capabilities are delivered based on a common metadata and security layer and are delivered via the same Web 2.0-style user interface. To create a query, users simply drag and drop data objects of interest from a business view of the data into a query builder window, with the option to add filters and parameter groups. The query results can then be laid out and formatted to produce a report that can be saved, shared, scheduled and distributed through the JasperServer platform.

Dashboards, including features like URL-based mashups and live refresh, can also be created through the same drag-and-drop interface. Flash charts and maps enhance the look and interactivity of the dashboards. Figure 4 shows an example of a dashboard with some of these features.

Jaspersoft offers a suite of analysis capabilities covering a broad range of user and organizational profiles, delivered chiefly within JasperAnalysis and in the integrated In-Memory Analysis tool within JasperServer. Big Data Analytics can be a challenging concept with multiple sources, a number of different ways to connect and users wanting insight faster than ever.

Jaspersoft has recently released Visualize.js, a JavaScript framework for advanced embedding of visualizations and reports in applications. The new framework, which is included in the newly released TIBCO Jaspersoft 7.1 product, delivers more control, simplicity and power for application developers by combining the power of the complete Jaspersoft analytic server with the simplicity and control of JavaScript. In addition to Visualize.js, Jaspersoft introduced a series of updates to its flagship platform. Version 7.1 of the Jaspersoft platform includes virtualized blending of relational and Big Data sources such as MongoDB, Hadoop and Cassandra, more powerful analytic calculations and visualizations, and new interactive reporting features.

Furthermore, the company now partners with Hortonworks, Amazon AWS and Cloudera to provide Analytics as a Service (AaaS), enabling data visualization at reduced cost.

Pentaho

Pentaho, which was founded in 2004 and bought in 2015 by Hitachi systems, is one the best-known open-source BI&A platforms in the market. Pentaho provides a breadth of functionality that can be considered the closest match to commercial offerings from companies such as Business Objects, Cognos or Oracle.

Pentaho's offerings include a simple reporting solution and the more comprehensive Pentaho BI&A suite, which also includes analysis, dashboard and data mining capabilities. For both offerings, customers can purchase a subscription. The subscription also enables access to additional BI&A platform functionality (for example, system auditing, performance monitoring as well as single sign-on or clustering), which is not available in the open-source version.

To round out the capabilities for the BI&A platform, Pentaho has acquired the assets and hired the lead developers of some complementing open-source projects, such as Mondrian or Kettle, for online analytical processing (OLAP) or extraction, transformation and loading (ETL) technology, respectively. The BI&A platform, which runs on most popular Windows versions, Linux distributions and even Mac OS, is based on server-side Java, a thin client Ajax front-end and an Eclipse-based design environment.

Pentaho provides a highly interactive and easy to use web-based design interface for the casual business user to create simple and *ad hoc* operational reports. Pentaho has also rich and highly interactive dashboards that help business users to easily identify the business metrics that are on track, and the ones that need attention. With no prior training, users can create personalized dashboards to turn organizational metrics into visual and interactive representations as seen in Figure 5.

Figure 5. Pentaho interactive dashboards
(source: https://www.hitachivantara.com/en-us/products/data-management-analytics/pentaho-platform.html)

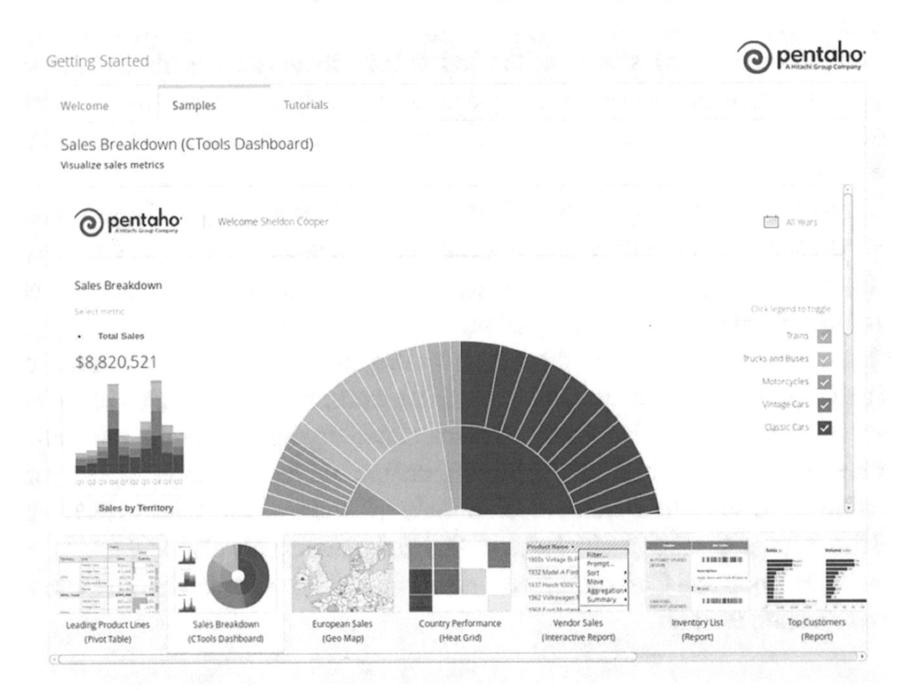

Pentaho 9.0, the last version of its business analytics and data integration platform, offers companies enhanced capabilities to scale up their Big Data operations and accelerates the delivery of value from any data by unleashing the power of highly scalable Big Data analytics.

The recent versions enable integration with SAP HANA and several Big Data systems such as Apache Spark and Hadoop, including Amazon Elastic MapReduce and YARN (MapReduce 2.0). Furthermore, it provides API to access Google Analytics' and enables business analytics on-demand by partnering with Amazon Web Services (AWS) in order to provide Pentaho in the cloud. These new functionalities empower organizations to rapidly blend and analyze the highest volumes of data, ultimately speeding the time to a 360-degree business view.

The Pentaho 9.0 platform enables companies, large or small, to take full advantage of Big Data without having to endure a lengthy, specialized process that often proves a barrier to entry to Big Data.

SpagoBI/Knowage

SpagoBI (SpagoBI, 2019) is an integrated platform for Business Intelligence and Analytics developed entirely in accordance with the free and open-source software idea which also integrates with proprietary solutions such as Microsoft Business Objects and Microsoft Analysis Services.

Created by several individuals and companies, this is a multiplatform that covers BI&A requirements both in terms of data analysis & management and administration & security, offering solutions for data reporting, multidimensional analysis, dashboarding and ad-hoc queries. The tool enables data extraction, transformation and loading (ETL) while supporting maintenance and management of analytical processes and implementing version control and workflow management.

SpagoBI follows a modular structure in which all modules are related to the core of the system, which ensures the harmony of the platform alongside its evolutions and improvements.

Figure 6 shows an image of the SpagoBI studio.

Figure 6. SpagoBI studio sample
(source: https://www.spagobi.org/).

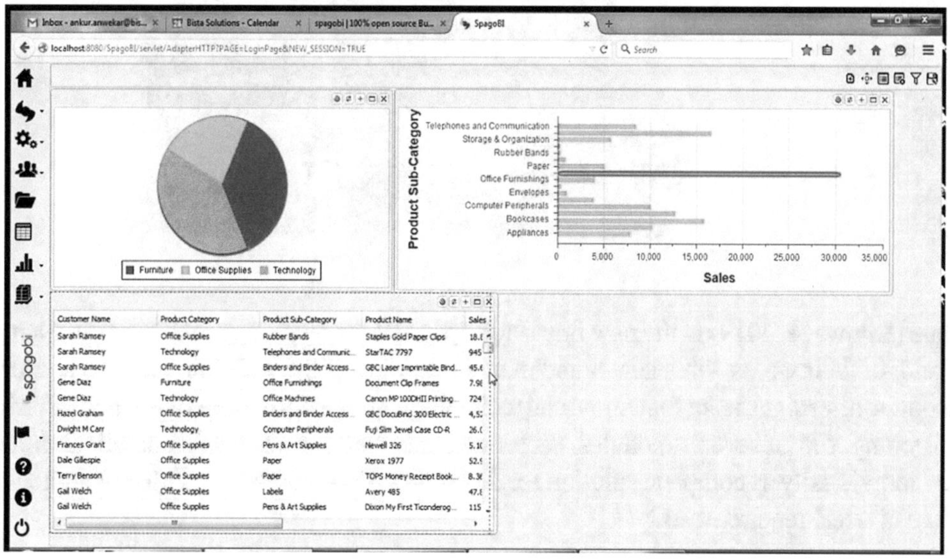

Developed by SpagoWorld and supported by an open source community, this tool consists of several modules such as:

- SpagoBI server represents SpagoBI's main module and offers the core and analytical capabilities of the application;
- SpagoBI studio allows the user to design and modify the analytical documents such as reports, dashboards, and data analysis processes. The interaction between this module and the server model is possible due to the SpagoBI SDK module;
- SpagoBI Meta is used for metadata management allowing users to edit and import external data from ETL and other tools, enlarging the metadata knowledge base from SpagoBI server, so that these can be easily queried by available tools;
- SpagoBI SDK is the dev kit used to integrate services provided by the server. This module allows the integration and publishing of analytical processes;
- SpagoBI Applications is a collection of analytical models developed using SpagoBI (Brandão *et al.*, 2016).

Figure 7 presents an example of a dashboard created on SpagoBI.

Figure 7. SpagoBI 6.0 (Knowage) sample
(source: https://www.spagobi.org/).

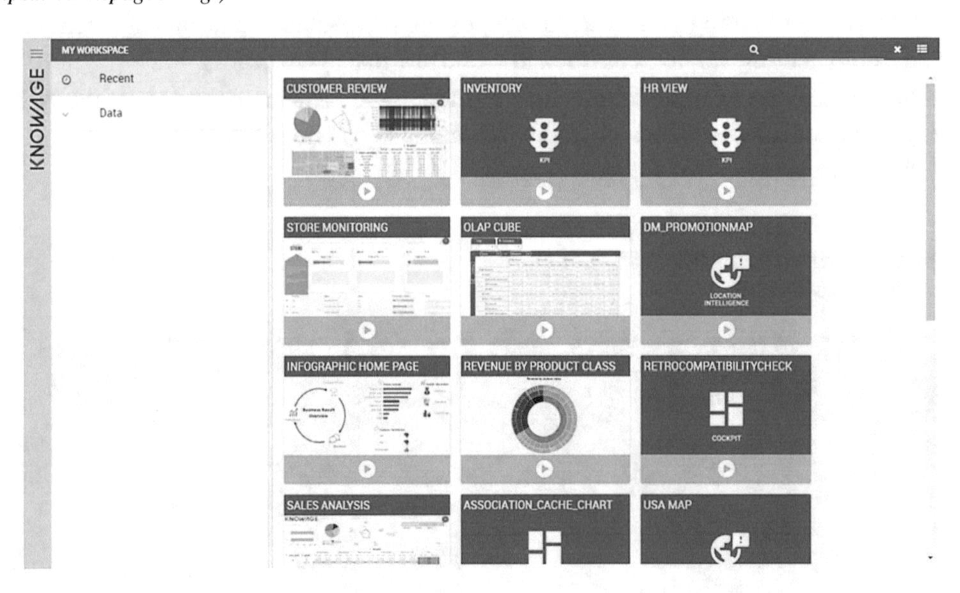

Knowage (Knowage, 2019) is the new brand for SpagoBI project – SpagoBI's 6.0 and latest version so forth – and, in their words, "this new brand marks the value of the well-known open source business intelligence suite after significant functional and technological transformations and a new offering model. The suite is composed of several modules, each one conceived for a specific analytical domain. They can be used individually as complete solution for a certain task or combined with one another to ensure full coverage of user' requirements."

This new version brings Big Data and NoSQL databases into SpagoBI, introduces self service capabilities, Location & predictive analytics and embedded analytics so that processing can be closest to data sources.

OSSPAL Methodology

To assertively evaluate the BI&A tools, we used OSSpal (Wasserman et al., 2017) assessment methodology, which combines quantitative and qualitative evaluation measures for software in several categories and determines which tool has the best score. The methodology is composed of seven categories:

- **Functionality**: How well does the software meet the average user's requirements;
- **Operational Software Characteristics**: How secure is the software; how well does the software perform; how well does the software scale; how good is the User Interface (UI) and how easy is it use by the end-users; how easy to install, configure, deploy, and maintain;
- **Support and Service**: How well is the software component supported; if there is any commercial and/or community support; are there people and organizations that provide training and consulting services onto this software;
- **Documentation**: Is there adequate tutorials and reference documentation for the software;
- **Software Technology Attributes**: How well is the software architected; how modular, portable, flexible, extensible, open, and easy to integrate it is; is the design, the code, and the tests of high-quality level; how complete and error-free are they;
- **Community and Adoption**: How well is the component adopted by community, market, and industry; how active and lively is the community for the software;
- **Development Process**: What is the level of professionalism onto the development pipeline and of the project organization as a whole.

This methodology is composed of four phases:

Phase 1: Within the first phase, it is necessary to identify a software component list to be analyzed, measure each component in relation to the evaluation criteria and removing from the analysis any software component that does not satisfy the use requirements;

Phase 2: Within the second phase, we assign weights for the categories and measures:

a. Assign a weight of importance to each category, totaling 100%;
b. For each measure within a category, it is necessary to rank the measurement in accordance with its weight;
c. To each measure within a category, there is the need to assign the importance by percentage, totaling 100% for all the measures.

Phase 3: In the third phase, data is gathered for each measure used in each category and calculated its weighting in a range between 1 to 5 (1 - Unacceptable, 2 - Poor, 3 - Acceptable, 4 - Very Good, 5 - Excellent);

Phase 4: In the fourth and final phase, we calculate the OSSpal final score.

The category 'Functionality' is calculated differently from the others. In this category is intended to analyze and evaluate the characteristics which the tools have or should have. The method to assess this category is as follows:

1. Set down the characteristics to analyze, scoring them from 1 to 3 (less important to very important);
2. Classify the characteristics in a cumulative sum (from 1 to 3);
3. Standardize the prior result to a scale from 1 to 5.

Therefore, the Functionality category will have the following scale:

- Under 65%, Score = **1** (Unacceptable)
- 65% - 80%, Score = **2** (Poor)
- 80% - 90%, Score = **3** (Acceptable)
- 90% - 96%, Score = **4** (Good)
- Over 96%, Score = **5** (Excellent).

Open Source Business Intelligence and Analytics Systems

In this section, we evaluate the Open Source BI&A systems based on the rankings presented in (Ferreira et al., 2017) and (Bernardino & Neves, 2019). We examine the open source tools described before: BIRT (Birt, 2019), Jaspersoft (Jaspersoft, 2019), Pentaho (Pentaho, 2019), and SpagoBI (2019).

Table 1 shows the ranking for Open Source BI&A vendors according to (Ferreira et al., 2017).

Table 1. Open Source Business Analytics Vendors – Stacked Rankings

Category	Score			
	BIRT	**Jaspersoft**	**Pentaho**	**SpagoBI**
Functionality	0.3	0.9	1.5	1.5
Operational software characteristics	0.84	0.9	0.82	0.7
Software technology attributes	0.48	0.51	0.46	0.36
Support and service	0.01	0.04	0.04	0.01
Documentation	0.04	0.05	0.05	0.05
Community and adoption	0.25	0.35	0.45	0.15
Development process	0.15	0.15	0.15	0.15
TOTAL	**2.07**	**2.9**	**3.47**	**2.92**

These are all top BI&A solutions that organizations can choose according to their dimension, requirements, and budget.

FUTURE RESEARCH DIRECTIONS

Within this section, we present the future research directions for the three most promising areas discussed in this paper. As such, we will detail Big Data research directions; Business Intelligence and Analytics research directions; and finally, Open Source Business Intelligence & Analytics Systems directions for future research.

Big Data

Many of today's enterprises are encountering the Big Data problems. A Big Data problem has four distinct characteristics (so called 4V features): the data volume is huge (Volume); the data type is diverse meaning a mixture of structured data, semi-structured data and unstructured data (Variety); the streaming data is very high (Velocity); and uncertain or imprecise data (Veracity). These 4V features pose a grand challenge to traditional data processing systems since these systems either cannot scale to the huge data volume in a cost-effective way or fail to handle data with variety of types (Abouzeid et al., 2008; Chattopadhyay, 2011).

With the increasingly large amount of data, building separate systems to analyse data becomes expensive and infeasible, caused by not only the cost and time of building the systems, but also the required professional knowledge on Big data management and analysis. Therefore, it is necessary to have a single infrastructure which provides common functionality of Big data management, and flexible enough to handle different types of Big data and Big data analysis tasks (Agrawal et al., 2014).

Although current technologies such as cloud computing provide infrastructure for automation of data collection, storing, processing and visualization, Big Data impose significant challenges to the traditional infrastructure, due to the characteristics of volume, velocity and variety.

Modern Internet and scientific research project produce a huge amount of data with complex interrelationship. These Big data need to be supported by a new type of infrastructure tailored for Big Data, which must have the performance to provide fast data access and process to satisfy users' just in time needs (Slack, 2012). Moreover, community standards for data description and exchange are also crucial (Lynch, 2008).

Nowadays, users are accessing multiple data storage platforms to accomplish their operational and analytical requirements (Horey et al., 2012). Efficient integration of different data sources is important (Devlin et al., 2012). For example, an organization may purchase storage from different vendors and need to combine data with different format stored on systems from different vendors (Slack, 2012). Data integration, which plays an important role for both commercial and scientific domains, combines data from different sources and provides users with a unified view of these data (Lenzerini, 2002). How to make efficient data integration with the 4V (volume, velocity, variety, and veracity) characteristics is a key research direction for the Big Data platforms.

Although Big data infrastructure and platform are critical, they can't create the same long-term value as various Big data analytics software which apply Big data to accelerate a market (Rouse, 2012). Big data analytics is the process of examining large amounts of data of various types to uncover hidden patterns, unknown correlations and other useful information (Analytics, 2014a). The Big data analytics algorithms are complex and far beyond the reach of most organization's capabilities. Moreover, there are too few skilled Big data practitioners available for every organization. Therefore, more and more organizations turn to Big Data Analytics Software-as-a-Service (Analytics-as-a-Service -AaS) to obtain the BI&A service that turns their unstructured data into an enhanced asset (Delen et al., 2013).

Big Data Analytics Software-as-a-Service exploits massive amounts of structured and unstructured data to deliver real-time and intelligent results, allowing users to perform self-service provisioning, analysis, and collaboration. Big Data Analytics Software-as-a-Service is typically Web-hosted, multi-tenant and use Hadoop based environment, NoSQL, and a range of pattern discovery and machine learning technologies (Analytics, 2014b). Users would typically use ELT (Extraction Loading and Transformation) systems, execute scripts and queries that big data engineers, programmers, data analysts and data scientists developed for them to generate reports and visualizations (EMC, 2012). Various Big Data analytics approaches can be implemented and encapsulated into services. By this way, users will be able to interact with Web-based analytics services easily without worrying about the underlying data storage, management, and analyzing procedures.

Considering the 4-V characteristics of Big Data (Roos et al., 2011), and the requirements for enterprises' application development, in order to overcome the challenges Big Data caused, and to help enterprises seize the opportunities Big Data brings, these platforms are desired to provide the following functions:

- **Scalability**. Data storage should always be considerately designed in data-intensive application development. Traditionally, we always preferably choose a relational database management system (RDBMS) in that several decades of successful application have proved that the system is reliable, effective and robust. Although it could be scaled up or scaled out to some extent, there are also many bottlenecks on its scale and speed when processing trillions of data (Lee, 2004; Leavitt, 2010).
- **Multi-typed data supported**. Besides existing relational databases, documents, e-mails and attachment, photos and videos, and Internet search indexes, log files and social media become the primary data sources that need to be stored, processed and utilized by enterprises. It is difficult for those rigidly-defined and schema-based existing approaches used by relational databases to quickly incorporate these unstructured and semi-structured data (Blumberg & Atre, 2003), which requires the products designed for unstructured data as well as structured data.
- **Business driven and agile development**. In order to adapt to rapidly changing business requirements, data-intensive applications are no longer complex and all-in-one, and this requires the products should be designed to be simpler and lighter than ever before and could be adjusted rapidly to meet the needs of business development.

Business Intelligence and Analytics

With the current rate of data growing, organizations have been experiencing several problems. Here, the challenges are not so much technical but rather human. There is the need to create effective data governance strategies that above all can ensure data quality an overall data driven mind that starts from C level all the way down to analysts and business developers in any company. Concerning data, organizations cannot conceal themselves into silos and departments, an effective data strategy must be in place across all departments of an organizations with no exceptions.

Therefore, the following challenges and research areas can be identified (Hindle et al., 2019):

- **Effective data management strategies**. Managing data is not an easy task – organizations are facing not only huge amounts of data but also different formats and frequency of data arrival. Thence, if they want to keep up with competitors, effective data management strategies need to be the focus for not only the IT department but for the overall organization;
- **Big Data Quality**. Implementing quality is difficult – In data science and business analytics, there is a sentence widely known and used: "garbage-in, garbage-out!". BI&A supports several other initiatives such as visual analytics and data science, so this means that if data quality is not controlled, the data used as input for data visualization or data science will generate wrong results and ultimately not result in any true value;
- **Ethics**. Ethics is an important subject concerning data – GDPR recalled that ethics is one of the most critical and sensitive research areas at this moment and in the near future. A balance between personalized services and discretized, bundled information is going to be a huge challenge in the next few years;
- **Data Science and Machine Learning**. Machine learning and data science do not solve all the problems. Nowadays there is the idea that Data science and machine learning solves all the problems. Although this is simply not true. In fact, to be able to use data science and machine learning effectively, a strong analytics practice must be in place, which includes data governance, data management and data quality issues.

Open Source Business Intelligence and Analytics Systems

The increasing focus on open-source software has reached the mainstream BI&A market. As some organizations are looking to reduce costs in their deployments, they are hoping that open source gives them greater leverage for their business strategies. Other open-source BI&A deployments are initiated by application developers that are looking for a way to embed BI&A functionality into their applications. Similarly, companies often cannot afford to roll out BI&A technology to hundreds or maybe thousands of users, even from their preferred vendor, because of steep licensing costs and are therefore considering an open-source solution to complement the current infrastructure.

There are about a dozen vendors or projects that can be considered offering open-source BI&A, although quite a few of those same companies also provide commercial versions of the software, often with significant enhancements over the "free" version.

As future research, we identify 7 different areas of interest (Technologyadvice, 2020; SelectHub, 2020):

- **Real-time integration**. Real time data is becoming a necessity and ultimately a requirement as businesses need to take decisions quickly to cope with nowadays market;
- **Data management incorporation**. One of most critical issues related to data is not entirely technical but rather human based. Data management is a very important issue that needs to be on the roadmap of any BI&A system as this helps to enforce data strategy and quality which is the basis for more advanced data and analytics techniques such as machine learning;
- **Improved machine learning algorithms integration**. BI&A cannot afford to be quoted as ("only") reporting systems anymore. As nowadays' organizations require machine learning (ML) algorithms, these systems need either to natively provide the implementation of such algorithms or have a tight integration with third party software which fully implement ML algorithms over data in a scalable and manageable way;

- **Native integration with IoT devices**. While it is true that organizations are experiencing huge amounts of data, it is also true that large chunks of that data come from sensors and other IoT devices (Informatica, 2020). As such, open source BI&A seriously need to consider the native integration with machine data and IoT devices while providing ways to act upon that data and extract information from there.
- **Improved data visualization techniques**. Data is only information if it is exploited, but information is only valuable if it can be understood. Data visualization plays an important role in this field and BI&A need to take it very seriously. Data visualization integration with IoT devices and real time data is becoming the general demand across all organizations as it is starting to be part of their business strategy;
- **Scalability and flexibility**. Being able to process and present only proofs of concepts is not enough as the amounts of data being used nowadays can easily consume cluster resources. Scalability, flexibility and easy manageable interfaces is really something that Open source BI&A needs to considerate;
- **Cloud based Solutions**. As BI&A systems are starting to be much more complex and scalability may be an issue, Cloud based systems start to make much more senses as it transfers (depending on the cloud model used) responsibility to the cloud provider while the big data engineers need only to worry about the end-results rather than the overall platform. Moreover, cloud-based solutions can significantly reduce the total cost of ownership (TCO) of implementing open source BI&A systems (Amazon, 2020).

CONCLUSION

Big Data represents a fundamental shift in business decision-making. Organizations are used to analyzing internal data – sales, shipments, and inventory. Now they are increasingly analyzing external data too, gaining new insights into customers, markets, supply chains and operations: the perspective of "outside-in view". We believe it is Big Data and the outside-in view that will generate the Biggest opportunities for differentiation over the next years. The key to winning in the Information Age is making decisions that are consistently better and faster than the competition. Using Big data for decision-making will lead to better decisions, better consensus, and better execution.

Business intelligence and analytics is an approach to managing business that is dedicated to providing competitive advantage through the execution of fact-based decision-making. At a tactical level, business analytics allows to achieve this goal by applying a decision-making cycle of analyzing information, gaining insight, taking action, and measuring results. At a strategic level, business analytics allows to use the results of analysis to create superior corporate strategies that outsmart competitors. BI&A essentially means putting relevant information at the fingertips of decision makers at all levels of the organization – functional areas, business units, and executive management. Technologies exist today to make this possible for all companies – large and small.

The BI&A technologies presented in this paper become valuable only when they are used to positively impact organization behaviour. Successful BI&A solutions provide businesspeople with the information they need to do their jobs more effectively. Using the OSSpal methodology, which combines quantitative and qualitative evaluation measures for software in several categories and determines which tool has the best score, Pentaho is the best open source solution so far. However, SpagoBI and Jaspersoft are also very good options as well. The lowliest option in comparison with these is BIRT.

In our opinion, Open Source Business Intelligence and Analytics tools are matured from the functional and business model point of view to become a solid option to meet and exceed the business analytics needs of an organization, especially among SMEs to enter into the new market and operate on a national and global scale.

Data is now the fourth factor of production, as essential as land, labour and capital. It follows that tomorrow's winners will be the organizations that succeed in exploiting Big Data, for example by applying advanced predictive analytic techniques in real time using Business Intelligence and Analytics tools.

Nowadays with the new techno-economic paradigm, the SMEs can also improve their competitiveness using open BI&A as an open innovation strategy. Thus, the use of open source BI&A systems in organizations can trigger immense possibilities of accelerating knowledge acquisition, innovation, intensifying entrepreneurship development and improving business skills, therefore, leading to business sustainability.

REFERENCES

Abouzeid, A., Bajda-Pawlikowski, K., Abadi, D. J., Rasin, A., & Silberschatz, A. (2009). HadoopDB: An Architectural Hybrid of MapReduce and DBMS Technologies for Analytical Workloads. *Proceedings of the VLDB Endowment International Conference on Very Large Data Bases*, *2*(1), 922–933. doi:10.14778/1687627.1687731

Agrawal, D. (2014). *Challenges and opportunities with big data. Leading researchers across the United States, Tech. Rep*. Retrieved August 25, 2014, from http://www.cra.org/ccc/files/docs/init/bigdatawhitepaper.pdf

Albright, S., & Winston, W. (2016). *Business Analytics: Data Analysis & Decision Making* (6th ed.). Cengage Learning.

Amazon. (2020). *What is cloud computing?* Retrieved January 12, 2020 from https://aws.amazon.com/what-is-cloud-computing/

Analytics. (2014a). *Why big data analytics as a service? Analytics as a Service*. Retrieved August 25, 2014, from http://www.analyticsasaservice.org/why-big-data-analytics-as-a-service/

Analytics. (2014b). *What is Big Data? Analytics as a Service in the Cloud. Analytics as a Service.* Retrieved August 25, 2014, from http://www.analyticsasaservice.org/what-is-big-data-analytics-as-a-service-in-the-cloud/

Arsham, H. (2015). *Applied Management Science: Making Good Strategic Decisions*. Retrieved October 27, 2015, from http://home.ubalt.edu/ntsbarsh/opre640/opre640.htm

Bates, P., Biere, M., Weideranders, R., Meyer, A., & Wong, B. (2009). *New Intelligence for a Smarter Planet*. Academic Press.

Bernardino, J., Furtado, P., & Madeira, H. (2002). DWS-AQA: a cost effective approach for very large data warehouses. *Proceedings of the International Database Engineering & Applications Symposium, IDEAS'02*, 233-242. 10.1109/IDEAS.2002.1029676

Bernardino, J., & Madeira, H. (2001). Experimental evaluation of a new distributed partitioning technique for data warehouses. *Proceedings of the International Database Engineering and Applications Symposium, IDEAS*, 312-321. 10.1109/IDEAS.2001.938099

Bernardino, J., & Neves, P. C. (2016). Decision-Making with Big Data Using Open Source Business Intelligence Systems. In H. Rahman (Ed.), *Human Development and Interaction in the Age of Ubiquitous Technology* (pp. 120–147). IGI Global. doi:10.4018/978-1-5225-0556-3.ch006

Bernardino, J., & Neves, P. C. (2019). Decision-Making with Big Data Using Open Source Business Intelligence Systems. In Web Services: Concepts, Methodologies, Tools, and Applications (pp. 431-458). Hershey, PA: IGI Global. doi:10.4018/978-1-5225-7501-6.ch025

Bessen, J. (2006). Open Source Software: Free Provision of Complex Public Goods. In J. Bitzer & P. J. H. Schröder (Eds.), *The Economics of Open Source Software Development*. Elsevier B.V. doi:10.1016/B978-044452769-1/50003-2

Birt. (2019). *BIRT BI.* Retrieved December 31, 2019, from https://www.eclipse.org/birt/

Blumberg, R., & Atre, S. (2003). The problem with unstructured data. *DM Review*, *13*(2), 42–49.

Burstein, F., & Holsapple, C. W. (2008). *Handbook on Decision Support Systems.* Springer Berlin Heidelberg.

Camm, J., Cochran, J., Fry, M., Ohlmann, J., & Anderson, D. (2020). Business Analytics (3rd ed.). Cengage Learning, Inc.

Chamoni, P., & Gluchowski, P. (2004). Integration trends in business intelligence systems-An empirical study based on the business intelligence maturity model. *Wirtschaftsinformatik*, *46*(2), 119–128. doi:10.1007/BF03250931

Chattopadhyay, B., Lin, L., Liu, W., Mittal, S., Aragonda, P., Lychagina, V., Kwon, Y., & Wong, M. (2011). Tenzing a SQL Implementation on the MapReduce Framework. *Proceedings of the VLDB Endowment International Conference on Very Large Data Bases*, *4*(12), 1318–1327. doi:10.14778/3402755.3402765

Cheng, R., Zhang, X. (2018). *Business Intelligence & Analytics (BI&A) Training: The Guidelines for Improving the Effectiveness of BI&A Systems Training for Business Users*. Academic Press.

Chesbrough, H. (2003). Open Innovation: How Companies Actually Do It. *Harvard Business Review, 81*(7), 12-14.

Chesbrough, H. & Crowther, A. K. (2006). Beyond high tech: early adopters of open innovation in other industries. R*&D Management, 36*(3), 229-236.

Dahlander, L., & Magnusson, M. G. (2005). Relationships between open source software companies and communities: Observations from Nordic firms. *Research Policy, 34*(4), –493.

Davenport, T. H., & Harris, J. G. (2007). *Competing on analytics: The new science of winning*. Harvard Business School Review Press.

Delen, D., Demirkan, H. (2013). *Data, information and analytics as services*. Academic Press.

Devlin, B., Rogers, S., & Myers, J. (2012). *Big data comes of age. Tech. Rep*. Retrieved August 25, 2014, from http://www.9sight.com/pdfs/Big_Data_Comes_of_Age.pdf

Dresner Advisory Services. (2014). *Wisdom of Crowds Business Intelligence Market Study*. From http://businesssintelligence.blogspot.com/2011/05/released-2011-edition-of-wisdom-of.html

Evans, J. (2019). *Business Analytics* (3rd ed.). Pearson.

Evelson, B., & Norman, N. (2008). *Topic overview: business intelligence*. Forrester Research. Retrieved August 23, 2014, from http://www.forrester.com/Topic+Overview+Business+Intelligence/fulltext/-/E-RES39218

Ferreira, T., Pedrosa, I., & Bernardino, J. (2017). Evaluating Open Source Business Intelligence Tools using OSSpal Methodology. *Proceedings of the 9th International Joint Conference on Knowledge Discovery, Knowledge Engineering and Knowledge Management*, 1, 283-288. 10.5220/0006516402830288

Gantz, J., & Reinsel, D. (2018). *The Digitization of the WorldFrom Edge to Core*. https://www.seagate.com/www-content/our-story/trends/files/idc-seagate-dataage-whitepaper.pdf

Hindle, G., Kunc, M., Mortensen, M., Oztekin, A., & Vidgene, R. (2018). *Business Analytics: Defining the field and identifying a research agenda*. Retrieved December 30, 2019, from https://www.journals.elsevier.com/european-journal-of-operational-research/call-for-papers/business-analytics-defining-the-field-and-identifying-a-rese

Horey, J., Begoli, E., Gunasekaran, R., Lim, S.-H., & Nutaro, J. (2012). Big data platforms as a service: challenges and approach. *Proceedings of the 4th USENIX conference on Hot Topics in Cloud Computing, HotCloud'12*, 16–16.

Informatica. (2020). *IoT Data Management: The Rise of Industrial IoT and Machine Learning*. Retrieved January 12, 2020, from https://www.informatica.com/resources/articles/iot-data-management-and-industrial-iot.html#fbid=GTwRdF2a3Xb

Jaspersoft. (2019). *Jaspersoft Business Intelligence*. Retrieved December 31, 2019, from https://www.jaspersoft.com/

Knowage. (2019). *Knowage Business Intelligence*. Retrieved January 4, 2020 from https://knowage.readthedocs.io/en/latest/

Lakhani, K. R., & von Hippel, E. (2003). How open source software works: "Free" user-to-user assistance. *Research Policy*, *32*(6), 923–943. doi:10.1016/S0048-7333(02)00095-1

Lapa, J., Bernardino, J., & Figueiredo, A. (2014). A comparative analysis of open source business intelligence platforms. *International Conference on Information Systems and Design of Communication, ISDOC 2014. ACM International Conference Proceeding Series*, 86-92. 10.1145/2618168.2618182

Leavitt, N. (2010). Will NoSQL databases live up to their promise? *IEEE Computer*, *43*(2), 12–14. doi:10.1109/MC.2010.58

Lee, R. (2004). *Scalability report on triple store applications.* Retrieved August 25, 2014 from http://simile.mit.edu/reports/stores/

Lenzerini, M. (2002). Data integration: A theoretical perspective. *Proceedings of the 21st ACM SIGMOD-SIGACT-SIGART Symposium on Principles of Database Systems*, 233–246.

Lin, Y. H., Tsai, K. M., Shiang, W. J., Kuo, T. C., & Tsai, C. H. (2009). Research on using ANP to establish a performance assessment model for business intelligence systems. *Expert Systems with Applications*, *36*(2), 4135–4146. doi:10.1016/j.eswa.2008.03.004 PMID:32288333

Loshin, D. (2011). *The Analytics Revolution 2011: Optimizing Reporting and Analytics to Optimizing Reporting and Analytics to Make Actionable Intelligence Pervasive*. Knowledge Integrity, Inc.

Luhn, H. P. (1958). A Business Intelligence System. *IBM Journal of Research and Development*, *2*(4), 314–319. doi:10.1147/rd.24.0314

Lynch, C. (2008). Big data: How do your data grow? *Nature*, *455*(7209), 28–29. doi:10.1038/455028a PMID:18769419

Marinheiro, A., Bernardino, J. (2015). Experimental Evaluation of Open Source Business Intelligence Suites using OpenBRR. *IEEE Latin America Transactions, 13*(3), 810-817.

Mearian, L. (2010). Data growth remains IT's biggest challenge, Gartner says. *Computerworld*. https://www.computerworld.com/s/article/9194283/Data_growth_remains_IT_s_biggest_challenge_Gartner_says

OpenText. (2019). Retrieved December 31, 2019, from https://www.opentext.com/products-and-solutions/products/opentext-product-offerings-catalog/rebranded-products/actuate

Pentaho. (2019). *Pentaho Open Source BI.* Retrieved August 31, 2019, from https://www.hitachivantara.com/en-us/products/data-management-analytics/pentaho-platform.html

Pollock, R. (2009). Innovation, Imitation and Open Source. *International Journal of Open Source Software & Processes, 1*(2), 114-127.

Power, D. J. (2007). *A Brief History of Decision Support Systems, version 4.0.* DSSResources.COM. http://DSSResources.COM/history/dsshistory.html

Roos, D., Eaton, C., Lapis, G., Zikopoulos, P., & Deutsch, T. (2011). *Understanding Big Data: Analytics for Enterprise Class Hadoop and Streaming Data*. McGraw-Hill Osborne Media.

Rouse, M. (2012). *Definition of big data analytics.* Retrieved August 25, 2014, from https://searchbusinessanalytics.techtarget.com/definition/big-data-analytics

Santos, R. J., Bernardino, J., & Vieira, M. (2011). A survey on data security in data warehousing: Issues, challenges and opportunities. *EUROCON 2011 - International Conference on Computer as a Tool - Joint with Conftele 2011.*

SelectHub. (2020). *The Future of Business Intelligence (BI) in 2020.* Retrieved January 12, 2020 from https://www.selecthub.com/business-intelligence/future-of-bi/

Sharda, R., Delen, D., & Turban, E. (2014). *Business Intelligence and Analytics: Systems for Decision Support* (10th ed.). Prentice Hall.

Skok, M. (2008). *The Future of Open Source: Exploring the Investments, Innovations, Applications, Opportunities and Threats.* North Bridge Venture Partners.

Slack, E. (2012). *Storage infrastructures for big data workflows.* Storage Switzerland White Paper. Retrieved August 25, 2014, from https://iq.quantum.com/exLink.asp?8615424OJ73H28I34127712

Spago, B. I. (2019). Retrieved January 04, 2020 from https://www.spagobi.org/

TDWI. (2019). *The Data Warehousing Institute.* https://tdwi.org/Home.aspx

Technologyadvice. (2020). *Business Intelligence Software - Compare over 100 vendors.* Retrieved January 12, 2020 form https://technologyadvice.com/business-intelligence/

Teradata. (2019). *Teradata University Network.* Retrieved December 30, 2019, from https://www.teradatauniversitynetwork.com/

The 451 Group. (2008). *Open Source Is Not a Business Model.* The 451 Commercial Adoption of Open Source (CAOS) Research Service.

Torres, R., Sidorova, A., & Jones, M. C. (2018). Enabling firm performance through business intelligence and analytics: A dynamic capabilities perspective. *Information & Management, 2018*(7), 822–839. Advance online publication. doi:10.1016/j.im.2018.03.010

Turban, E., King, D., Lee, J. K., & Viehland, D. (2004). *Electronic Commerce 2004: A Managerial Perspective* (3rd ed.). Prentice Hall.

Tzu, S. (1963). *The Art of War. United Sates of America.* Oxford University Press.

Wang, S., Yeoh, W., Richards, G., Wong, S. F., & Chang, Y. (2019). Harnessing business analytics value through organizational absorptive capacity. *Information & Management, 56*(7), 103–152. doi:10.1016/j.im.2019.02.007

Wasserman, A. I. (2017). OSSpal: Finding and Evaluating Open Source Software. In *Springer International Publishing* (pp. 193–203). doi:10.1007/978-3-319-57735-7_18

Watson, H. J., Wixom, B. H., & Goodhue, D. L. (2004). Data Warehousing: The 3M Experience. In H. R. Nemati & C. D. Barko (Eds.), *Org. Data Mining: Leveraging Enterprise Data Resources for Optimal Performance* (pp. 202–216). Idea Group Publishing. doi:10.4018/978-1-59140-134-6.ch014

West, J., & Gallagher, S. (2006). Patterns of Open Innovation in Open Source Software. In H. Chesbrough, W. Vanhaverbeke, & J. West (Eds.), *Open Innovation: Researching a New Paradigm* (pp. 82–106). Oxford University Press.

West, J., & Lakhani, K. (2008). Getting Clear About the Role of Communities in Open Innovation. *Industry and Innovation, 15*(2), 223–231. doi:10.1080/13662710802033734

Wit, B., & Meyer, R. (2010). *Strategy: Process, Content, Context* (4th ed.). Cengage Learning EMEA.

ADDITIONAL READING

Al-Aqrabi, H., Liu, L., Hill, R., & Antonopoulos, N. (2015, February). Cloud BI: Future of business intelligence in the Cloud. *Journal of Computer and System Sciences, 81*(1), 85–96. doi:10.1016/j.jcss.2014.06.013

Arefin, S., Hoque, R., & Bao, Y. (2015). The impact of business intelligence on organization's effectiveness: An empirical study. (2015). *Journal of Systems and Information Technology, 17*(3), 263–285. doi:10.1108/JSIT-09-2014-0067

Arora, S., Nangia, V. K., & Agrawal, R. (2015). Making strategy process intelligent with business intelligence: An empirical investigation. *International Journal of Data Analysis Techniques and Strategies, 7*(1), 2015. doi:10.1504/IJDATS.2015.067702

Back, W. D., Goodman, N., & Hyde, J. (2013). *Mondrian in Action: Open source business analytics* (2nd ed.). Manning Publications.

Bell, J. (2020). *Machine Learning: Hands-On for Developers and Technical Professionals* (2nd ed.). John Wiley & Sons Inc. doi:10.1002/9781119642183

Bulusu, L. (2012). *Open Source Data Warehousing and Business Intelligence* (1st ed.). CRC Press. doi:10.1201/b12671

Chang, Y., Hsu, P., & Wu, Z. (2015). Exploring managers' intention to use business intelligence: The role of motivations. *Behaviour & Information Technology, 34*(3), 2015. doi:10.1080/0144929X.2014.968208

Grant, B., Bassens, A., & Krohn, J. (2019). *Deep Learning Illustrated: A Visual, Interactive Guide to Artificial Intelligence*. O'Reilly.

Grossmann, W., & Rinderle-Ma, S. (2015). *Fundamentals of Business Intelligence*. Springer Berlin Heidelberg. doi:10.1007/978-3-662-46531-8

Höpken, W., Fuchs, M., Keil, D., & Lexhagen, M. (2015, June). Fuchs, M.; Keil, D. & Maria Lexhagen (2015). Business intelligence for cross-process knowledge extraction at tourism destinations. *Information Technology & Tourism, 15*(2), 101–130. doi:10.100740558-015-0023-2

Minelli, M., Chambers, M., & Dhiraj, A. (2013). *Big Data, Big Analytics: Emerging Business Intelligence and Analytic Trends for Today's Businesses* (1st ed.). Wiley. doi:10.1002/9781118562260

Moro, S., Cortez, P., & Rita, P. (2015, February 15). (2025). Business intelligence in banking: A literature analysis from 2002 to 2013 using text mining and latent Dirichlet allocation. *Expert Systems with Applications*, *42*(3), 1314–1324. doi:10.1016/j.eswa.2014.09.024

O'Reilly Media Inc. (2016). *Big Data Now: 2016 Edition* (6th ed.). O'Reilly Media.

Orgaz, G. B., Barrero, D. F., R-Moreno, M. D., & Camacho, D. (2015). R-Moreno, M. D. & Camacho, D. (2015). Acquisition of business intelligence from human experience in route planning. *Enterprise Information Systems*, *9*(3), 303–323. doi:10.1080/17517575.2012.759279

Provost, F., & Fawcett, T. (2013). *Data Science for Business: What you need to know about data mining and data-analytic thinking* (1st ed.). O'Reilly Media.

Roldán, M. C. (2017). *Learning Pentaho Data Integration 8 CE* (3rd ed.). Packt.

Sabherwal, R., & Becerra-Fernandez, I. (2013). *Business Intelligence: Practices, Technologies, and Management* (1st ed.). Wiley.

Shan, C., Chen, W., Wang, H., & Song, M. (2015). *The Data Science Handbook: Advice and Insights from 25 Amazing Data Scientists*. Data Science Bookshelf.

Sharda, R., Delen, D., Turban, E., & King, D. (2013). *Business Intelligence: A Managerial Perspective on Analytics* (3rd ed.). Prentice Hall.

Sherman, R. (2014). *Business Intelligence Guidebook: From Data Integration to Analytics* (1st ed.). Morgan Kaufmann.

Shollo, A. (2011). Using Business Intelligence in IT Governance Decision Making. Governance and Sustainability in Information Systems, 3-15.

Su, S. I., & Chiong, R. (2011). Business Intelligence. Encyclopedia of Knowledge Management, 72-80.

Turban, E., Sharda, R., & Delen, D. (2010). *Decision Support and Business Intelligence Systems* (9th ed.). Prentice Hall.

KEY TERMS AND DEFINITIONS

BI&A Tools: Represents the tools and systems that play a key role in the strategic planning process of the organization. These systems allow a company to gather, store, access and analyze corporate data to aid in decision-making. Generally, these systems will illustrate business analytics in the areas of customer profiling, customer support, market research, market segmentation, product profitability, statistical analysis, and inventory and distribution analysis to name a few.

Big Data: Big data has four distinct characteristics (so called 4V features): the data volume is huge (volume); the data type is diverse meaning a mixture of structured data, semi-structured data and unstructured data (variety); the streaming data is very high (velocity); and uncertain or imprecise data (veracity).

Business Intelligence and Analytics: Business intelligence and analytics is a broad category of strategies, applications, and technologies for gathering, storing, analyzing, and providing access to data to help enterprise users make better business decisions. BI&A applications include the activities of decision support systems, query and reporting, statistical analysis, forecasting, and data mining. BI&A refers to a management philosophy and tool that help organizations to manage and refine business information to make effective decisions.

Cloud: Cloud consist in a distributed environment that uses resource virtualization to unify and abstract resource management and provide several high performance, available and redundant services in a pay-as-you-go model that allows cloud service provider's clients not to worry about IT concerns and focus mainly in their business needs. Among the services provided in the cloud are the Infrastructure as a Service (IaaS), which allow clients to purchase resources on demand; Platform as a Service (PaaS) where platform instances are hired as a service, trusting the cloud provider to maintain servers; Software as a Service (SaaS), which is used to deploy software in a scalable and available way; Big Data as a Service – a specification of PaaS – that provides cloud's customers a way to store, manage and process big volumes of data; Analytics and Business analytics as a Service (AaaS and BAaaS) enables users to mine data to find interesting patterns, correlations and trends.

Data Warehousing: Analytical databases focused on providing decision support information and deriving business analytics for enterprises. According to Kimball definition, "a data warehouse is a copy of transaction data specifically structured for query and analysis". This is a functional view of a data warehouse. Typically, a data warehouse is a massive database (housed on a cluster of servers, or a mini or mainframe computer) serving as a centralized repository of all data generated by all departments and units of a large organization. Advanced data mining software is required to extract meaningful information from a data warehouse.

Decision Support Systems (DSS): Decision support systems are a specific class of computerized information system that supports business and organizational decision-making activities. A properly designed DSS is an interactive software-based system intended to help decision makers compile useful information from raw data, documents, personal knowledge, and/or business models to identify and solve problems and make decisions.

Information and Communications Technology (ICT): ICT refers to technologies that provide access to information through telecommunications. It is similar to Information Technology (IT), but focuses primarily on communication technologies. This includes the Internet, wireless networks, cell phones, and other communication mediums. In the past few decades, information and communication technologies have provided society with a vast array of new communication capabilities. For example, people can communicate in real-time with others in different countries using technologies such as instant messaging, voice over IP (VoIP), and video-conferencing.

Information Technology (IT): Set of tools, processes, and methodologies (such as coding/programming, data communications, data conversion, storage and retrieval, systems analysis and design, systems control) and associated equipment employed to collect, process, and present information. In broad terms, IT also includes office automation, multimedia, and telecommunications. It refers to anything related to computing technology, such as networking, hardware, software, the Internet, or the people that work with these technologies. Many companies now have IT departments for managing the computers, networks, and other technical areas of their businesses. IT jobs include computer programming, network administration, computer engineering, Web development, technical support, and many other related occupations. Since we live in the "information age," information technology has become a part of our everyday lives.

Open Source: Open source doesn't just mean access to the source code. The distribution terms of open-source software must comply with the following criteria: 1. Free Redistribution; 2. Source Code; 3. Derived Works; 4. Integrity of The Author's Source Code; 5. No Discrimination Against Persons or Groups; 6. No Discrimination Against Fields of Endeavor; 7. Distribution of License; 8. License Must Not Be Specific to a Product; 9. License Must Not Restrict Other Software; 10. License Must Be Technology-Neutral. For the complete definition see https://opensource.org/docs/osd.

Open Source Software (OSS): Open source software refers to software that is developed, tested, or improved through public collaboration and distributed with the idea that the must be shared with others, ensuring an open future collaboration. The collaborative experience of many developers, especially those in the academic environment, in developing various versions of the UNIX operating system, Richard Stallman's idea of Free Software Foundation, and the desire of users to freely choose among a number of products - all of these led to the Open Source movement and the approach to developing and distributing programs as open source software.

Chapter 11
E-Government Policy Implementation in Thailand:
Success or Failure?

Mergen Dyussenov
Ministry of Culture and Sports of the Republic of Kazakhstan, Kazakhstan

Lia Almeida
Universidade Federal do Tocantins, Brazil

ABSTRACT

This chapter investigates the current e-government and ICT policy in Thailand from an actor-centered perspective. It reviews existing literature on e-government implementation, while looking into the interaction of government institutions and citizens. It seeks to answer questions, such as the following: What are the key actors in driving the implementation of e-government policies in Thailand? How do Thai citizens perceive e-government efforts and ICT policy implementation especially in the context of present military government power? What are some of the risk factors typically embedded in e-government initiatives and policies implemented in Thailand? Some scholars specifically emphasize the key role of central government institutions in driving the e-government and ICT policy implementation pointing at its readiness to transform toward E-Government 4.0. These observations notwithstanding, issues related to the often-omitted crucial role of citizens and local customers in driving policy implementation and the problem of digital divide remain across much of the developing world.

1. INTRODUCTION

Thailand, along with other nations of the South East Asian region e.g. Indonesia, largely remains at an early stage both of e-government policy implementation (e.g. Mirchandani et al. 2008; Sagarik et al. 2018; Hassan 2019) and engaging the public in policymaking and service delivery processes in general (OECD/ADB 2019). Thailand, however, still appears relatively better developed vis-à-vis some of its ASEAN neighboring states, such as the Philippines and Indonesia, in terms of basic infrastructure preparedness to facilitate e-government policy implementation efforts (Bukht & Heeks 2018), as outlined in Table 1.

DOI: 10.4018/978-1-7998-5849-2.ch011

Table 1. Thailand's key indicators of ICT development vis-à-vis ASEAN states

Country	ICT Development	Global IDI Ranking	Total Mobile Penetration	Unique Mobile Subscribers	3G + 4G	Smartphone
Thailand	4.8	74	122%	85.47%	82.47%	58.98%
Indonesia	3.8	108	126%	58.43%	40.48%	40.37%
Singapore	7.9	19	145%	71.52%	63.14%	78.16%
Malaysia	5.2	64	142%	76.1%	60.3%	64.63%
Philippines	4.0	98	117%	65.09%	44.74%	40.9%
Vietnam	4.1	102	152%	49.66%	36.48%	27.84%

Source: DTAC (n.d.), as cited in Bukht & Heeks (2018).

As Kawtrakul et al. (2011) suggest, in order to boost Thailand's competitiveness in the context of ASEAN economic integration, the next step should focus on the implementation of more concrete policy initiatives, such as Internal Smart with E-Government, International Smart with intergovernmental processes and overcoming language barriers (Kawtrakul et al. 2011), as well as e-Auction, ID Smart Card, and Government Fiscal Management Information Systems (Lorsuwannarat 2006). In an effort to make an initial move in transforming the nation towards becoming a smarter society, the E-Government national policy has been largely implemented since 2000 with the primary goal to improve government services and interactions with citizens and businesses alike (Kawtrakul et al. 2011). Furthermore, even earlier the 2001-2006 Thaksin Shinawatra's government was determined to employ e-government policy as a tool to facilitate the Thailand's transformation towards becoming a knowledge-based society. However, problems and limitations remain that need to be addressed, especially related to the policy formulation stage of policymaking (Lorsuwannarat 2006).

This paper attempts to contribute to scholarly knowledge primarily by evaluating the current e-government policy implementation in Thailand based on a review of existing literature and specifically in terms of local citizens' perception including regular citizens (Mirchandani et al. 2008, Kawtrakul et al. 2011), taxpayers (Rotchanakitumnuai 2008), business communities (Sutanonpaiboon & Pearson 2006, Kawtrakul et al. 2011), students (Sukasame 2004) and navy military staff (Vathanophas 2008 et al.). It also seeks to analyze the key actors involved in implementation processes, predominantly the role of government agencies in implementing e-government and information technology (IT) policy initiatives in Thailand (e.g. Sagarik et al. 2018, Varavithya & Esichaikul 2003, Keretho et al. 2015 etc.), and the role of the public and citizenry (e.g. Mirchandani et al. 2008, Sukasame 2004). It then concludes with a summary of remaining issues, along with policy recommendations.

What follows below is a brief review of policy implementation theories in general, as well as e-governments models of implementation. Next, we detail the literature review carried out, and how theoretical viewpoints and evidence in the existing literature collaborated to identify and analyze the key actors involved and their roles in e-government implementation and as well as the citizen perceptions of the issue. Then, we move on to data analysis, where we first present a contextualization of Thailand's cases through a concrete case and present the data analysis and discussion of the findings. We then conclude with a summary of remaining issues for future research along with policy recommendations.

2. THEORETICAL BACKGROUND

2.1 The Evolution of Policy Implementation Theories

By around the 1990s, three distinct generations of policy implementation research had largely developed (Goggin et al. 1990; Pülzl & Treib 2006). The initial interest to implementation studies manifested itself in the 1970s in the United States. The first generation of implementation research, around the mid-1970s, mainly focused on the effectiveness of public policies, with numerous cases related to implementation failure (e.g. Derthick 1972; Pressman & Wildavsky 1973; Mazmanian & Sabatier 1983 as cited in Pülzl & Treib 2006). First-generation scholars developed so-called "top-down" theories that emphasized the central role of national governments in implementing policy goals based on hierarchical control and supervision (Stewart, Hedge & Lester 2007, Pülzl & Treib 2006).

Particularly, in their original publication, Pressman & Wildavsky (1973) viewed implementation as an "interaction between the setting of goals and actions geared to achieve them" (p. xv). This implies a linear relationship between goals and their implementation. Thus, this necessitates the public agencies involved in implementing policy goals to possess sufficient resources at their disposal, along with a hierarchical control to monitor the implementation process (Pülzl & Treib 2006). This is a vital observation: indeed, as Chutimaskul (2002) notes below with regard to the implementation of e-government development projects in modern Thailand, the national government employs centralization and distribution functions in order to effectively coordinate the activities of ministries in pursuing stated goals.

In response to the "top-down" approach to analyzing implementation, bottom-up theories had been largely developed by the early 1980s. This marked the third generation of implementation research, preceded by the intermediary second generation of hybrid implementation theories (see Fig. 1 below). These new theories (e.g. Lipsky 1971, 1980; Elmore 1980, as cited in Pülzl & Treib 2006) sought to question an often-assumed causal relationship between initial policy goals and outcomes. The bottom-up school advocated the need to analyze "what was happening on the recipient level" (Pülzl & Treib 2006, p. 92)

Figure 1. An overview of evolutionary policy implementation research theories
Source: Pülzl & Treib (2006)

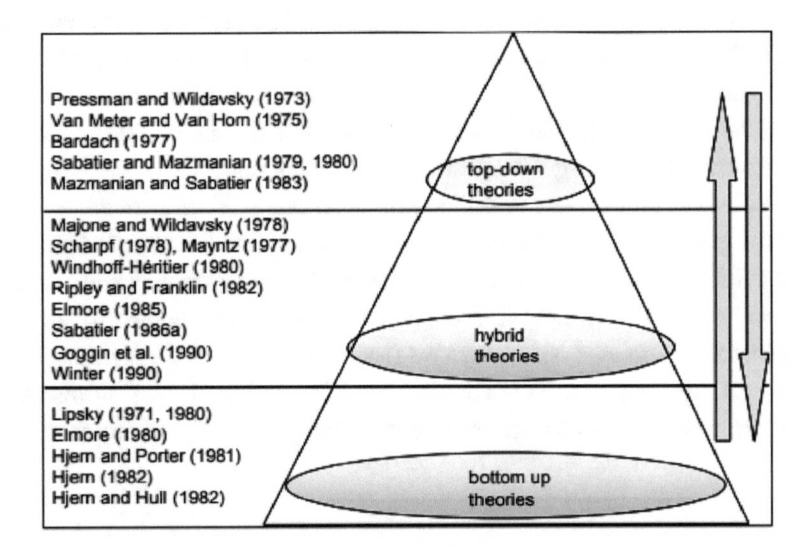

thus rejecting the notion that implementation policies should be decided at the central government level. Unlike the top-down approach, the bottom up approach assumes the primary need to identify a range of actors involved in delivery of local services and define their goals, activities and contacts (Stewart, Hedge & Lester 2007). The contacts are then employed to identify a range of "local, regional, and national actors involved in the planning, financing and execution of the relevant governmental and non-governmental programs" (Sabatier 1986, p. 32). These actors can include civil servants (bureaucracy), the legislature, political executives, pressure groups, community organizations (Stewart, Hedge & Lester 2007), and the private sector (Sabatier 1986).

Among the bottom up camp scholars, Lipsky (1971, 1980) developed the notion of "street-level bureaucrats" referring to public service workers, such as teachers, police officers, social workers etc. to emphasize their often crucial role in policy implementation. This is because their daily work tends to benefit regular people. Hudson (1989, cited in Pülzl & Treib 2006) further elaborated the notion of "street-level bureaucrats" noting that their authority encompasses not only their regular control of local citizens' behavior but also a significant degree of autonomy and discretion in decision-making. Another prominent "bottom up" scholar, Elmore (1980), coined the concept of "backward mapping" that implies the primary need to analyze a specific problem followed by an examination of local actors' actions to address the policy issue as part of an implementation process.

Thus, the period from the 1970s to 1980s witnessed a gradual transformation from top-down to bottom up approaches to studying implementation research, especially in the western world. The "bottom up" camp scholars criticize the top-down approach as it fails to account for the needs of "street-level bureaucrats" (Lipsky 1971, 1980), and by the use of hierarchical control it largely restraints the autonomy of local-level civil servants in decision-making and policy implementation processes (Hudson 1989, as in Pülzl & Treib 2006). Furthermore, according to Sabatier (1986), a major flaw embedded in the top-down approach is the failure to incorporate strategies employed by street-level bureaucrats in implementing policy goals. Thus, implementation processes should be analyzed by looking into subsystems as a unit of analysis, as contrasted with organizations or programs, since "...there is seldom a single dominant program at the local/operational level" (Sabatier 1999, p. 119). Instead, what is often observed is a range of programs that start at different levels of public administration that local actors then attempt to employ in their own interest (e.g. Hjern & Porter 1981; Sabatier 1986, as cited in Sabatier 1999).

2.2 E-government Models

E-government development can be implemented in two major ways: through the means of centralization, i.e. central planning and monitoring, and distribution (Varavithya & Esichaikul 2003; Chutimaskul 2002). The former suits those agencies less experienced in implementing e-government projects, while the latter seems to work for better funded and competent governments that possess significant e-government project experience (Chutimaskul 2002).

The major models of e-government project development can be categorized as follow:

- Individual e-government project development. Under this model, each government ministry develops its own information system aimed to provide specific public services, e.g. in areas of transportation, construction, foreign services etc. The model does not assume a centralized system maintained by the central government (ibid).

- Individual e-government project development with the national IT plan in place. Aimed to address the problem of a lacking national IT plan as related to the previous category, this model implies that the government should introduce both short-term and longer-term IT national plans to enforce local agencies to implement the e-government development policy (ibid).
- Workgroup e-government project development. Under this model, projects are developed for/by a number of national ministries that share similar requirements. It assumes centralization by one agency (among a group of ministries), and the success of project development is dependent on staff experience (ibid).
- Workgroup e-government project development with the national IT plan in place. This model differs from the previous one by incorporating the national IT plan to ensure better oversight over government operation. However, as a possible shortcoming, the integration of information systems from various ministerial workgroups may lead to problems related to standardization and compatibility of system requirements (ibid).
- Hybrid e-government project development. This assumes the presence of a central IT agency (e.g. the National IT agency or the Ministry of information etc.), integrating centralization and distribution of e-government development projects in a state. The key function of the IT agency is to coordinate activities of various governmental workgroups and units to streamline the implementation of e-government projects. A key risk factor involved under this model is free riding on the part of one of a few agencies involved. Indeed, in the words of Chutimaskul (2002), "Poor involvement of any agency will reduce the reliability, completeness, consistency, accuracy, and creditability on both government and e-Government as seen in many countries. Thailand has adopted this model for e-Government development" (p. 116). Specifically, under the national IT plan, the Thai government utilized both distribution and centralization tools under the authority of the National IT Agency to drive the e-government development implementation process (Varavithya & Esichaikul 2003).

Figure 2. A typology of major e-government models
Source: Chutimaskul (2002)

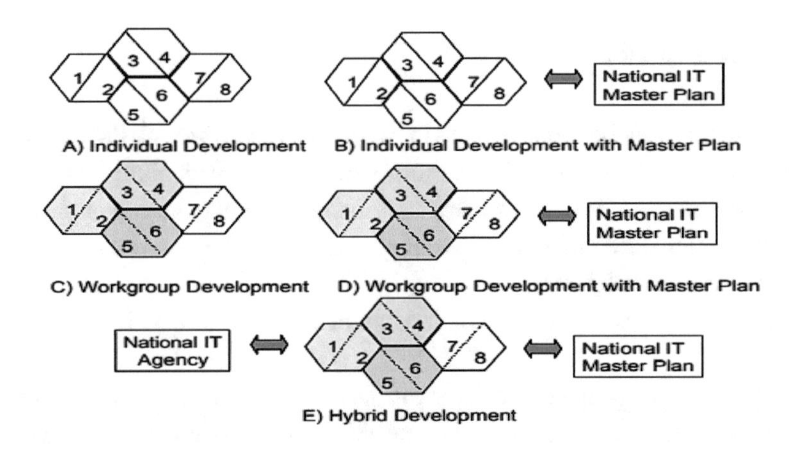

3. LITERATURE REVIEW ON E-GOVERNMENT POLICY IMPLEMENTATION IN THAILAND

The articles collected by the literature review were systematized according to the research objectives: first, to identify the key actors involved in implementation processes, predominantly the role of government agencies in implementing e-government and information technology (IT) policy initiatives in Thailand, as well as the role of the public and citizenry. Second, outline and discuss the local citizens' perception toward e-government policy implementation, including regular citizens, taxpayers, business communities, students and navy military staff.

It is worth mentioning that this literature review was not exhaustive, we focused on the most representative works that would help to understand the categories analyzed in the context of Thailand, considering the space available and the purpose of this article. The total body of literature sources (listed in Appendix) seeks to analyze e-government development policy and program implementation in modern Thailand. In so doing, part of the selected literature mainly examines the role of key actors involved in implementation processes (e.g. Sagarik et al. 2018, Mirchandani et al. 2008, Keretho et al. 2015 etc.), while another part of scholarly sources reviews the efficiency of specific Thai e-government programs and projects such as the Digital Economy Plan (Bukht & Heeks 2018, Sagarik et al. 2018), the Thai Smart ID Card project (Gunawong & Gao 2017, Lorsuwannarat 2006), and the e-Auction project and GFMIS (Lorsuwannarat 2006).

In the analysis of the role of actors in implementing Thai e-government development projects, after identifying the main actors according to the researched literature, we analyze how this literature presents the role of government agencies (ranging from a more positive continuum to criticism and negative sentiment), for this we classify the analyzes using the sentiment analysis metric. Regarding the participation of citizens, we classified the works seeking to understand two categories: vital success and risk factors perceived by citizens.

3.1 The Role of Actors in Implementing Thai e-Government Development Projects

As part of this literature review, it is vital to analyze the major actors involved in implementing Thai e-government development policies and IT initiatives. These include the role of government agencies in Thailand-based implementation processes (Sagarik et al. 2018, Gunawong & Gao 2017, Bukht & Heeks 2018, Varavithya & Esichaikul 2003, Kawtrakul et al. 2013, Kawtrakul et al. 2011, Khampachua & Wisitpongphan 2015, Funilkul et al. 2011, Keretho et al. 2015, Lorsuwannarat 2006), the role of the public and local citizens (Mirchandani et al. 2008, Funilkul et al. 2011, Gunawong & Gao 2017, Khampachua & Wisitpongphan 2015, Wangpipatwong et al. 2009, Wangpipatwong et al. 2008, Vathanophas et al. 2008, Sukasame 2004), distantly followed by the role of international organizations e.g. ASEAN (as in Kawtrakul et al. 2011, Keretho et al. 2015), and business communities (e.g. small- and medium-sized businesses as in Sutanonpaiboon & Pearson 2006). Table 2 summarizes key points.

Thus, as Table 2 suggests, significant part of scholars emphasize the role of government agencies in driving the implementation of e-government development projects in Thailand. This largely resembles the top-down approach to studying implementation processes as described above (section 2.1). This is a vital observation: while much of the developed world has moved away from top-down to bottom-up approaches, the Thai policy research tradition appears to continue emphasizing the relevance of

Table 2. The summary of key actors in Thai e-government policy implementation

Key Actors Involved in Implementing e-Government Development Projects and Goals			
The Government (Ministries)	**The Public and Citizens**	**International Organizations (ASEAN)**	**Business Communities**
Sagarik et al. 2018, Bukht & Heeks 2018, Varavithya & Esichaikul 2003, Gunawong & Gao 2017, Kawtrakul et al. 2013, Kawtrakul et al. 2011, Khampachua & Wisitpongphan 2015, Funilkul et al. 2011, Keretho et al. 2015, Lorsuwannarat 2006	Mirchandani et al. 2008, Funilkul et al. 2011, Gunawong & Gao 2017, Kawtrakul et al. 2013, Khampachua & Wisitpongphan2015, Wangpipatwong et al. 2009, Wangpipatwong et al. 2008, Vathanophas et al. 2008, Rotchanakitumnuai 2008, Sukasame 2004	Kawtrakul et al. 2011, Keretho et al. 2015	Sutanonpaiboon& Pearson 2006, Kawtrakul et al. 2013.

Source: The author's own analysis of existing literature review

government dominance through the use of hierarchical control (or 'central control' as in Sagarik et al. 2018) and monitoring under the authority of the National IT Agency (Varavithya & Esichaikul 2003, Chutimaskul 2002).

Furthermore, among those scholars emphasizing the role of Thai government agencies, some describe government activities in rather supportive and positive sentiments (Bukht & Heeks 2018, Sagarik et al. 2018, Varavithya & Esichaikul 2003), while others raise cautious remarks and possible risk factors as related to the government activities in implementing e-government development goals (Kawtrakul et al. 2011, Keretho et al. 2015, Sagarik et al. 2018, Gunawong & Gao 2017, Funilkul et al. 2011, Lorsuwannarat 2006). Thus, Bukht & Heeks (2018) commend Thai national government's efforts for developing e-government policy with "dedicated policies and governance structures for the digital economy" in place (p. 1), while Varavithya & Esichaikul (2003) note that in the face of problems related to IT use in Thai public agencies, the government developed the first national IT Master Plan in 1996employing strategic approaches. Sagarik et al. (2018), however develop rather mixed insights into government efforts. On the one hand, they note that the state has accumulated long experience in implementing e-government policies and frameworks, especiallyrecent IT transformations through central control. On the other, the number of key actors involved in e-government implementation processes is rather large (including the Digital Economy Promotion Agency, the National Statistical Office, the Meteorological Department, the Telephone Organization of Thailand, and the Communications Authority of Thailand – all under the coordination of the Ministry of Digital Economy and Society), which raises questions over the likelihood of successful implementation (ibid).

Furthermore, while Sagarik et al. (2018) commend the government for its efforts to upgrade the IT infrastructure in public agencies, Kawtrakul et al. (2011) remain skeptical noting the slow progress in government IT transformation efforts, "for six main reasons: lack of national data standards and standard governance body, lack of clear understanding about common processes across all involved stakeholders, lack of best practices and knowledge sharing in implementation, lack of data quality and data collection resources, lack of laws and regulations in data sharing and absence of a proactive mindset" (p. 183), while Kawtrakul et al. 2013 add another factor that is lack of taking customers' needs into consideration. Keretho et al. (2015) observe that while ASEAN pushes Thailand to implement ASEAN-wide common e-government development policies to facilitate regional trade, the major challenge facing Thai agencies is to conduct sound and objective evaluation of e-government development projects. Gunawong & Gao 2017 advocate the need for the Thai government to adopt an actor-network approach instead of completely monopolizing the e-government implementation power, Khampachua&Wisitpongphan2015 similarly call for the inclusion of a range of government agencies, industrial sectors and citizens in e-government implementation processes. Funilkul et al. (2011) further point to a number of problems in local e-government policy implementation, including IT privacy and ethics issues, network infrastructure quality, lack of clarity in rules and regulations, underdeveloped security policy for public agencies' websites, misunderstanding among government staff regarding e-government development policy, and a lack of citizen trust toward government information and IT use. Finally, Lorsuwannarat (2006), referring to the 2001-2006 Thaksin Shinawatra government, notes that despite the intent to implement a range of e-government policies (e.g. e-Auction, Smart Card, and GFMIS), the government mainly focused on physical technology and developing system capabilities at the expense of contextual factors, the actual readiness of government staff and agencies to implement these projects, local people's behavior and culture. Finally, Gunawong & Gao (2017) analyze factors that explain the failure of the Thai Smart ID Card project, which include a lack of an actor network in managing the project, failure to adopt an open principle in establishing key objectives, and readiness of ICT system design. Table 3 summarizes the above-described sentiment analysis.

Table 3. The sentiment analysis of sources emphasizing the role of Thai government

Positive Sentiments	Mixed Sentiments	Cautious/Negative Sentiments
Bukht & Heeks 2018, Varavithya & Esichaikul 2003	Sagarik et al. 2018	Kawtrakul et al. 2011, Kawtrakul et al. 2013, Keretho et al. 2015, Gunawong& Gao 2017, Khampachua, T., & Wisitpongphan 2015, Funilkul et al. 2011, Lorsuwannarat 2006

Source: The author's own analysis of existing literature review

Followed by the government, a number of scholars also emphasize the role of the public and citizens in implementing Thai e-government policy initiatives. Vathanophas et al. (2008) analyze the perceptions ofThai navy finance officers in terms of accepting new e-government development policies. Their study found a number of external factors that influence their perceptions, such as prior experience in using e-government platforms and tools, relevance to their job, commitment, trustworthiness, and autonomy, while other vital factors being training and infrastructure problems. Mirchandani et al. (2008) attempt to assess citizen perceptions (across Thailand and Indonesia) toward the use of e-government services and websites, and find that as perceived by Thai citizens, there is a positive correlation between the importance of financial transaction services and the importance of website efficiency and a positive correlation between the importance of local information services and the importance of citizen identification with the e-government website. Similarly, both Wangpipatwong et al. 2009 and Wangpipatwong et al. 2008 analyze perceptions of Thai citizens toward enhanced use of e-government websites. Wangpipatwong et al. 2009 find three key factors – information quality, system quality and service quality – enhance the continued use of e-government websites by citizens, while Wangpipatwong et al. 2008 concludes that perceived usefulness and perceived ease of use of e-government websites, and citizens' computer self-efficacy enhanced continued intention to use the websites. Rotchanakitumnuai (2008) analyzes perceptions of personal income tax payers towards e-government service quality, and found that the taxpayers tend to value service design, website design, technical support and the quality of client support. Furthermore, the taxpayers point to vital risk factors such as performance reliability, privacy issues in using e-government services and financial audit risk (ibid). Next, Sukasame (2004) analyzes the perceptions of undergraduate students toward the use of e-services through the government web portal (www.thaigov.net) and finds the following key factors correlated with statistical significance: reliability of data, content, linkage (defined as the number and quality of links offered on a web site [Santos 2003, as in Sukasame 2004]), ease of use, and self-service (Table 4).

Table 4. Correlation analysis of key factors influencing student perceptions

Variable		Pearson Correlation	
Dependent	**Independent**	*r*	*p*-**Value**
E-service	H1: Content	.709**	.000
	H2: Linkage	.690**	.000
	H3: Reliability	.784**	.000
	H4: Ease of use	.588**	.000
	H5: Self-service	.705**	.000

** Correlation is significant at the 0.01 level.
Source: Sukasame (2004)

Finally, while the previous scholars (Vathanophas et al. 2008, Mirchandani et al. 2008, Sukasame 2004, and Rotchanakitumnuai 2008) emphasize the importance of analyzing the perceptions of different strata of the Thai public and citizens in assessing the importance and relevance of e-government platforms and systems, Funilkul et al. 2011 critically analyze a range of possible problems in Thai e-government implementation noting that citizens face the followingchallenges when using e-government platforms: a lack in understanding of tools and IT knowledge, access to government provided networks and the quality of networks, citizen awareness of e-services, and a lack of trust toward government IT.

3.2 Understanding Thailand-based e-Government Implementation in Context: The Case of Smart ID Cards

Thailand has begun implementing economic transformation policy aimed to boost growth by means of transforming its economy from industry intensive to high-tech economy. To this regard, in 2016 the government adopted the 2017-2021 Thailand Digital Government Development Plan (see Appendix) and the Thailand 4.0 policy (Kuncinas 2017, as cited in Bukht & Heeks 2018). The Thailand 4.0 strategy seeks to transform the nation from a middle-income toward a value-based economy (Yoon 2016, as in Bukht & Heeks 2018), by promoting innovation, technology-driven economic development, and a move from products-based to service-oriented economy. The key components of the Digital Thailand Plan reflect vital technological trends and include virtual reality, advanced information systems, big data, artificial intelligence, cloud computing, and block chain among others (EABC 2019).

However, as the above section (3.1) demonstrates, scholars criticize the government-led transformation efforts, instead pointing to the need to account for the greater role of citizens (e.g. Sukasame 2004, Khampachua & Wisitpongphan 2015) in driving implementation efforts as related to e-government policy in Thailand. Furthermore, this is triangulated with earlier observations in implementation theory evolution (section 2.1), as international scholars (such as Elmore 1980 and Hjern 1982, as cited in Pülzl & Treib 2006; and Lipsky 1971, Lipsky 2010) call for the need to analyze implementation processes at the level of citizens and customers, thus taking a bottom-up approach and introducing the third generation of implementation research.

It is, therefore, unsurprising that Thai scholars increasingly raise warning signs and point to specific cases of failed e-government projects (e.g. Sagarik et al. 2018). One such case in point is the Smart ID Card project the Shinawatra government attempted to implement beginning 2003. This project arguably falls under the Big Data component of the 2017-2021 Digital Thailand Plan (as detailed in Appendix).

The initial vision of Prime Minister Thaksin Shinawatra was that every Thai citizen would possess smart identification cards, facilitating further efficiency and transparency (Gunawong & Gao 2017). The smart card includes a microchip that allows individual data storage and further connecting individual user databases to other databases (Lorsuwannarat 2006). The Thai government developed a 4–phase goal to implement the Smart ID Card project, as follows (Department of Provincial Administration 2004, as cited in Gunawong & Gao 2017).

Phase 1, 2004. The Smart ID Card was set to substitute free medical treatment cards, social security- and ATM cards. Smart ID Card users could access public and private e-services via electronic machines or websites.

Phase 2, 2005. It was set to substitute debit, credit, and phone cards.It further could be used for recording home address change and voting in referenda.

Phase 3, 2006. It was set to substitute passports and driver's licenses.

Phase 4, 2007. Smart ID Cards would offer alternative uses such as 'dual contact' (Gunawong & Gao 2017).

The initial budget for the Smart Card project was an equivalent of 37 million UK pounds (as in Gunawong & Gao 2017). "There were 5 ministries, 6 organizations and 18 subprojects" under this project (ibid, p. 779). This dovetails with another vital observation (Chutimaskul2002) who noted the free-riding problem as applied to e-government project implementation that involves a large number of government agencies. A major implementation problem related to the ID card project was observed around March 2004 with a conflict among the Smart ID Card government committee about the security of ID card manufacturing (Gunawong & Gao 2017). This particular observation is also reflected in the works of Funilkul et al. (2011), who referred to the unreliability of Thai e-government websites, and Lorsuwannarat (2006) who noted that a key problem in ID card implementation was a lack of government capacity to ensure security standards. Another problem appeared in July 2006 when an auction for ID card production revealed "the possibility of corruption in the process" (as in Gunawong & Gao 2017, p. 779). Corruption in Thailand, more generally, remains widespread, due to the state and regulatory capture that involves top government officials (Mutebi 2008, as cited in Dyussenov 2017). Other problems in ID card implementation include the need to rely on imported microchip technology, lack of citizen participation in policy formulation, lack of agencies' capacity in terms of data reliability, personnel development and public relations (Lorsuwannarat 2006). As a final note, Gunawong & Gao 2017 note that the Thai government administration system continues to operate through a top down approach, where "public servants are dependent on what the government tells them to do" (p. 782).

4. DISCUSSION OF KEY FINDINGS

As the above detailed analysis shows, the two major actors (that should be) involved in Thai e-government development project implementation processes are government agencies, and the public and citizenry (see summary tables 2 and 3). Specifically, on the one hand some scholars (Bukht & Heeks 2018, Varavithya & Esichaikul 2003, Sagarik et al. 2018) praise government efforts or emphasize the key role attributed to government organizations in driving Thai e-government project implementation. On the other hand, however, a bulk of scholars point to the need to include a range of other actors in implementation processes, primarily including the public and local citizens (e.g. Funilkul et al. 2011, Rotchanakitumnuai 2008, Kawtrakul et al. 2013 etc.), but also the business community (Sutanonpaiboon & Pearson 2006) and industry sector representatives (Khampachua & Wisitpongphan 2015, Kawtrakul et al. 2013).

In terms of citizen perceptions toward e-government policy implementation, the following key important value factors are observed: service quality (Wangpipatwong 2009, Rotchanakitumnuai 2008) although Kawtrakul et al. 2013 note recent improvement in service design and innovation; content, reliability of data, linkage, ease of use, and self-service (Sukasame 2004); availability of financial transaction services and local information on websites (Mirchandani et al. 2008); prior experience, job relevance, commitment and trust (Vathanophas et al. 2008), perceived usefulness and ease of use (Wangpipatwong 2008). Furthermore, citizens perceive the following risk factors embedded in e-government project implementation, which should form the basis for policy recommendations: compromised performance reliability, privacy (Rotchanakitumnuai 2008); poor quality of network infrastructure (Vathanophas et al. 2008, Funilkul et al. 2011) and poor user training (Vathanophas et al. 2008), low accuracy of information, website security and ethics, andthe public's unawareness of services offered by e-government websites (Funilkul et al. 2011). Key factors as perceived by citizens are summarized in Table 5.

Table 5. Vital success and risk factors as perceived by citizens in e-government project implementation

Key Success Factors	Major Risk Factors
Service quality (Wangpipatwong 2009, Rotchanakitumnuai 2008, Kawtrakul et al. 2013); content, reliability of data, linkage, ease of use, and self-service (Sukasame 2004); financial transaction and local information services (Mirchandani et al. 2008); prior experience, job relevance, commitment and trust (Vathanophas et al. 2008), perceived usefulness and ease of use (Wangpipatwong 2008)	Performance reliability, privacy (Rotchanakitumnuai 2008); poor quality of network infrastructure (Vathanophas et al. 2008, Funilkul et al. 2011) and poor user training (Vathanophas et al. 2008), low accuracy of information, website security and ethics, and the public's unawareness of services (Funilkul et al. 2011).

Source: The author's own analysis of existing literature review

The above citizen-perceived observations regarding the need to incorporate the public and citizens in driving the implementation of e-government policies are further supported with an analysis of implementation theory evolution. Indeed, as research moved to its third generation beginning the early 1980s, scholars (e.g. Elmore 1980 and Hjern 1982, as cited in Pülzl & Treib 2006) increasingly called for developing bottom up approaches to implementation analyses, thus focusing on the role of citizens in driving policy implementation processes.

All these tentative findings are further triangulated with the literature review of Thai e-government policy implementation (tables 2 and 3). First, the summary in Table 2 may suggest that scholars mainly emphasize greater roles of government agencies (e.g. Bukht & Heeks 2018) and the public (e.g. Funilkul et al. 2011). However, further sentimental analysis (summarized in Table 3) points that the bulk of Thai scholars lament the government dominance in implementing e-government projects (e.g. Gunawong & Gao 2017, Lorsuwannarat 2006, Khampachua & Wisitpongphan 2015 etc.) and call for a more inclusive, bottom up approach to account for the needs of the public (e.g. Vathanophas et al. 2008, Sukasame 2004), the business community (Sutanonpaiboon & Pearson 2006, Kawtrakul et al. 2013) and industry sectors (Khampachua & Wisitpongphan 2015, Kawtrakul et al. 2013).

Thus it does not seem surprising that scholars (e.g. Sagarik et al. 2018, Gunawong & Gao 2017) remain rather cautious when assessing the implementation efforts of e-government policies in Thailand in terms of success or failure. A strong case in point is the Smart ID Card project under the Shinawatra government. Despite benign intent of the government, the project largely ended as failure due to problems such as coordination of a large number of stakeholders (Gunawong & Gao 2017) and free-riding (Chutimaskul 2002), failure to sustain security (Gunawong & Gao 2017, Lorsuwannarat 2006, Funilkul et al. 2011), dependence on imported microchips, lack of citizen participation in policy formulation, lack of agencies' capacity in terms of data reliability, personnel development and public relations (Lorsuwannarat 2006).

Finally, other remaining issues related to Thai e-government program implementation generally include digital divide not only in Thailand (Sagarik et al. 2018) but across much of the developing world (e.g. Margetts et al. 2016); remaining corruption among top-level political actors (Mutebi 2008 as cited in Dyussenov 2017, Gunawong & Gao 2017); and continuing concentration of political power at the central government level at the expense of capacity-building of local governments (Sagarik et al. 2018) and other actors (Gunawong & Gao 2017), such as industry sectors (Khampachua & Wisitpongphan 2015, Kawtrakul et al. 2013).

5. CONCLUDING REMARKS

The implementation of major e-government and ICT development policy in Thailand has been driven by central government authorities, through the means of centralization (i.e. central planning) and distribution of resources (Varavithya & Esichaikul 2003; Chutimaskul 2002). This largely resembles a top-down approach based on hierarchical control and subordination (Stewart, Hedge & Lester 2007, Pülzl & Treib 2006).

This paper finds that the two major actors that should drive Thailand-based e-government and ICT policies are the government and the citizenry, while the role of other actors, e.g. international organizations and business communities, appears less pronounced (Table 2). Although much of the developed world has evolved away from top-down toward either bottom-up or 'hybrid' approaches to e-government and ICT project implementation (Figure 1), the Thai policy implementation context continues to emphasize government dominance in this area.

In terms of citizen and scholarly perceptions, the government is yet to improve its own performance with regard to more efficient policy implementation. Thai citizens tend to point to the problems associated with a lack of adequate training to employ new ICT facilities and infrastructure issues (Vathanophas et al. 2008), low performance reliability, breach of privacy in using e-government services (Rotchanakitumnuai 2008). Some scholars, furthermore, remain skeptical of Thai e-government implementation efficiency pointing to a lack of national data standards, lack of knowledge-sharing in implementation (Kawtrakul et al. 2011), problems related to coordination of multiple government agencies involved (Sagarik et al. 2018), IT privacy and ethical issues, network infrastructure quality (Funikul et al. 2011), and inadequate training for government staff in terms of e-government and ICT project implementation skills (Lorsuwannarat 2006). Thus, key policy recommendations include the need to provide comprehensive training both for citizens in using e-government facilities and for government staff in acquiring skills to better implement e-government and ICT policies, build and put in action e-government privacy protection and ethical policy measures (e.g. Code of ethics), develop common data standards, and motivate best practice and knowledge-sharing among different government units, e.g. by establishing inter-governmental committees involving local citizenry and media.

This research has only focused on an evaluation of Thailand-based e-government and ICT development policy implementation. Further studies should look into more comprehensive comparative analyses of ICT policy evaluation among a number of ASEAN countries, beyond Mirchandani et al.'s (2008) attempt to compare citizen perceptions of using e-government services in Thailand and Indonesia. Yet others might focus on specific 'success' cases in e-government policy implementation in Thailand, analyze enabling factors and attempt to scale them onto a wider policy span with specific recommendations.

REFERENCES

Bukht, R., & Heeks, R. (2018). *Digital Economy Policy: The case example of Thailand*. Development Implications of Digital Economies.

Chutimaskul, W. (2002). E-government Analysis and Modeling. *Knowledge Management in e-Government KMGov-2002, 7*, 113-123.

Dyussenov, M. (2017). Who Sets the Agenda in Thailand? Identifying key actors in Thai corruption policy, 2012–2016. *Journal of Asian Public Policy, 10*(3), 334–364. doi:10.1080/17516234.2017.1326448

Elmore, R. F. (1980). Backward Mapping: Implementation Research and Policy Decisions. *Political Science Quarterly, 94*(4), 601–616. doi:10.2307/2149628

European Association for Business and Commerce. (2019). Retrieved from: https://www.eabc-thailand.org › download

Funilkul, S., Chutimaskul, W., & Chongsuphajaisiddhi, V. (2011, August). E-government Information Quality: A case study of Thailand. In *International Conference on Electronic Government and the Information Systems Perspective* (pp. 227-234). Springer. 10.1007/978-3-642-22961-9_18

Goggin, M. L., Bowman, A. O., Lester, J., & O'Toole, L. (1990). *Implementation Theory and Practice: Toward a Third Generation*. Harper Collins.

Gunawong, P., & Gao, P. (2017). Understanding E-government Failure in the Developing Country Context: A process-oriented study. *Information Technology for Development, 23*(1), 153–178. doi:10.1080/02681102.2016.1269713

Hassan, N. (2019, Oct 12). Getting Digital IDs Right in Southeast Asia: Countries in Southeast Asia are racing to implement digital IDs, but governments must proceed with caution. *The Diplomat*. Retrieved from: https://thediplomat.com/2019/10/getting-digital-ids-right-in-southeast-asia/

Kawtrakul, A., Mulasastra, I., Khampachua, T., & Ruengittinun, S. (2011). The Challenges of Accelerating Connected Government and Beyond: Thailand perspectives. *Electronic Journal of E-Government, 9*(2), 183.

Kawtrakul, A., Pusittigul, A., Ujjin, S., Lertsuchatavanich, U., & Andres, F. (2013, June). Driving connected government implementation with marketing strategies and context-aware service design. In *Proceedings of the European Conference on e-Government* (p. 265). Academic Press.

Keretho, S., Lent, B., Suchaiya, S., & Naklada, S. (2015). Evaluation of National e-Government Development Levels in Thailand. *Evaluation*.

Khampachua, T., & Wisitpongphan, N. (2015, June). Implementing Successful IT Projects in Thailand Public Sectors: A Case Study. In *European Conference on e-Government* (p. 403). Academic Conferences International Limited.

Lipsky, M. (1971). Street-level Bureaucracy and the Analysis of Urban Reform. *Urban Affairs Quarterly, 6*(4), 391–409. doi:10.1177/107808747100600401

Lipsky, M. (2010). *Street-level Bureaucracy: Dilemmas of the individual in public service*. Russell Sage Foundation.

Lorsuwannarat, T. (2006, November). Lessons Learned from E-government in Thailand. *Proceeding of Seminar on 'Modernising the Civil Service in Alignment with National Development Goals', Eastern Regional Organization for Public Administration (EROPA)*.

Margetts, H., John, P., Hale, S., & Yasseri, T. (2015). *Political turbulence: How social media shape collective action*. Princeton University Press. doi:10.2307/j.ctvc773c7

Mirchandani, D. A., Johnson, J. H. Jr, & Joshi, K. (2008). Perspectives of citizens towards e-government in Thailand and Indonesia: A multigroup analysis. *Information Systems Frontiers, 10*(4), 483–497. doi:10.100710796-008-9102-7

Napitupulu, D., & Sensuse, D. I. (2014, October). Toward Maturity Model of e-Government Implementation based on Success Factors. In *2014 International Conference on Advanced Computer Science and Information System* (pp. 108-112). IEEE. 10.1109/ICACSIS.2014.7065887

OECD/ADB. (2019). *Government at a Glance Southeast Asia 2019*. OECD Publishing., doi:10.1787/9789264305915-

Pressman, J. L., & Wildavsky, A. (1973). *Implementation. Barkley and Los Angeles*. University of California Press.

Pülzl, H., & Treib, O. (2006). Implementing Public Policy. In Handbook of Public Policy Analysis. Theory, Politics and Methods. CRC Press.

Rotchanakitumnuai, S. (2008). Measuring E-government Service Value with the E-GOVSQUAL-RISK model. *Business Process Management Journal, 14*(5), 724–737. doi:10.1108/14637150810903075

Sabatier, P. (1986). *Top-down and Bottom-up Approaches to Implementation Research: A critical analysis and suggested synthesis*. Retrieved from: http://0www.jstor.org.opac.sfsu.edu/stable/pdfplus/3998354.pdf?acceptTC=true

Sabatier, P. (1999). *Theories of the Policy Process*. Westview Press.

Sagarik, D., Chansukree, P., Cho, W., & Berman, E. (2018). E-government 4.0 in Thailand: The role of central agencies. *Information Polity*, *23*(3), 343–353. doi:10.3233/IP-180006

Stewart, J. Jr, Hedge, D. M., & Lester, J. P. (2007). *Public policy: An evolutionary approach*. Nelson Education.

Sukasame, N. (2004). The Development of e-Service in Thai Government. *BU Academic Review*, *3*(1), 17–24.

Sutanonpaiboon, J., & Pearson, A. M. (2006). E-commerce Adoption: Perceptions of managers/owners of small-and medium-sized enterprises (SMEs) in Thailand. *Journal of Internet Commerce*, *5*(3), 53–82. doi:10.1300/J179v05n03_03

Varavithya, W., & Esichaikul, V. (2003, September). The Development of Electronic Government: A case study of Thailand. In *International Conference on Electronic Government* (pp. 464-467). Springer. 10.1007/10929179_85

Vathanophas, V., Krittayaphongphun, N., & Klomsiri, C. (2008). Technology Acceptance toward E-government Initiative in Royal Thai Navy. *Transforming Government: People. Process and Policy*, *2*(4), 256–282. doi:10.1108/17506160810917954

Wangpipatwong, S., Chutimaskul, W., & Papasratorn, B. (2008). Understanding Citizen's Continuance Intention to Use e-Government Website: A Composite View of Technology Acceptance Model and Computer Self-Efficacy. *Electronic Journal of E-Government*, *6*(1).

Wangpipatwong, S., Chutimaskul, W., & Papasratorn, B. (2009). Quality Enhancing the Continued Use of E-government Web Sites: Evidence from e-citizens of Thailand. *International Journal of Electronic Government Research*, *5*(1), 19–35. doi:10.4018/jegr.2009092202

ADDITIONAL READING

Bhattarakosol, P. (2003). IT direction in Thailand: Cultivating an e-society. *IT Professional*, *5*(5), 16–20. doi:10.1109/MITP.2003.1235317

Bhuasiri, W., Zo, H., Lee, H., & Ciganek, A. P. (2016). User Acceptance of e-government Services: Examining an e-tax Filing and Payment System in Thailand. *Information Technology for Development*, *22*(4), 672–695. doi:10.1080/02681102.2016.1173001

Chaijenkij, S. (2010). Success factors in E-Government policy development and Implementation: the e-Revenue Project in Thailand.

Chutimaskul, W., & Chongsuphajaisiddhi, V. (2004, May). A framework for developing local e-government. In *IFIP International Working Conference on Knowledge Management in Electronic Government* (pp. 335-340). Springer, Berlin, Heidelberg.

Gunawong, P., & Gao, P. (2010, December). Challenges of egovernment in developing countries: actor-network analysis of Thailand's smart ID card project. In *Proceedings of the 4th ACM/IEEE international conference on information and communication technologies and development* (pp. 1-9). 10.1145/2369220.2369235

Mahmood, Z. (Ed.). (2013). *E-government implementation and practice in developing countries.* IGI Global. doi:10.4018/978-1-4666-4090-0

Palvia, S. C. J., & Sharma, S. S. (2007, December). E-government and e-governance: definitions/domain framework and status around the world. In *International Conference on E-governance* (No. 5, pp. 1-12).

Rahman, H. (2010). Framework of e-governance at the local government level. In *Comparative e-government* (pp. 23–47). Springer. doi:10.1007/978-1-4419-6536-3_2

Rahman, H. (Ed.). (2012). *Cases on Progressions and Challenges in ICT Utilization for Citizen-centric Governance.* IGI Global.

Sorn-In, K., Tuamsuk, K., & Chaopanon, W. (2015). Factors affecting the development of e-government using a citizen-centric approach. *Journal of Science & Technology Policy Management.*

Suchaiya, S., & Keretho, S. (2014, September). Analyzing national e-Government interoperability frameworks: A case of Thailand. In *Ninth International Conference on Digital Information Management (ICDIM 2014)* (pp. 51-56). IEEE. 10.1109/ICDIM.2014.6991416

Sunalai, A., & Tubtimhin, J. (2009). Thailand e-government: A step forward into the right track. *Global e-governance-Advancing e-governance through innovation and leadership, 2*, 3-25.

Thompson, N., Mullins, A., & Chongsutakawewong, T. (2020). Does high e-government adoption assure stronger security? Results from a cross-country analysis of Australia and Thailand. *Government Information Quarterly, 37*(1), 101408. doi:10.1016/j.giq.2019.101408

Thuvasethakul, C., & Koanantakool, T. (2002, March). National ICT policy in Thailand. In *Africa-Asia Workshop: Promoting Co-operation in Information and Communications Technologies Development.*

Thuvasethakul, D., & Pooparadai, K. (2003). ICT Human Resource Development within Thailand ICT Policies Context. In *Proceedings of the Second Asian Forum for Information Technology (2nd AFIT), Ulaabaatar, Mongolia* (pp. 1-11).

Waehama, W., McGrath, M., Korthaus, A., & Fong, M. (2014). ICT Adoption and the UTAUT Model. In *Proceedings of the International Conference on Educational Technology with Information Technology* (Vol. 17, pp. 24-30).

Wagner, C., Cheung, K., Lee, F., & Ip, R. (2003). Enhancing e-government in developing countries: Managing knowledge through virtual communities. *The Electronic Journal on Information Systems in Developing Countries, 14*(1), 1–20. doi:10.1002/j.1681-4835.2003.tb00095.x

Wescott, C. G. (2001). E-Government in the Asia-pacific region. *Asian Journal of Political Science*, *9*(2), 1–24. doi:10.1080/02185370108434189

KEY TERMS AND DEFINITIONS

E-Government Policy: Can be defined as a tool employed by central government authorities to facilitate the Thailand's transformation towards becoming a knowledge-based society (Lorsuwannarat 2006).

ICT: Information and communications technology is a term that stresses the role of unified communications and integration of telecommunications and computers that allow users to access, store and share information (Wikipedia). ICT can be defined as a facilitating factor to propel national economic, social development and to improve competitiveness as applied to Thai context (Thuvasethakul & Koanantakool 2002).

IT National Plan: Is broadly defined as a central government document that seeks to enforce local agencies to implement the e-government development policy (Chutimaskul 2002).

Policy Implementation: Can be viewed as an "interaction between the setting of goals and actions geared to achieve them" (Pressman & Wildavsky 1973, p. xv). Implementation can be defined as one of the key stages in policymaking usually followed by evaluation.

Thai ID Card: A national government project aimed at issuing official government documents to Thai citizens aged between 7 and 70 (Wikipedia). It is largely viewed as an e-government failure case by scholars (e.g., Gunawong & Gao 2017).

Thailand 4.0 Strategy: Is the keystone of Thailand 4.0 policy that seeks to transform the nation from a middle-income toward a value-based economy (Yoon 2016, as in Bukht & Heeks, 2018), by promoting innovation, technology-driven economic development, and a move from products-based to service-oriented economy.

APPENDIX

"Key Components of the Thailand Digital Government Development Plan 2017-2021": Strategic Technology Trends for Digital Government

- **Virtual Reality / Augmented Reality:** Application of Virtual Reality (VR) and Augmented Reality (AR) technologies in simulating environment or situations for the purpose of public safety management, telemedicine, and new formats of education and tourism
- **Advanced Geographic Information System:** Application of Advanced Geographic Information System technology in geographical data management, as well as its applications in management of agricultural resources, transportation system and other areas
- **Big Data:** Processing big data and make forecasts and estimations in business environment, using Internet of Things (IoT) and Smart Machine technologies to perform real time analysis and responses with users
- **Open Any Data:** Disclose informative data to users through refurbishment of database and website to allow wider public access and promote linkage of those disclosed data with other entities
- **Smart Machine / Artificial Intelligence:** Application of Smart Machine technology to enable management and responses of automated services - the Smart Machine system will gradually evolve and consequently be able to evaluate and address problems throughout the service supply chain
- **Cloud Computing:** Application of Cloud Computing technology for data storage to reduce complication in system installation, reduce system maintenance cost, and save network establishment investment
- **Cyber Security:** Addressing cyber security issues by setting cyber security standards, revising related regulations to make them more updated and flexible, as well as reforming the mindsets in handling cyber security issues
- **Internet of Things:** Using the Internet of Things (IoT) technology to facilitate the transformation of government services into digital formats, and at the same time, the IoT technology can also support government's works in communication, utilization of mobile technology, analyzing big data, and cooperation with private business sector
- **Block Chain / Distributed Ledger Technology:** Application of Block Chain technology in data storage and utilization of the network for the purpose of verification and reduction of intermediaries under reliable security environment

Source: https://www.eabc-thailand.org › download

Chapter 12
An Evaluation of Measuring the Publicness Level of Interiors in Public Building Design:
Visual Graph Analysis (VGA) Approach

Pelin Aykutlar
https://orcid.org/0000-0002-5157-9919
İzmir Kavram Vocational School, Turkey

Seçkin Kutucu
https://orcid.org/0000-0002-7035-2656
Yasar University, Turkey

Işın Can-Traunmüller
İzmir Institute of Technology, Turkey

ABSTRACT

This study examines the publicness level of the interior spaces of public buildings. As a method, VGA (visual graph analysis) is used for analyzing the early design phases of selected municipal service buildings. In this study, the authors utilized from VGA for quantifying the publicness level of the two selected architectural competitions of municipality buildings. The method allows us analyzing the floor plans of each project in obtaining an eventual assessment of permeability and accessibility which give an idea of the levels of publicness comparatively. Subsequently, representation parameters are compared under two main criteria: connectivity and integration. The aim of the study is to understand the level of publicness and efficiency of spatial settings for the users circulating in the public buildings, which have dissimilar plan schemes. This method would be used by the designers for early design stage and provide useful feedback for understanding the level of accessibility and permeability of the structures and adjust their schemes accordingly.

DOI: 10.4018/978-1-7998-5849-2.ch012

1. INTRODUCTION AND MOTIVATION

Municipal service buildings, which reflect public structure, identity and the society's periodic ideological stance, represent an important type in these public administration structures. Therefore, public administration buildings and within these, municipal service buildings, assume an important role of visual mediation between the public and the administration. Functional and formal maturity is simply not sufficient by itself for a representative aura of municipal service buildings. This is why municipal service building designs represent an important type in terms of examining the concept of publicness and publicness value. Additionally, public buildings are defined as a public domain, where everybody can use the space in an equal way, and which does not belong to a particular person, affinity group or foundation. In this context, public usage becomes crucial.

In the late 20th century, public realm discussions increased in the global architectural agenda after Hannah Arendt, Jürgen Habermas and Richard Sennett published their explorations on public space. In 1984, in Turkey, after the drastic social and political changes of 1980, one of the most significant liberal changes introduced was the sovereignty of local municipalities in the development of a master plan, instead of a central government and The Ministry of Public Works. The aim of this this study is to determine the changes in the design of public space use of service buildings. In order to do that, this research focuses on analyzing selected architectural competitions of municipality service buildings from 1984 to 2013. This study focus on the selected architectural competitions of municipality buildings for design experimentation and simulation that can be used at an early design stage. In pursuing this specific goal, the following questions constitute the core of this study:

- How can we interpret the publicness level in the spatial layout of municipality service building design of architectural competitions?
- How does the publicness in the spatial layout of the architectural design competition projects differentiate?
- How can we understand the forms of publicness through examining the physical structure of architectural designs?
- Can Visibility Graph Analysis (VGA) method be used as a quantitative tool to determine the differences in the publicness levels of the projects in terms of permeability and integration?

This study's content is based on selected municipality buildings with criteria to determine the publicness level through permeability. The criteria are consisted of being chosen from the national competitions, which concern the architectural programs of municipality service buildings. The selected projects were picked on the basis of the precondition that they are smaller than 20.000 m^2 and they do not serve for any other public utilities and neither contain any function that belongs to the buildings of other types. This filtration process provided a shortlist of two selected projects.

In this study, however, only plan schemes have been used, because this study does not only analyze public measures according to real structures and spaces, but also examines from the perspectives of the designers and their design reports. These levels of publicness belong to closed spaces of the selected municipality service building projects. The study focuses exclusively on the publicness levels of closed spaces, as opposed to open public spaces. In this regard, it provides a new frame of measuring levels of publicness for these locations, which can be measured by integration and connectivity measures of visibility graph analysis.

1.1 Method of the Study

Architectural design competitions are the main sources of creative novelty in the architectural history of a country. Jury reports, competition contracts, jury criteria for the winning projects are the sources that transfer information tools. In this context, architectural design competition projects become the repository for architectural understanding over long time-ranges. Statistics of architectural design competitions indicate that these are mostly opened for public buildings, especially for municipality buildings in Turkey. The method of this study is based on space organization analysis of selected architectural competitions of municipality building designs' layouts through their level of publicness. This measuring is based on the conceptions of the selected projects. Interpretation of results stand on the measure of integration and the correlation between most used common spaces and their functions, correlation of public areas and the cores and way-finding. Therefore, this study investigates the relationship between the spatial layout of the selected architectural competitions of municipality buildings and their public usage. Additionally, the level of publicness in the selected floors of the buildings are examined through the original design phase of each building. Project's floor plan layouts are analyzed by the visibility graph analyses. Subsequently, the results of the analyses are normalized to compare with each other in consequence of having different plan type.

2. SPACE SYNTAX AS A QUANTITATIVE APPROACH IN SPATIAL ANALYSIS

Space syntax is a theory and a method to analyze spaces both in urban and building scale. The method was developed by the research team led by Bill Hillier in the University Collage London (UCL) since late 1970s. As mentioned in the book of 'The Social Logic of Space' written by Hillier and Hanson in 1984, space syntax method is used to understand the relation of social life and space. The theory based on that social life comes out from the space's physical organization (Hillier and Hanson, 1984). There are studies on space syntax that have analyzed the relation between spatial layout and movement (Koohsari, M. et al., 2019), (Turner, A. et al., 2001), (Othman, F. et al., 2019), communication (Bafna, 2003), personnel encounters (Goldfarb, M., Donegan, L., 2017), co-awareness and way-finding (Hedhoud, A., et al. 2014). Space syntax method is used as an analytical method, which scholars have used to describe spatial arrangements on permeability that refer to physical environments with spatial behaviors (Hillier, 1984).

Space is a vital environment that is comprised of bringing individuals together, allowing them to perform their actions and detectable limits. There are lots of definition of space until today (Lefebvre, 1991), (Soja, 1996), (Hasol, 1998). Kuban (1998) mentioned that space within a building is a phenomenon that created by restricted space and common elements of the limits together and it is impossible to define it with only its limits and volume. Lefebvre (1991) defines space as a social product or a complex system that based on values and social production which affects spatial perceptions and practices (Lefebvre, 1991). It can be described in accordance with geometrical rules or rational aspects and perceived by anyone who moves in space as a subject different, emotional and irrational (Ataç, 1990). Benedikt (1979) describes space as;

"Historically psychologists and architects have shared a vital interest in the nature of space. Coinciding with the birth of modern experimental psychology, it was the late nineteenth century when space was first propounded as being of the essence in the experience of architecture" (Benedikt, 1979, p.20).

Physical components have a great impact in the formation of space together with the influences of human behaviors and relationships. Social and cultural structure are the predictors of human's behaviors to the contrary physical spaces. There are two main subject of a spatial form. These are the residents, who live within the space and the relationships of the residents with visitors who are coming from outside (Hillier and Hanson, 1984).

According to Hillier (1993); space is one of the primary means by which the ascent of a building as cultural transmission to architecture and a theoretical intent is made. This means that one aspect of the abstract comparability of forms in architecture centers on spatial form, which implies space as an objective property of buildings. Hence, buildings and cities stand for us in two different ways; as the physical forms that we see, and as the spaces that we use and move through. After the late nineteenth century and onwards, architecture began to represent theories about space. During the twentieth century, space was increasingly articulated as a dimension of architectural expression. By the end of this century, most architectural and urban theories include a chapter related with space (Hillier, 2005).

On the other side, Hillier (1993) explains what spatial forms carry in the book of *Space is the Machine* as below:

"It is because this is so that spatial organization through buildings and built environments becomes one of the principle ways in which culture is made real for us in the material world, and it is because this is so that buildings can, and normally do, carry social ideas within their spatial forms. To say this does not imply determinism between space to society, simply that space is always likely to be structured in the spatial image of a social process of some kind" (Hillier, 1993, p. 52).

According to Hillier and his colleagues (Hillier and Hanson, 1984), the relation between social structure and space have a mutual interaction. Space is a product that is affected by the society and social structure, as well as it affects the society (Hillier and Hanson, 1984). The concept of spatial enclosure defines the space by reference to the physical forms. (Hillier, 1993). Hence, space syntax focuses on creating a platform for space and society to give a spatial nature to society as well as giving a social dimension to space (Karimi,1997). Today spatial approach is benefited from architecture, urban design, planning, transport and interior architecture to archeology, information technology, urban and human geography, anthropology, landscape architecture and informatics. This method is not only being used to understand the city's physical components and the relationships between them, but also aims to understand the social, economic and conceptual components and the relationship between the physical components. Space syntax, including housing and urban scale is used to analyze the spatial organization of different scales (Steadman, 2014).

Hillier and his colleagues (1987, p.217) define space syntax as "a model for representation, analysis and interpretation". Building entrances have a role in forming the relation between the inside and the outside as well as the residents and the outsiders. The understanding of the method stands on how buildings gather together and define a continuous open system (Hillier et al., 1987).

Space is not the background to human activity, but more than that it is the background to the movement (Hillier, 2005). So there is more an intricate relationship between space and movement. Space is as a fundamental state of everything human beings do (See Figure 1.). In this sense, there are three types of spaces. Linear space, as people move through space, convex space as people interact with each other in a space, and isovist space as people have different point of view from a particular point in space. How we use or experience space is described by each of these geometric organizations. For this reason, how we

Figure 1. Movement and Space
(Hillier, 2005)

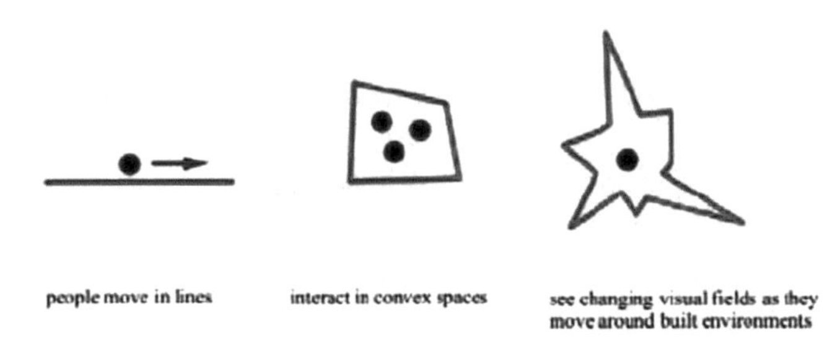

create, use or understand spatial configuration stands on how the buildings and cities are organized in terms of these geometric ideas. Geometric language reflecting human behavior and experience creates the language of the city (Hillier, 2005).

Space syntax methodology identifies relational characteristic of space as spatial configuration and proposed the idea that it is its characteristic forms which define human behavior and social knowledge. Space syntax aims to develop strategies for the description of configuring inhabited spaces with underlying social meanings. Effects of spatial configuration on various social or cultural variables let the research field to develop more practical explanations. Thus, the methodology seeks to understand not only the configured space itself, but also its developmental process and its social meaning (Bafna, 2003).

2.1 Public Space Studies in Space Syntax

Space syntax covers the properties as spatial and physical characteristics of space, accessibility of movement networks (pedestrians, bus, cars etc.), pattern of land-use attractors and quality of public realm. Besides, space syntax has pioneered the development of new techniques for the public space. There are various studies examining the accessibility of public squares within cities via using space syntax method. These studies either assessed how integrated those public spaces within the urban grain of the city or examined their location for the possibility of use by various residents or visitors (Ruben, T., 2012), (Knöll, M. et al., 2015), (A. Barry, G. Thomas et al., 2012). Hillier (1984) studied on the performance of public spaces that a successful urban square depends on the correct balance between static and moving people, whereas the number of people choosing to stop and make informal use of the public space. This value is calculated by the sum of integration values of all lines. (Hillier, 1984).

The most important point in space, which comes together and creates a meaningful whole is relational structure. In order to understand relational structures of space, *morphological studies* should be carried out. Morphology, in the most general sense is known as physical form or structure formation, -morph meaning shape, fom and –ology meaning knowledge (Hamawand, 2011), in "Architectural Morphology "(1983), it is mentioned that past and present design mainly deals with the composition and shape of architectural elements and bringing together the two-dimensional space and its elements. In order to determine the process of bringing the spaces together, it is emphasized that spatial relationships should be understood and efforts should be made to solve the structure formation (Steadman, 1983). The concept of building and space analysis in architecture is defined by the structured physical environment. This definition occurs according to different criteria as a result of the architectural design and building construction. Configuration on space syntax techniques defines the abstract relational order of the structure's characteristic forms (Hillier and Hanson, 1997). It does not only define the simple relationship but also the complex relationship between each element.

In many areas such as; architecture, urban planning, interior design, landscape architecture, transport and IT, space syntax methods are used. In parallel with, representation graph varieties and scales belonging to "Space Syntax" and "VGA" method, application areas and units of measure are discussed. VGA focuses on building scale. Amongst the classification of the spatial analysis in the research, VGA will be used as a method to understand the spatial configurations and linkages between physical environments. Besides, VGA is identified as a method for providing social understanding of buildings in terms of way-finding. The reason for application of this scientific method is the numerical and statistical data desired to be obtained in the changes of the selected public projects for further correlations between integration, connectivity and circulation. Mapping of the space organization and the accessibility of the changed situation through forms of publicness is examined with this method. It helps to enunciate the social meaning behind the confirming inhabited spaces. When trying to find measurable indicators for each of the themes of publicness, it is understood that a crucial aspect related to a space's publicness is its accessibility. The accessibility of a public space can be judged by its own connections to its surroundings both physical and visual. A successful public space is visible both from a distance and up close as well as easy to get to get and get through (Whyte, 2000). Based on the small-scale buildings and public building type studies, VGA method is used.

2.2 Visibility Graph (ISOVIST)

The application of visibility graph analysis to building environments was first introduced as early as 1980 by Braaksma and Cook (Turner, 2001). They calculated the co-visibility of various units within an airport layout, and produce an adjacency matrix to represent these relationships, placing a "1" in the matrix, where two locations are mutually visible, and a "0" where they are not. From this matrix, they present a measure to compare the number of existing visibility relationships with the number which could possibly exist, in order to quantify how usefully a plan of an airport satisfies a goal of total mutual visibility of locations (Tahar and Brown, 2003).

Visibility analysis and the use of isovists are introduced in the form of analysis, isovists and isovists fields by Benedikt (1979), in which he defines isovists as 'the set of all points visible from a given vantage point in space and with respect to an environment' (Benedikt, 1979, p.49). Measures of isovists,

Figure 2. The isovist and isovist field
(Turner, 2004)

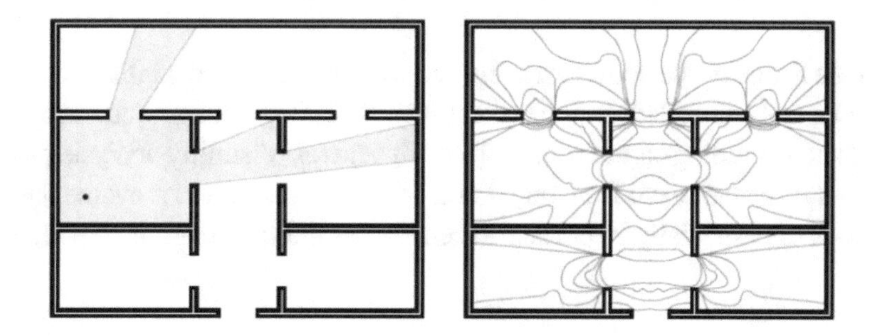

such as their areas, perimeters, circularity and radials variance and skewness can be used to compare the quality of different spatial experiences (See Figure 2).

Established on the idea from Benedikt (1979) that involves the experience of a space is related to the interplay of isovists, Turner and his colleagues (Turner et al., 2001) developed the technique of *visual graph*. This is used to determine how visible any point in the spatial configuration is from any other point. Based on the technique, they developed the software which is called "Depthmap", that divides any given plan into a grid, whose size can be determined by the user. All mutually visible points across the grid are connected. The resulted visibility graph has two sets of elements, the set of vertices and the set of edge connections joining pairs of vertices. The properties of isovist are represented in several different measures based on (according to the) the number of vertices and edges (Turner et al., 2001). Visible points can be transferred into accessible points in the context of program. All mutually accessible points across the grid are connected as well. Accessibility analysis regards glass walls and ponds as blocks considering accessibility whereas these are visible in visibility analysis based on the aim of the study. All mutually accessible points across the grid are connected as well. Based on the aim of the study, visibility analysis and accessibility analysis are used. This method is to understand how visual characteristics at locations are related and one that has a potential `social' interpretation. Graph based representations used in social theories of networks lead us to use isovists to derive a visibility graph of the environment, the graph of mutually visible locations in a spatial layout. Through movement and occupation of the environment that the graph represents, the effects of spatial structure on social function in architectural spaces become defined (Turner et al., 2001). In this sense, the visibility graph is a tool with which we can begin consciously to explore the visibility and permeability relations in spatial systems. The relation between visibility and permeability is a vital component of how systems work spatially and are experienced by their occupants (Tahar and Brown, 2003). In invisibility graph analysis, it is aimed to analyze the organization of movement in complex structures such as museums, hospitals and exhibition centers, the analysis of important parameters to improve the quality and sharing of structures especially like museums, pre-construction choice of location for any structure or activity, estimate of the effects of the newly added structure to the urban scape in the context of organization of movement, determination of the interior organization of architecture, investigation of indoor use in the historical development of traditional architecture and to investigate the effect of space organization in the variations of the typology of structure (Su & Sailer, 2015).

Visibility Graph Analysis (VGA), one of the spatial analysis techniques that particularly emphasize the role of visual information on space syntax, concerns the effect of the visual information on the choice of movement routes. In order to investigate the relationship between spatial layout, the delivery of social network spaces, and space use patterns are directly observed from Visibility Graph Analysis (Turner, 2001). Subsequently, the predicted movement of sighted persons in the same spaces is determined by the use of the Depthmap software. From the results provided by this software we utilized the connectivity and integration values.

Depending on the nature of the boundaries, the accessibility, i.e. permeability, and visibility between inside and outside can be controlled. Both permeability, where you can go and visibility what you can see directly, affects how buildings are xxx organized? in general. Visibility analysis provides that visual fields have their own form that result from the interaction of geometry and movement and that the shape and size of the isovist is especially important in relation to the information provided to the observer (Güney, 2007).

Visibility Graph Analysis is used to calculate the visual integration and connectivity of each building in a technical way. Apart from storage areas, elevators and staircases, all spaces were included in the analysis.

2.3 Syntactic Measures of Space Syntax

The spatial relationship between the spatial elements such as boundaries, convex spaces, axial lines and units can then be described by several measures. By applying these measures to the description of building form, space syntax scholars believe they can capture the spatial and functional differences in different plans (Chai, 2012). Hillier and colleagues (1983) consider each place as a system in which each point has two dimensions. The first dimension is the immediate relationship of the point with its environment (local dimension), the second dimension is the point's place in the overall system (global dimension). These two features come together to create patterns with different characteristics (Hillier, 1983). In other words, spaces create sub-spaces with different degrees of integration and perception. Thus, analysis of the graph is divided into two types as global and local measures. Global measure refers to be constructed using information from all the vertices in graph whereas local measures refer to be constructed information from the immediate neighborhood of each vertex in the graph. The user may elect to perform both or either of these types of measure by selecting from the program according to the aim of the research (Turner, 2001).

Briefly, connectivity and integration are reviewed as the representation of spatial elements in the building layouts of this study. Remarkable result with respect to way-finding and usability issues is the correlation of integration with connectivity. Hence, this study utilizes from the two space syntax measures; *integration* and *connectivity.*

2.3.1 Connectivity

Connectivity is a local spatial property that refers to how many immediate neighbors each node can see. It refers to the degree of direct visual connection (Turner, 2004). Connectivity or degree of a node captures the amount of space directly visible or accessible from another space. It measures the depth between spaces and refers to the degree of intersection.

In the analyses routes with high connectivity values are shown with warm colors. Spaces, which are connected to more spaces are respectively shown in red, orange and yellow colors means that these spaces are more permeable and connected to other spaces in the configuration. Additionally, the correlation between local and global values is called 'intelligibility'. The spaces, which have the higher value of local and global means that these spaces have the high potential for meeting points.

2.3.2 Integration

Integration value is a global measurement. It refers to a measurement of the depth of a space within the existing relationship. If the integration value is low, it indicates that the space is shallow, means that the space is not integrated to the whole spatial organization. (Hillier, 2007). In other words, less integrated spaces mean they are the places visited less by the people. Integration measures how many turns and changes one has to make in order to access from one space to the other in the whole system. It discloses how a space is related with the whole in terms of integratedness and segregatedness (Can, 2011). A place has the value of being a strong option means choice, when it has the shortest transport routes, it is linked to all places in the system and pass through facilities have been provided. These differences manage the effect of space of movements within the system. Less deep space will attract more movement and the deeper space will attract less movement (Hillier, 2001). This depth will give the most important formation related to the whole which is the integration value. This states clearly the inverse relationship between the value of depth and integration values. It is a measure which indicates the encounter rate and intensity of use. If more accessible places are considered as syntactic centers, integration values are used as criteria for the comparison of different sized systems. The higher the integration value of a site line the shallower the place, the lower the integration value the deeper the place (Hillier, 1983). In other words, the higher the integration value the more accessible the place, the lower the value the more difficult it is to access the space in the system. Integrated places have the potential to bring together all the people that live in a place or who are there for any reason. The most integrated spaces are places where you may even pass through to go somewhere else. These places are called integrated cores. Integration value; is calculated by calculating the average depth value for each line for all lines in the system (n). This is the global integration (R-n) value for the whole area (Hillier, 1984).

Integration value is the main criteria for space shaping parameters. According to this, accessibility rates are expressed respectively with red, orange and yellow color lines (Gündoğdu, 2014).

Spaces of the whole system can be ranged according to their integration levels. Space which stands in the middle of the whole system, the spaces around it get increased then it shows that the space is integrated. It shows that the space can be accessible according to its integration level. According to Hillier (2005), the closeness of each element to all others is in fact the integration value of a space, we can color from red for high integration through to blue for low in order to understand the degree of accessibility (Hillier, 2005). They are represented as a spectral range from indigo for low values through blue, cyan, green, yellow, orange, red to magenta for high values (Turner, 2001).

3. CASE STUDIES: PUBLICNESS CONVERGENCE OF MUNICIPALITY BUILDINGS

The selected projects on municipality service buildings in Turkey are examined by VGA analysis. This method allows the selected floor plans of each project to be analyzed in purpose of obtaining an eventual assessment of permeability, which is the level of publicness. Collected results in the form of thematic maps and mathematical values that reflect permeability levels are then compared with the jury reports of each competition. Chosen from the national design competitions which concern the architectural programs of municipality service buildings, the selected projects were picked on the basis of the pre-condition that they are smaller than 20.000 m^2, they do not serve for any other public utilities and neither contain any function that belongs to the buildings of other types to make them comparable with each other. This filtration process provided a shortlist of two selected projects. The data for the analysis was prepared by conducting a CAD file drawing of each project plan. This data was then specifically prepared on the basis of Depthmap requirements in order to be made ready for the eventual VGA analysis.

Towards analyzing the integration and connectivity relations of specific municipality service buildings of architectural design competitions, a VGA is implemented on each building's chosen floor on the basis of its relation with the ground level and public usage. All spaces except storage areas are included in the analysis. It is important to emphasize that the set grid parameter in the Depthmap program was selected to be a standard 0.5 at all projects since this allowed the specific identification of permeability and integration levels in the study to be defined on a common basis of publicness level. Visibility representation parameters are compared under two main criteria: *Connectivity* and *Global Integration*. The adjacency matrix specifies the relationship between the locations by allowing '1' to indicate the mutually-integrated locations and '0', to the non-integrated locations. In consequence of this, the VGA integration measure was found to be a highly significant discriminator between the preferred and non-preferred locations in terms of privacy. The integration and connectivity maps obtained from the analyses are interpreted according to a color chart that exhibits a range from red to blue. The chart basically indicates that the red color represents a high level of publicness and connectivity, whereas purple or the darkest blue represents a high level of privacy and lower level of connectivity (See Figure 3.).

Figure 3. Color range of Depthmap

Since the building layouts comprise of different scales and involve detached buildings on the same floor, the integration and connectivity values obtained from each building were normalized through dividing these specific values into the number of the grid cells of the nodes. This calculation procured an integration and connectivity value for each node, ensuring different buildings to become comparable on a standard plain. Total visual node counts were taken from Depthmap. Subsequently, high integration and connectivity values belonging to each of the building layouts were taken from Depthmap column properties for calculating the selected visual node counts with high integration and connectivity values. In order to calculate the percentage of areas with high levels, the selected visual node counts with high integration and connectivity values were proportioned to the total visual node count of each layout. These values and percentages were evaluated with circulation percentage of the projects. This is how the circulation areas reveal public usage and accessibility. Each of the projects is assessed on the basis of four criteria, listed below:

- Remarks in the report that reveal jury's evaluations on publicness and public use of the project.
- Remarks in the report that reveal competitors' interpretations on publicness and public use of the project.
- VGA analysis and its results; integration and connectivity levels, critical access points (their relation between the core – staircases and elevations), the most and the least accessible and integrated public functions of the layouts.
- Comparison of the relative percentage of high connectivity and integration levels with the relative percentage of circulation areas of each project.

In order to provide a series of sample buildings which can be studied through comparative structural analysis, the below criteria has been followed:

- The examined architectural competitions were limited only with national architectural competitions in Turkey.
- Architectural Design Competitions were selected if only their program covers administrative and public functions of municipality service buildings. The competitions, which also include mixed used functions were excluded in order to provide programmatic homogeneity of the samples.
- The size of the selected samples was narrowed down to a maximum of 20.000 sqm. in closed space. Because bigger municipality building's spatial layout changes as its corporate structure changes.

Within these framework, nine projects were selected in the thesis whereas in this study two projects are discussed regarding plan type as; fragmental and compact: Gaziantep Municipality Service Building project and Karabük Municipality Service Building project.

3.1. Architectural Competition of Gaziantep Municipality Service Building Design (1986)

Gaziantep Municipality Service Building Design Competition was held by Gaziantep Metropolitan Municipality in 1986. Hasan Özbay and Tamer Başbuğ were awarded by the jury. The project is approximately 20.000 m^2 (Circulation, elevators and staircases etc. included) (See Fig. 4., 5.).

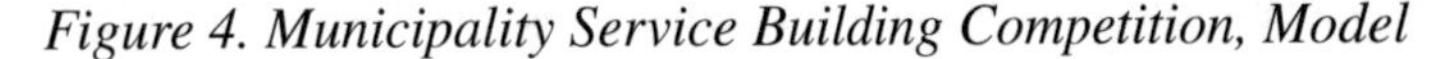

Figure 4. Municipality Service Building Competition, Model

3.1.1. Design Approach of the Designer

According to the designers, the main idea of design relies entirely on the 'council chamber'. The platform that rises through the eastern and western directions ends by the council chamber; constituting the main entrance. At the very center of the geometry, the Council Block (Figure 4. and 5., Building SG2) possesses important functions. Its ground floor includes the main entrance, its sub-ground floor the dining hall and its basement floor, the kitchen and depots; all of which are the common spaces used by all of the units of the building (See Figure 4. and 5.). The council chamber structure around the other four

Figure 5. Gaziantep Municipality Service Building Competition, First Prize, Site plan

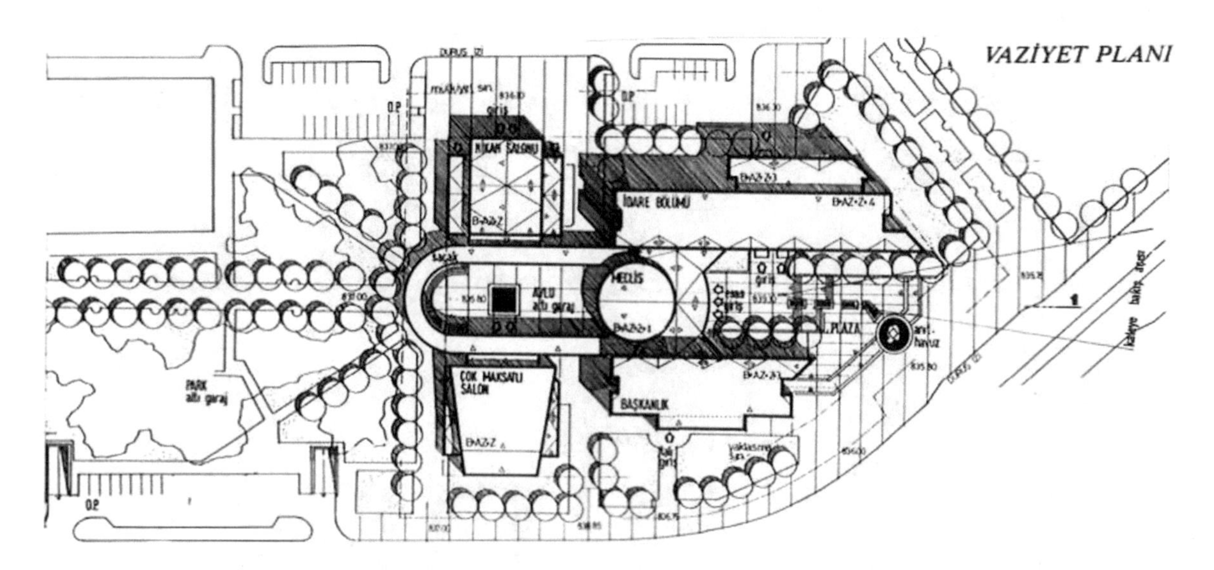

masses constitutes the president's office and relevant offices. The other side of the president's office is inhabited by administrative offices. The remaining two blocks are located around the yard, incorporating the wedding hall and a multi-purpose hall. (See Figure 4. And 5.) Thus, each single unit is defined as a proper functional group and the function groups altogether complete a whole.

3.1.2. Jury's Evaluations on the Project

The jury members have found the overall composition formed by the scaled outdoor spaces on the background of the outside blocks from the city centre compliant and positive. The continuity of the entrance axis that develops through the east and west; the controlled entrances located at the northern and southern directions and the continuity of the spaces that remain between the outdoor spaces and the blocks have also been found favorable. It is furthermore emphasized in the report that the indoor and vertical circulations in the project bear an excellent relation. The jury deemed that there exists sufficient wideness in the indoor circulation and that the milestones of this circulation, as well as the direction choices in the general planning, the publicness function of the yard, the relation of the wedding hall with the car park and the entrances are all well-defined. The plastic of the outdoor and indoor spaces in addition to the mass in scale of the indoor and outdoor spaces were especially found successful. The building's public use was given importance by the jury and specifically mentioned in the jury report. On the other hand, the wedding and meeting halls were found small in terms of size by the jury. It is reported that the lack of visual connections of spaces and the invisibility of the entrances from the highest plato are found inadequate in means of accessibility (Specifications, 2015).

3.1.3. VGA Analysis and Results

According to the results put forth by the integration and connectivity maps of the sub-ground floor, the highest connectivity and integration value belongs to the part of the foyer at the wedding hall entrance

Figure 6. Gaziantep Municipality Service Building Competition, First Prize, Connectivity map of sub-ground floor plan

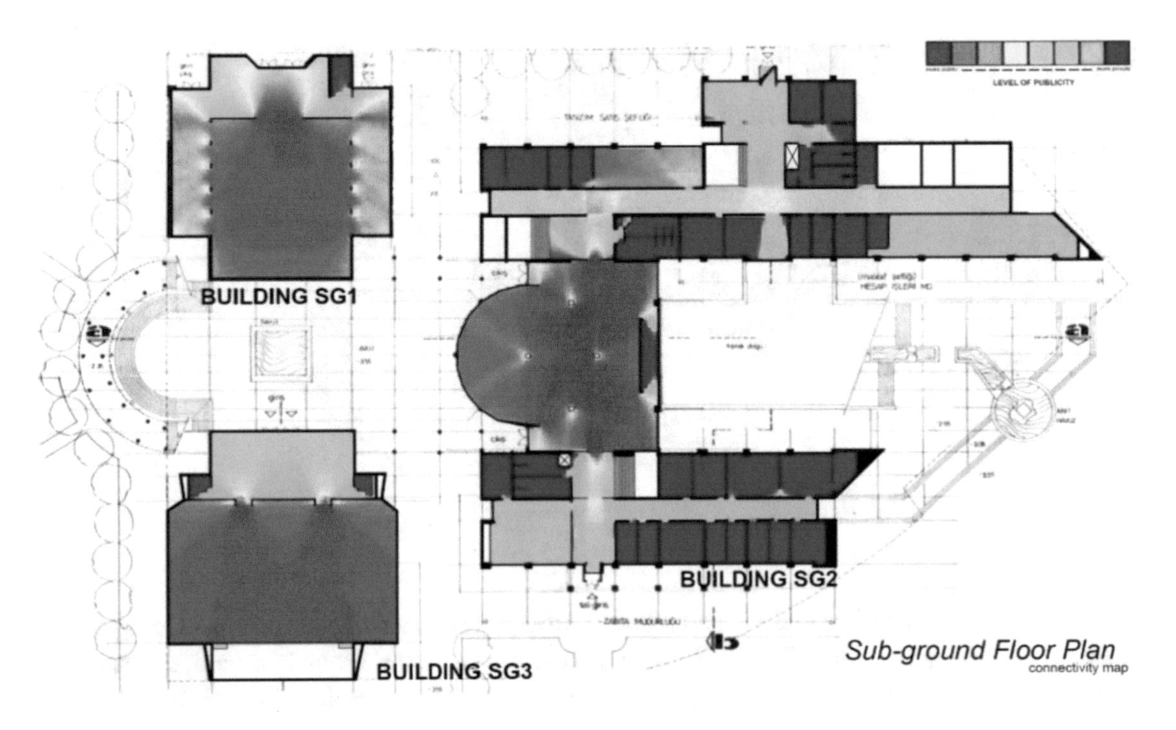

Figure 7. Gaziantep Municipality Service Building Competition, First Prize, Integration map of sub-ground floor plan

in Building SG1 (HH Value:107.475, Connectivity Value:2298). The lowest connectivity and integration value meanwhile, is identified at the staircase that is close to the entrance. (Connectivity Value: 83, Integration Value: 9.04) It can be figured out from the layout of the Building SG1 that the wedding hall has been planned as the main public space (See Figure 6.).

The integration and connectivity maps of Building SG 2 on Sub-ground plan indicate that the highest integration value belongs to the dining hall (HH Value: 11.73, Connectivity Value: 2146) Spaces located at the northern and southern part of the plan scheme, such as the offices, are the deepest spaces of the layout. (Connectivity Value: 29, Integration Value: 2.37) (See Figure 8. and 9.) (See Table 1.).

The multi-purpose hall is the space with the highest connectivity and integration value at Building SG3 (Connectivity Value: 1804, Integration Value: 55.15). The lowest connectivity and integration value meanwhile, belongs to the service places on the eastern and west side of the foyer in Building SG3 (Connectivity Value: 10, Integration Value: 4.77) (See Figure 8. and 9.).

Figure 8. Gaziantep Municipality Service Building Competition, First Prize, Connectivity map of ground floor plan

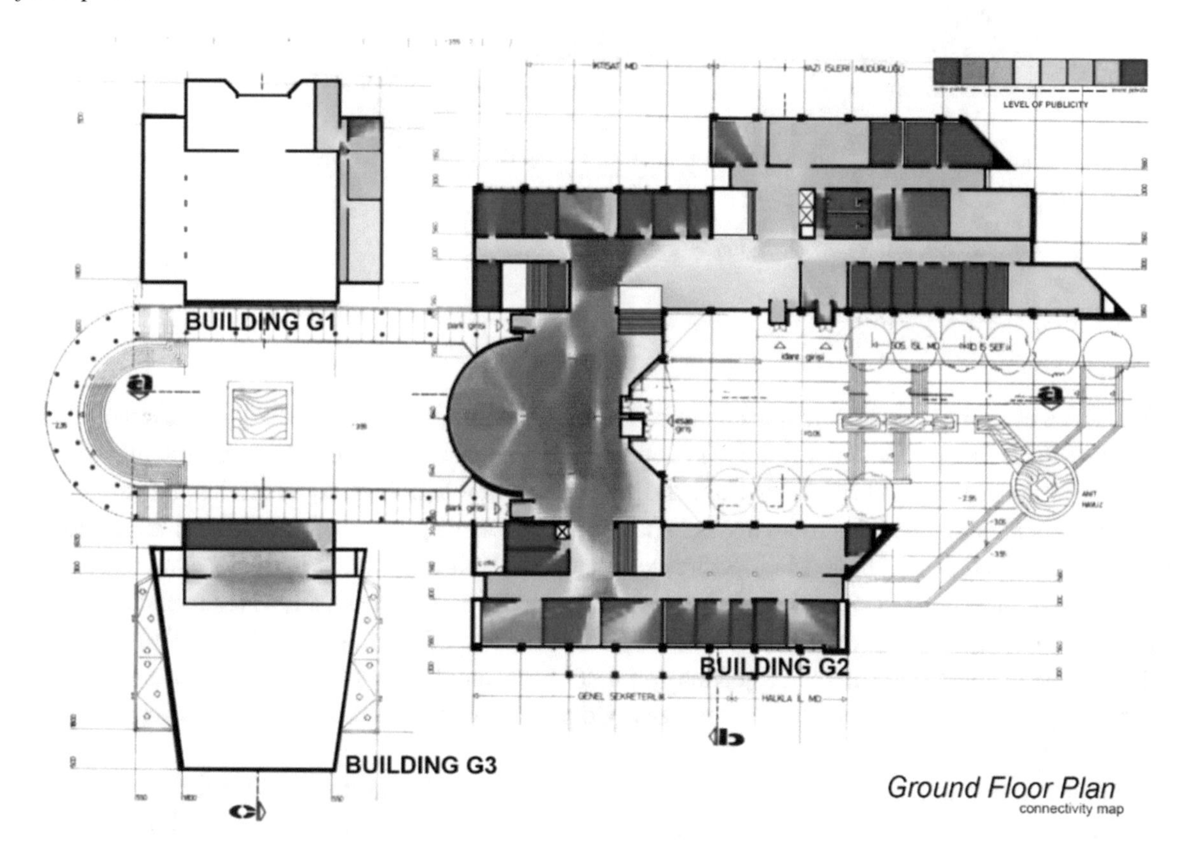

Ground Floor integration and connectivity map results set forth that the corridor intersection area is the most integrated and connected public space in Building G1 (Connectivity Value: 52, Integration Value: 52.48). The deepest space meanwhile, is the office next to the staircase. (Connectivity Value:1902, Integration Value: 6.60) In Building G2, the corridor in the northern side which connects the exhibition

Figure 9. Gaziantep Municipality Service Building Competition, First Prize, Integration map of ground floor plan

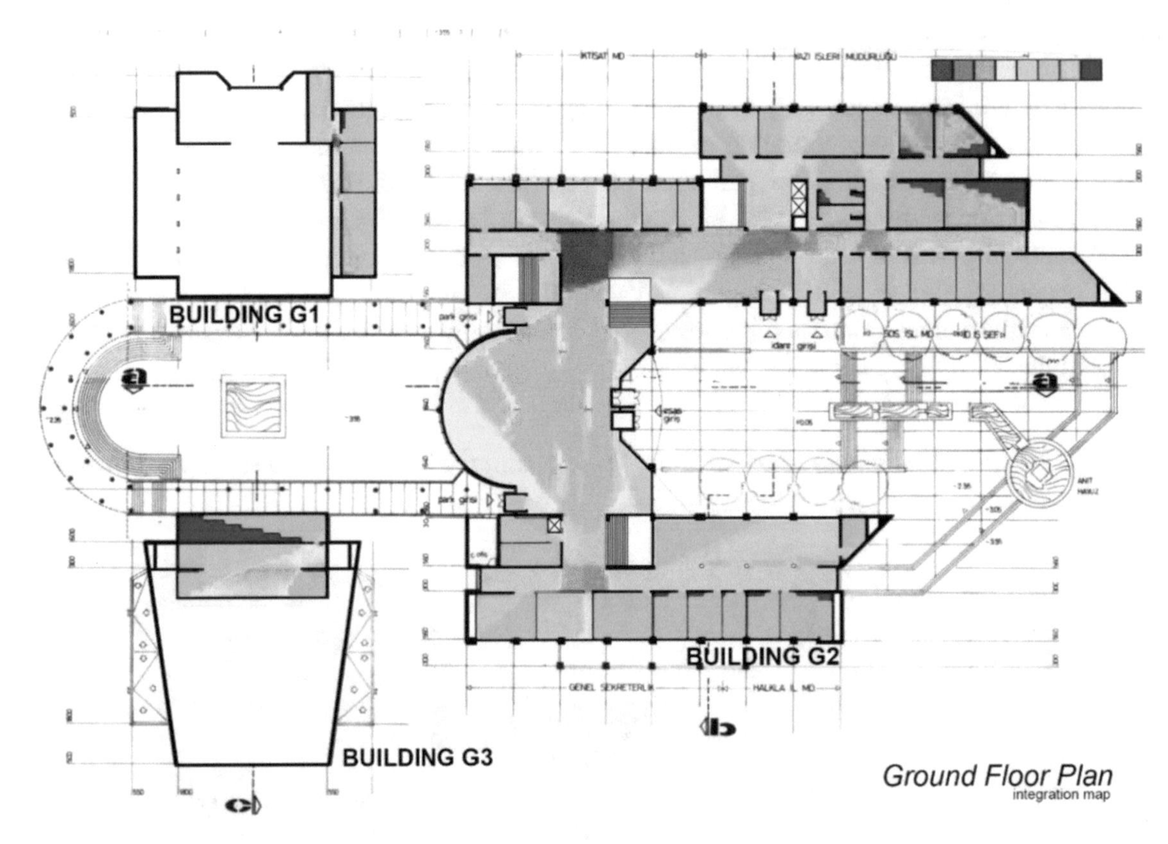

hall and the economy department appears to be the most integrated and connected public space (Connectivity: 2031, Integration Value: 90.777). It opens to more spaces than the spaces at the southern part. The deepest space is the archive which bears the lowest level of integration, connectivity and publicness. This result is natural since this room needs to have a high level of privacy (Connectivity: 209, Integration Value: 2.25) (Figure 10. and 11.). Staircases located in Building G2 are both close to the corridor connection points and to the most integrated parts. The middle part of the balcony is the most integrated and connected part of the layout of Building G3 (Connectivity:2359, Integration Value: 91.64). In this building, the deepest space is the projection room. The reason of this is that the projection room is a technical space which does actually need privacy (Connectivity: 9, Integration Value: 5.95) (See Figure 10. and 11.).

The integration and connectivity maps of Building FF1 exhibit that the corridor that connects the chamber council with the foyer is the most integrated and connected public space in Building F1. (Connectivity: 2414, Integration Value: 9.70). This is why this specific intersection part opens up to more spaces than the southern intersection part does, as the First Floor connectivity map shows (Figure 10. and 11.). Archive room at the north part has the lowest level of integration in Building F1 (Connectivity: 5, Integration Value: 2.03). The analyses show that these are private spaces (See Figure 10. and 11.). Staircases in Building FF1 are both close to the corridor connection points and to the most integrated and connected parts. Main routes are highly integrated and connected (See Figure 10. and 11.).

Figure 10. Gaziantep Municipality Service Building Competition, First Prize, Connectivity map of first floor plan

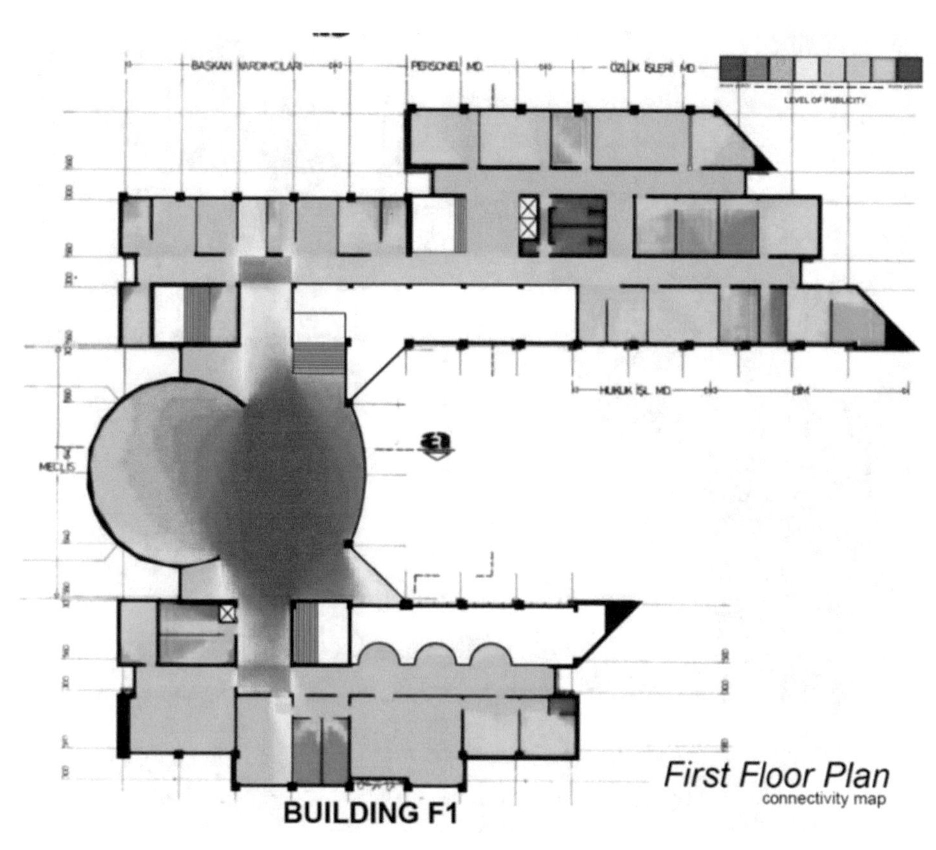

According to the VGA results of the project (see Table 1.), the integration value after the normalization process to compare with each other is 0.0084 and the connectivity value is 0.343. Building SG1 and SG2 are at the sub-ground floor while Building G1 and G2 reveal a higher integration and connection level than the average because of having larger spaces with public use (Table 1.). The relative percentage of circulation areas to the total building area indicates that the circulation areas have a high integration and connectivity value. 7.9% of Building G2 consists of circulation areas. Against this low circulation rate, there exists a high integration and connectivity value; representing that Building G2 contains spaces with a high level of connection and integration (see Table 1.). This is why the main function of Building G2 is defined with the exhibition hall and offices.

Staircases and lifts are located in close proximity with the spots that indicates high connectivity and integration values on the sub-ground, ground and first floor. The main routes of the sub-ground and ground floors seem to be shallow. This can be associated with the purpose of providing ease of way-finding to the visitors (See Figure8., 9., 10. and 11.).

Figure 11. Gaziantep Municipality Service Building Competition, First Prize, Integration map of first floor plan

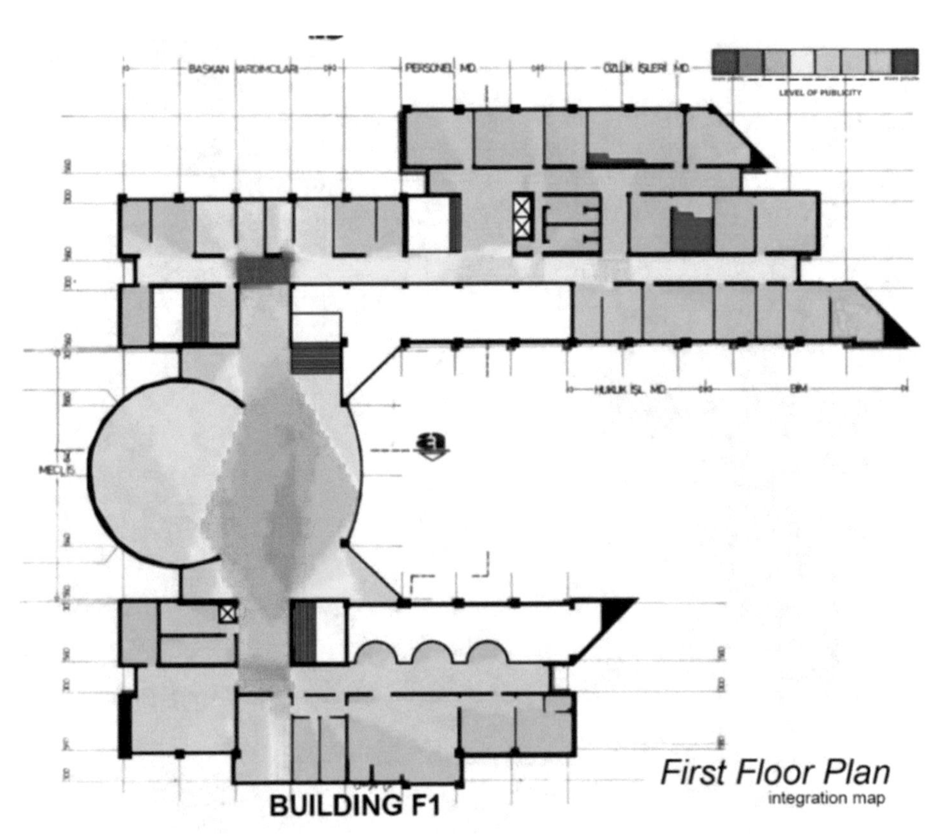

Table 1. VGA Results and Relative Percentage of Circulation Areas of Architectural Competition of Gaziantep Municipality Building Design

Architectural Competition of Gaziantep Municipality Building Design (1986)									
	Building Names	Mean Value of Connectivity	Mean Value of Integration	Visual Node Count	Integration Value after Normalization	Connectivity Value after Normalization	Relative Percentage of Areas with High Integration Level	Relative Percentage of Areas with High Connectivity Levels	Relative Percentage of Circulation Areas to Total Areas
Sub-ground Floor	Building SG1	1741.87	36.084	2.495	0.0140	0.698	30%	33%	28.6%
	Building SG2	1651.28	30.581	2.519	0.0120	0.655	73%	51.6%	21.2%
	Building SG3	697.498	6.495	7.422	0.0008	0.093	33%	25.2%	28.1%
Ground Floor	Building G1	98.746	5.178	399	0.0120	0.247	12%	44.3%	24.2%
	Building G2	301.521	11.450	597	0.0190	0.505	25%	84%	7.9%
	Building G3	763.156	5.336	7.767	0.0006	0.098	37%	27.1%	45.9%
First Floor	Building FF1	800.182	5.152	7.368	0.0006	0.108	32%	27.3%	32.8%
				AVARAGE:	0.0084	0.343	34.5%	41.7%	26.9%

3.2 Architectural Competition of Karabük Municipality Service Building Design (2005)

Karabük Municipality Service Building Design Competition was held by Karabük Municipality in 2005. Architects Erkin Mutlu and his team were awarded by the jury. The project is approximately 16.500 m^2 (Circulation, elevators and staircases included) (See Fig. 3.9 and 3.10). The project has compact plan scheme.

Figure 12. Karabük Municipality Service Building Competition, First Prize, Model (Anonymous, 2015)

Figure 13. Karabük Municipality Service Building Competition, First Prize, Site plan (Anonymous, 2015)

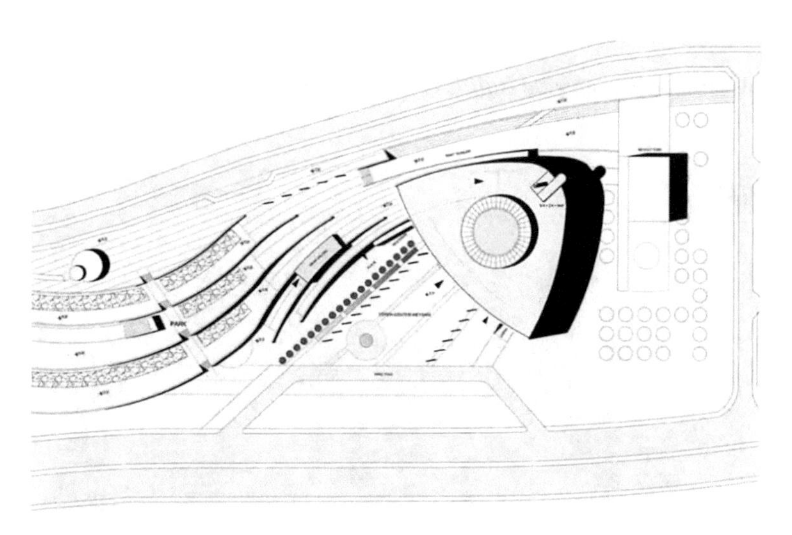

3.2.1. Design Approach of the Designer

Benefiting the opportunity of opening the building to the city both through upper and lower levels has been the main idea of the project. As a result, there are two entrances from two different levels to preserve the existing topography. The main entrance of the building is designed as a space for ceremonies. The continuity of the pedestrian flow which approaches from the upper and lower level is ensured by the interior void and this generated a visual perception in the interior (Anonymous, 2015).

3.2.2. Jury's Evaluations on the Project

The jury assessed the competing projects on the basis of the three following criteria:

- Fire escapes should be easily accessible.
- Service places and their relation with other places, priority of access to the kitchen, should be well-designed.
- Offices should not be organized as cellular offices but as a single open office.

3.2.3. VGA Analysis and Results

Integration and connectivity analyses set forth that the highest connectivity at Building G1 ground floor belongs to the front side of the restaurant which intersects with the exhibition hall. The highest integration on the other hand, belongs to the greeting room by the restaurant (Connectivity Value: 4760, Integration value: 9.780, see table 3.2). In truth, the exhibition hall and the greeting room constitute the main function of the Building's ground floor. The main staircases are far away from this area. The least connected space is the WC which is close to the wedding hall on the west side of the layout (Connectivity Value: 5, Integration Value: 1.950) (See Figure 14. and 15.).

Figure 14. Karabük Municipality Service Building Competition, First Prize, Connectivity map of ground floor plan

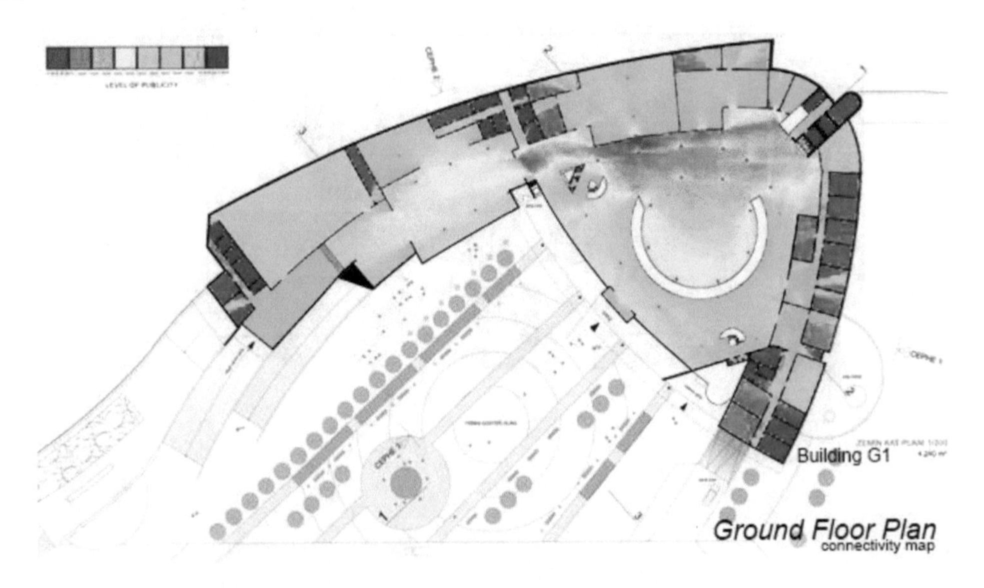

Figure 15. Karabük Municipality Service Building Competition, First Prize, Integration map of ground floor plan

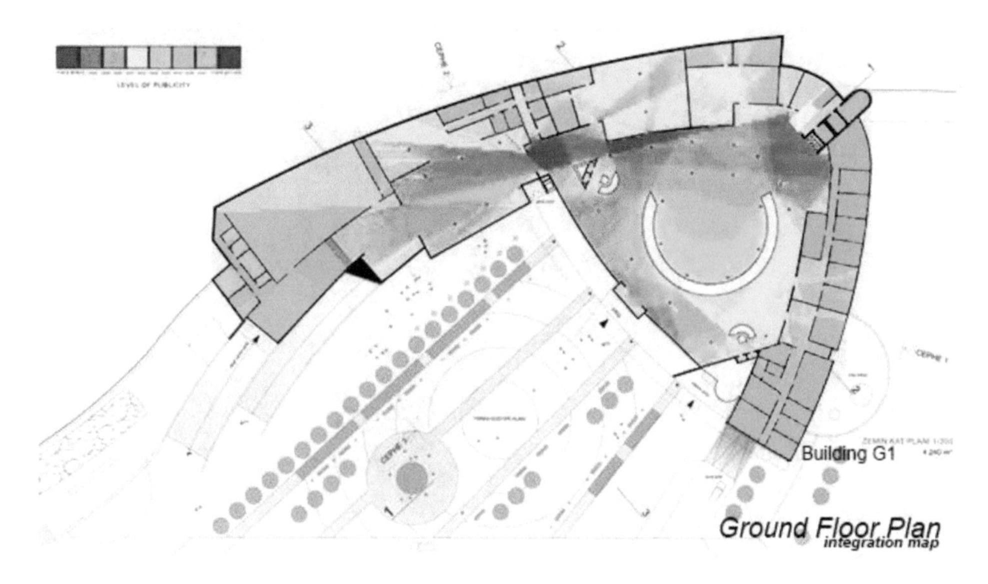

Figure 16. Karabük Municipality Service Building Competition, First Prize, Connectivity map of first floor plan

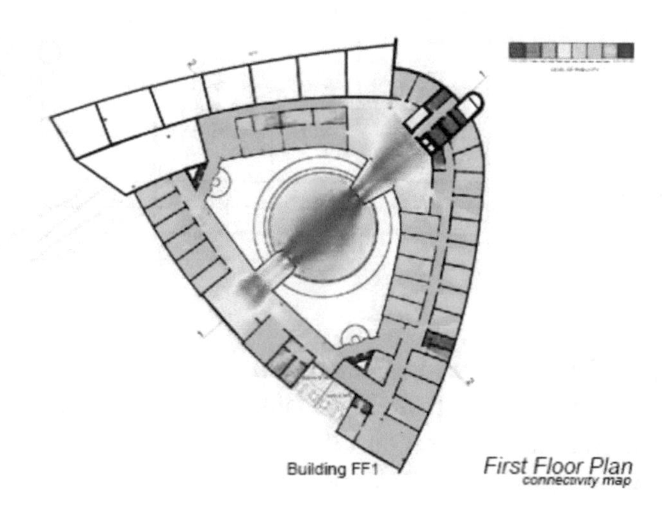

According to the results of integration and connectivity analyses, the highest value of connectivity at Building FF1 first floor is owned by the corridor which opens to the council chamber (Connectivity Value: 1975). The highest value of integration meanwhile, is of the municipal council room which is close to the council chamber (Integration value: 7.466) Besides, the main staircases are close to corridor leading to the council chamber whereas far away from the borough council. The least connected and integrated space was legal affairs on the layout. (Connectivity Value: 1, Integration Value: 2.127) (See Figure 16. and 17.).

Figure 17. Karabük Municipality Service Building Competition, First Prize, Integration map of first floor plan

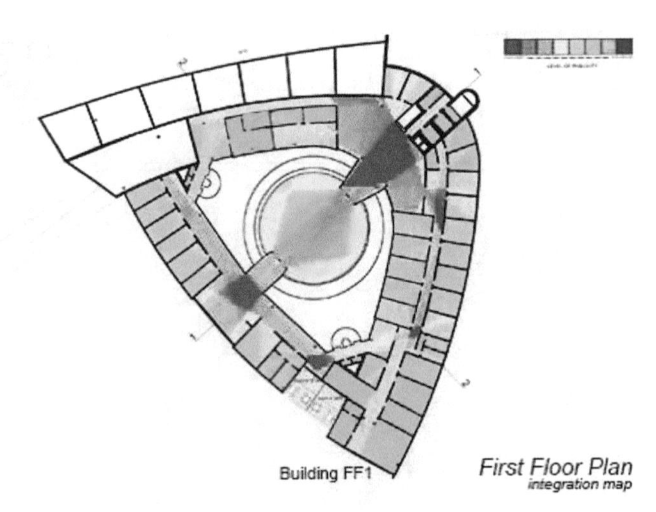

Table 2. VGA Results and Relative Percentage of Circulation Areas of Architectural Competition of Karabük Municipality Building Design

Architectural Competition of Karabük Municipality Building Design (2005)										
	Building Names	Mean Value of Connectivity	Mean Value of Integration	Visual Node Count	Integration Value after Normalization	Connectivity Value after Normalization	Relative Percentage of Areas with High Integration Level	Relative Percentage of Areas with High Connectivity Levels	Relative Percentage of Circulation Areas to Total Areas
Ground Floor	Building G1	1553.45	5.941	16052	0.0003	0.0967	37.6%	24.2%	45.4%
First Floor	Building F1	570.327	4.785	8244	0.0005	0.0691	40.6%	29.5%	26.4%
				AVARAGE:	0.0004	0.0829	39.12%	26.8%	35.9%

According to the VGA results, the integration value after normalization is 0.0004, while the connectivity value after normalization is 0.0829 (see table 2.). Building G1 at its ground floor possesses a higher integration and connectivity level than the average, whereas Building F1 at its first floor has a lower integration and connectivity value than the average. This is because Building G1 has more spaces than Building F1. In spite of the presence of a high rate of relative circulation, the relative percentage of the areas with a high level of relative connectivity and integration is low in Building F1 (Table 2.). Staircases and lifts are positioned in close proximity to the areas that bear the highest connectivity and integration values at the ground floor. However, the same cannot be said for the first floor since neither the staircase nor the lifts are close to the areas with the highest connectivity and integration values here. The main routes of the ground floor and the first floor appear to be shallow. This can be interpreted to the purpose of promoting ease of way-finding (See Figure 14., 15., 16. and 17.).

4. FUTURE RESEARCH

Further research on this topic can involve further three dimensional data regarding common public areas such as courtyards, gallery voids, etc. In this study, the publicness level of the building interior spaces was determined. Publicness levels of exterior spaces of buildings must also be measured at the urban scale. If these are supported by VGA, surveys and observation techniques the results may be more rewarding. The analysis conducted in the study was oriented to measure the level of publicness of only the plans of the drafts. Thus, in the VGA analyses obtained it is beneficial to measure the qualifications of the designs made quantitatively. Also if VGA is cross-correlated with other scientific methods better information about the results are given. The most accurate results can be applied as the building is used and with user's feedbacks. This study includes only assessment that can be made early in the design phase. Finally, publicness evaluation via VGA can be done in the early design phases by the correlation of the relative percentage of circulation areas with the relative percentage of areas with high integration and connectivity levels.

5. CONCLUSION

This study utilizes from VGA analysis in order to determine the publicness level of the floor plans of selected municipal service buildings. Approaches regarding public use were assessed by jury reports and competitor's objectives of each of the competitions. Places with high integration and connectivity values have been examined whether being open to public use or not. As a result of the analysis, correlation tables have been created. These tables helped us to review the ratio of the areas with higher average integration and connectivity values in each project in comparison to the whole area. Additionally, the ratio of circulation areas with high permeability is considered as common areas in comparison to the whole area. Moreover, in each project the proximity of the core to places with high integration and connectivity is taken into account. In terms of plan schemas, projects are divided into two; sprawled programmatic solution in two or more buildings (fragmental schema), and compact programmatic solution in single buildings. Amongst the projects which are considered as sprawled programmatic solution in two or more buildings are considered as compact programmatic solution in a single building (see Figures 18 and 19). The reason for this is that the layout of sprawled programmatic solution in two or more buildings the interior setup of circulation areas has more accessibility and permeability levels.

Figure 18. Distribution of Connectivity Value of Each Project after Normalization

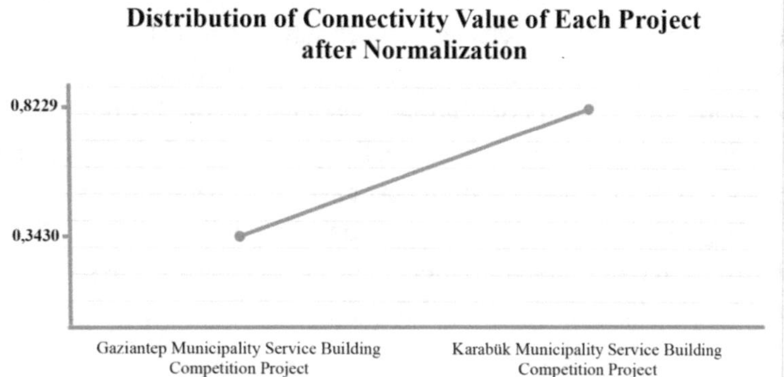

Figure 19. Distribution of Integration Value of Each Project after Normalization

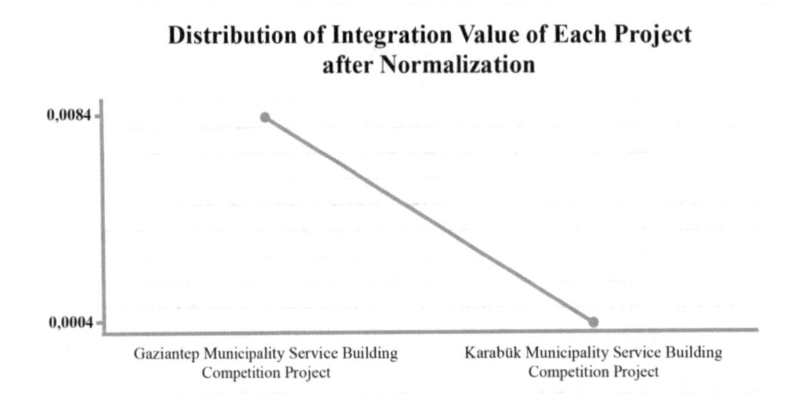

Various overlapping methods might at least contribute to a better understanding of determining publicness levels within the architectural design competitions projects. Regarding to the research questions, this study had explored the results and their correlation of closed space organization and social interaction in two different architectural competition of municipality service design building through using VGA method approach.

Each case is evaluated and compared by objective graphs and mathematical relative percentage of permeability and connectivity to understand how the publicness level is changed. By correlating and overlapping the results of each one, a more holistic comprehension is achieved. Space syntax models can be used for research without modification as well for design experimentation and simulation. Through using an analytical representation technique of space syntax, VGA, we can extend both isovist and graph–based analyses of architectural space in order to form a new methodology for the investigation and configurational relationships. The measurements of local and global characteristics of the graph is a growing interest from an architectural perspective. These measurements allow us to describe a configuration with reference to accessibility. Also they can be compared from location to location within system and systems with different geometries can be compared too (Turner et al., 2001). Though, Space syntax, especially VGA methodology can be used at an early design stage which can prove beneficial feedback in assessing the strengths and weaknesses of publicness level of municipality service buildings. Through VGA analysis method, publicness level can be measured according to their connectivity and integration levels.

The interpretation of publicness levels through architectural design competitions is a statistical process. In order to obtain the data needed, selected architectural projects have been investigated. Secondly, jury reports via designer and team reports have been analyzed if aims and intentions match with the evaluations of the jury. Then floor plans of each selected projects have been analyzed by VGA method. There is a point of view which is the limitation of the study includes only the designers and jury attitudes. Lastly, the results have been implemented and correlations have been made. Within this correlation, relative percentage of areas with high integration and connectivity values of the projects and relative percentage of circulation areas have been studied. Comments have been made about the project in accordance with these values. This gives us interesting results about the publicness level. Projects which have a high rate of relative percentage of circulation area, their publicness level associated with connectivity and integration levels are also high. It is accepted that if the relative percentage of circulation areas are high and relative percentage of areas with high connectivity and integration levels are low, public use is low. And if relative percentage of circulation areas are low, and relative percentage of areas with high connectivity and integration levels are high then the layout is more open to public use.

Public administration buildings and within these, municipal service buildings, assume an important role of visual mediation between the public and the administration. Functional and formal maturity is simply not sufficient by itself for a representative aura of municipal service buildings. These buildings are ideally rather expected to reflect the administration's philosophy and ideology to the public, communicate with the people and use in this very context their publicness as a tool. This is why municipal service building design competitions represent an important type in terms of examining the concept of publicness and its value. In this context, examining the publicness level and comparing to each other is important in this study.

Reference to the visibility and accessibility, VGA properties may give the clues to interpret manifestations of spatial perception such as way-finding, movement and space use within a building. Based on the case studies, in terms of way-finding, it is possible to define meeting points. By the help of it, proper place for notice boards can be defined on this public building. The hierarchy of public and private spaces are defined in a logical manner, helping to maintain confidentiality and publicness.

REFERENCES

Akkar, M. (2005). The changing 'publicness' of contemporary public spaces: A case study of the Grey's Monument Area, Newcastle upon Tyne. *URBAN DESIGN International*, *10*(2), 95–113. doi:10.1057/palgrave.udi.9000138

Ataç, İ. (1990). Typological Analysis of the Concept of Space. *Design*, *5*, 84–87.

Bafna, S. (2003). Space syntax: A brief introduction to its logic and analytical techniques. *Environment and Behavior*, *35*(1), 17–29. doi:10.1177/0013916502238863

Barry, A., Thomas, G., Debenham, P., & Trout, J. (2012). Augmented Reality in a Public Space: The Natural History Museum, London. *Computer*, *45*(7), 42–47. Advance online publication. doi:10.1109/MC.2012.106

Benedikt, M. L. (1979). To take hold of space: Isovist and fields. *Environment & Planning B*, *6*(1), 47–65. doi:10.1068/b060047

Can, I. (2011). *In-Between Space and Social Interaction: A Case Study of Three Neighborhoods in Izmir* (PhD Thesis). University of Nottingham.

Chai, H. (2012). *Making ''Invisible Architecture Visible'': A Comparative Study Nursing Unit Typologies in The United States and China* (PhD Thesis). Georgia Institute of Technology.

Davis, M. (2011). *Fortress Los Angeles, The Militarization of Urban Space, Cultural Criminology: Theories of Crime*. Academic Press.

Desyllas, J., & Duxbury, E. (2001). Axial Maps and Visibility Graph Analysis A comparison of their methodology and use in models of urban A comparison of their methodology and use in models of urban pedestrian movement. *3rd International Space Syntax Symposium*, Atlanta.

Donegan, L., & Goldfarb, M. (2017). New Building for A New Way? New School Pedagogic and its building spatial configuration in Paraíba (Brazil). *Proceedings of the 11th Space Syntax Symposium.*

Dursun, P. (2007). Space Syntax in Architectural Design. *6th International Space Syntax Symposium*, İstanbul.

Ehsan Abshirini, D. K. (2013). Visibility Analysis, Similarity, and Disimilarity in General Trends of Building Layouts and Their Functions. *Ninth International Space Syntax Symposium*, Seoul.

Gündoğdu, M. (2014). Mekan Dizimi Analiz Yöntemi ve Araştırma. *Art-Sanat*, 252-274.

Güney, Y. İ. (2007). *Analyzing Visibility Structures in Turkish Domestic Spaces. 6th International Space Syntax Symposium*, İstanbul.

Hamawand, Z. (2011). Morphology in English: Word Formation in Cognitive Grammar. *Continuum.*

Hasol, D. (1998). *Ansiklopedik Mimarlık Sözlüğü*. YEM.

Hedhoud, A., Foudil, C., & Amina, B. (2014). *Way-finding based on space syntax, Proceedings* (Vol. I). Space Syntax Today.

Hillier, B. (1993). *Specificially Architectural Theory: A Partial Account of the Ascent from the Building as Cultural Transmission*. The Harvard Architecture Review.

Hillier, B. (2005). The art of place and the science of space. World Architecture, 96-102.

Hillier, B. (2007). Space is the machine. Academic Press.

Hillier, B., Grajewski, T., Jones, L., Jianming, X., & Greene, M. (1990). *Broadgate Spaces Life in Public Places, Report of Research into the use of public spaces in the Broadgate Development. Unit for Architectural Studies*. Bartlett School of Architecture and Planning, University College London.

Hillier, B., & Hanson, J. (1984). *The Social Logic of Space*. Cambridge University Press. doi:10.1017/CBO9780511597237

Hillier, B. H. (1983). Space syntax: a different urban perspective. *Architect's Journal, 30.*

Hillier, B., Burdett, R., Peponis, J., & Penn, A. (1987). Creating Life: Or, does architecture determine anything? *Arch.&Comport Arch. Behav.*, *3*(3), 233–250.

Jin, S. K., & Sailer, K., (n.d.). Seeing and being seen inside a museum and a department store A comparison study in visibility and co-presence patterns. *Proceedings of the 10th International Space syntax Symposium.*

Karimi, K. (1997). The Spatial Logic of Organic Cities in Iran and the United Kingdom. In *Proceedings of the First International Symposium on Space Syntax*. London: University College London.

Kim, M., & Choi, J. (2009). Angular VGA and Cellular VGA An Exploratory Study for Spatial Analysis Methodology Based on Human Movement Behavior. *Proceedings of the 7th International Space Syntax Symposium.*

Knöll, M., Li, Y., Neuheuser, K., & Rudolph-Cleff, A. (2015). Using space syntax to analyse stress ratings of open public spaces. *Proceedings of the 10th International Space Syntax Symposium.*

Kohsari, M., Oka, K., Owen, N., & Sugiyama, T. (2019). Natural movement: A space syntax theory linking urban form and function with walking for transport. *Cities (London, England).*

Kuban, D. (1998). *Mimarlık Kavramları, Tarihsel Perspektif İçinde Mimarlığın Kuramsal Sözlügüne Giriş*. İstanbul: YEM Yayın.

Kutucu, S., & Yılmaz, E. (2011). *Architectural Competitions as A Method of Building Production Process in Turkey. In 23rd International Building & Life Congress: The Milieu of the Architectural Profession* (p. 477). Bursa: Bursa Chamber of Architects Bursa Branch.

Larson, M. S. (1994). Architectural Competitions as Discursive Events. *Theory and Society, 23*(4), 469–504. doi:10.1007/BF00992825

Lerman, Y., Rofé, Y., & Omer, I. (2014). Using Space Syntax to Model Pedestrian Movement in Urban Transportation Planning. *Geographical Analysis, 46*(4), 392–410. doi:10.1111/gean.12063

Madanipour, A. (2003). *Public and Private Space of the City*. doi:10.4324/9780203402856

Madanipour, A. (2019). Rethinking public space: Between rhetoric and reality. *URBAN DESIGN International, 24*(1), 38–46. doi:10.105741289-019-00087-5

Othman, F., Mohd Yusoff, Z., & Abdul Rasam, A. R. (2019). Isovist and Visibility Graph Analysis (VGA): Strategies to evaluate visibility along movement pattern for safe space. *IOP Conf. Ser.: Earth Environ. Sci., 385.*

Parvin, A., Ye, A. M., & Jia, B. (2007). Multilevel Pedestrian Movement: Does Visibility Make Any Difference? *6th International Space Syntax Symposium*, İstanbul.

Peponis, J., & Wineman, J. (2002). Spatial Structure of Environment and Behavior. In R. B. Bechtel & A. Churchman (Eds.), *Handbook of Environmental Psychology*. John Wiley & Sons, Inc.

Rohloff, I. K., Psarra, S., & Wineman, J. (2009). Experiencing Museum Gallery Layouts through Local and Global Visibility Properties in Morphology an inquiry on the YCBA, the MoMA and the HMA. *7th International Space Syntax Symposium*, Stockholm.

Ruben, T. (2012). *Improving Pedestrian Accessibility to Public Space Through Space Syntax Analysis.* Conference: 8th Space Syntax Symposium, Santiago de Chile.

Soja, E. (1996). *Thirdspace: Journeys to Los Angeles and Other Real-and-Imagined Places.* Blackwell.

Specifications. (2015). Retrieved from Arkitera: https://sartname.arkitera.com/Main.php?MagID=63&MagNo=220

Steadman, P. (1983). *Architectural Morphology, An Introduction to the Geometry of Building Plans, London.* Pion.

Tahar, B., & Brown, F. (2003). The visibility graph: An approach for the analysis of traditional domestic M'zabite spaces. *4th International Space Syntax Symposium*, London.

Turner, A. (2001). Depthmap: A Programme to Perform Visibility Graph Analysis. *Proceedings of the Third International Symposium on Space Syntax*, 31.1.

Turner, A. (2004, June). *Depthmap 4 A Researcher's Handbook*. Academic Press.

Turner, A., Doxa, M., O'Sullivan, D., & Penn, A. (2001). From isovists to visibility graphs: A methodology for the analysis of architectural space. *Environment and Planning. B, Planning & Design*, *28*(1), 103–121. doi:10.1068/b2684

Whyte, W. H. (2000). The Social Life Of Small Urban Spaces, Common Ground? Readings and Reflections on Public Space. Routledge.

ADDITIONAL READING

Ascensão, A., Costa, L., Fernandes, C., Morais, F., & Ruivo, C. (2019). *3D Space Syntax Analysis: Attributes to Be Applied in Landscape Architecture Projects*. Urban Sci.

Bafna, S. (2003, January). Space Syntax, A Brief Introduction to Its Logic and Analytic Techniques. *Environment and Behavior*, *35*(1), 17–29. doi:10.1177/0013916502238863

Benedikt, M. L. (1979). To take hold of space: Isovist and fields. *Environment & Planning B*, *6*(1), 47–65. doi:10.1068/b060047

Hillier, B. (1996). *Space is the Machine*. Cambridge University Press.

Hillier, B., & Hanson, J. (1984). *Social Logic of Space*. Cambridge University Press. doi:10.1017/CBO9780511597237

Kim, G., Kim, A., & Kim, Y. (2019). A new 3D Space Syntax Metric Based on 3D Isovist Capture in Urban Space Using Remote Sensing Technology. *Computers, Environment and Urban Systems*, *74*, 74–87. Advance online publication. doi:10.1016/j.compenvurbsys.2018.11.009

Lu, Y., Gou, Z., Ye, Y., & Sheng, Q. (2019). Three-Dimensional Visibility Graph Analysis and its Application. *Environment & Planning B*, *46*(5), 948–962. doi:10.1177/2399808317739893

Morais, F. (2018). *DepthSpace3D: A New Digital Tool for 3D Space Syntax Analysis in Formal Methods in Architecture and Urbanism*. Cambridge Scholars Publishing.

Syntax, S. (2002). *"Tate Britain", Report on the Spatial Accessibility Study of the Proposed Layouts, July, 2002*. Space Syntax Limited.

Turner, A. (2010). *UCL Depthmap: Spatial Network Analysis Software (Version 10.10.16b)*. University College London, VR Centre of the Built Environment.

Turner, A., Doxa, M., O'Sullivan, D., & Penn, A. (2001). From Isovists to Visibility Graphs: A Methodology for the Analysis Of Architectural Space. *Environment and Planning. B, Planning & Design*, *28*(1), 103–121. doi:10.1068/b2684

KEY TERMS AND DEFINITIONS

Axial Map: Set of fewest and longest lines of sight or access that passing through all spaces of a system.

Clustering Coefficient: the number of edges between all the vertices in the neighbourhood of the generating vertex divided by the total number of possible connections with that neighbourhood size.

Connectivity: A local measurement that refers to how many immediate neighbours each node can see.

Depthmap: A program to perform visibility graph analysis.

Integration: A global measurement that refers how many turns and changes one has to make in order to access one space from another space in whole system.

Neighbourhood Size: The set of vertices immediately connected through an edge.

Visibility Graph (Isovists): The set of all points visible from a given vantage point in space and with respect to an environment.

Compilation of References

(2010, February). Liao, Tung-Shan & Rice, J. (2010). *Original Research Article Research Policy*, *39*(1), 117–125.

Best Coding Apps for Kids. (2019). https://www.educationalappstore.com/best-apps/10-best-coding-apps-for-kids

Abouzeid, A., Bajda-Pawlikowski, K., Abadi, D. J., Rasin, A., & Silberschatz, A. (2009). HadoopDB: An Architectural Hybrid of MapReduce and DBMS Technologies for Analytical Workloads. *Proceedings of the VLDB Endowment International Conference on Very Large Data Bases*, *2*(1), 922–933. doi:10.14778/1687627.1687731

Advanced Online Collaboration. (n.d.). *Where Creative Minds Meld.* Retrieved from http://www.yorku.ca/dzwick/what_is_octopz.htm

Agrawal, D. (2014). *Challenges and opportunities with big data. Leading researchers across the United States, Tech. Rep.* Retrieved August 25, 2014, from http://www.cra.org/ccc/files/docs/init/bigdatawhitepaper.pdf

Akkar, M. (2005). The changing 'publicness' of contemporary public spaces: A case study of the Grey's Monument Area, Newcastle upon Tyne. *URBAN DESIGN International*, *10*(2), 95–113. doi:10.1057/palgrave.udi.9000138

Akkermans, M., & Vollaard, B. (2015). *Effect van het WhatsApp-project in Tilburg op het aantal woninginbraken – een evaluatie*. Tilburg University. https://hetccv.nl/fileadmin/Bestanden/Onderwerpen/Woninginbraak/Documenten/Effect_van_het_WhatsApp-project_in_Tilburg_op_het_aantal_woninginbraken/tilburg-whatsapp_191015.pdf

Akrich, M. (1992). The De-Scription of Technical Objects. In W. E. Bijker & J. Law (Eds.), *Shaping Technology / Building Society: Studies in sociotechnical change* (pp. 259–264). MIT Press. https://mitpress.mit.edu/books/shaping-technology-building-society

AlAghi, I. (2012). KnowledgePuzzle: A Browsing Tool to Adapt the Web Navigation Process to the Learner's Mental Modle. *Journal of Educational Technology & Society*, *15*(3), 275–287.

Albright, S., & Winston, W. (2016). *Business Analytics: Data Analysis & Decision Making* (6th ed.). Cengage Learning.

Aldrich, H. E., & von Glinow, M. A. (1992). Business start-ups: the HRM imperative. In S. Birley & I. C. MacMillan (Eds.), *International Perspectives on Entrepreneurial Research* (pp. 233–253). North-Holland.

Al-Fahad, F. (2009). Students' attitudes and perceptions towards the effectiveness of mobile learning in king Saud University, Saudi Arabia. *The Turkish Online Journal of Educational Technology*, *8*(2). Retrieved November 18, 2012, from http://www.tojet.net/articles/8210.pdf

Amadieu, F., Salmerón, L., Cegarra, J., Paubel, P.-V., Lemarié, J., & Chevalier, A. (2015). Learning from Concept Mapping and Hypertext: An Eye Tracking Study. *Journal of Educational Technology & Society*, *18*(4), 100–112.

Amara, N., Landry, R., Becheikh, N., & Ouimet, M. (2008). Learning and novelty of innovation in established manufacturing SMEs. *Technovation*, *28*(7), 450–463. doi:10.1016/j.technovation.2008.02.001

Amazon. (2020). *What is cloud computing?* Retrieved January 12, 2020 from https://aws.amazon.com/what-is-cloud-computing/

Analytics. (2014a). *Why big data analytics as a service? Analytics as a Service*. Retrieved August 25, 2014, from http://www.analyticsasaservice.org/why-big-data-analytics-as-a-service/

Analytics. (2014b). *What is Big Data? Analytics as a Service in the Cloud. Analytics as a Service*. Retrieved August 25, 2014, from http://www.analyticsasaservice.org/what-is-big-data-analytics-as-a-service-in-the-cloud/

Andrejevic, M. (2007). *Ispy: Surveillance and Power in the Interactive Era*. University Press of Kansas. https://www.bookdepository.com/ISpy-Mark-Andrejevic/9780700616862

Andrejevic, M. (2002). The Work of Watching One Another: Lateral Surveillance, Risk, and Governance. *Surveillance & Society*, *2*(4). Advance online publication. doi:10.24908s.v2i4.3359

Anonymous. (1998). *Maximum security: a hacker's guide to protecting your internet site and network*. Techmedia.

Antonelli, C. (2005). Models of knowledge and systems of governance. *Journal of Institutional Economics*, *1*(1), 51–73. doi:10.1017/S1744137405000044

Arora, A. (2002). Licensing tacit knowledge: Intellectual property rights and the market for know-how. *Economics of Innovation and New Technology*, *4*(1), 41–59. doi:10.1080/10438599500000013

Arora, A., Fosfuri, A., & Gambardella, A. (2001). Markets for technology and their implications for corporate strategy. *Industrial and Corporate Change*, *10*(2), 419–450. doi:10.1093/icc/10.2.419

Arsham, H. (2015). *Applied Management Science: Making Good Strategic Decisions*. Retrieved October 27, 2015, from http://home.ubalt.edu/ntsbarsh/opre640/opre640.htm

Artemis. (2009). *Multi-Annual Strategic Plan and Research Agenda 2009*. ARTEMIS-I ARTEMIS-IRC-02-V061008-draft (MASP). ARTEMISIA.

Ataç, İ. (1990). Typological Analysis of the Concept of Space. *Design*, *5*, 84–87.

Autant-Bernard, C., Fadairo, M. & Massard, N. (2012). *Knowledge diffusion and innovation policies within the European regions: Challenges based on recent empirical evidence*. Original Research Article, Research Policy, In Press, Corrected Proof, Available online 10 August 2012. doi:10.1016/j.respol.2012.07.009

Baetens, J., & Truyen, F. (2013). Hypertext Revisited. *Leonardo*, *46*(5), 477–480. doi:10.1162/LEON_a_00644

Bafna, S. (2003). Space syntax: A brief introduction to its logic and analytical techniques. *Environment and Behavior*, *35*(1), 17–29. doi:10.1177/0013916502238863

Barnes, B. (2001). Practice as collective action. In T. R. Schatzki, K. Knorr Cetina, & E. von Savigny (Eds.), *The Practice Turn in Contemporary Theory* (pp. 25–36). Routledge.

Barnes, S. J., & Mattsson, J. (2016). Understanding current and future issues in collaborative consumption: A four-stage Delphi study. *Technological Forecasting and Social Change*, *104*, 200–211. doi:10.1016/j.techfore.2016.01.006

Barney, J., & Clark, D. (2007). *Resource-based Theory: Creating and Sustaining Competitive Advantage*. Oxford University Prass.

Baron, N. (2015). *Words onscreen: The fate of reading in a digital world*. Oxford University Press.

Barry, A., Thomas, G., Debenham, P., & Trout, J. (2012). Augmented Reality in a Public Space: The Natural History Museum, London. *Computer*, *45*(7), 42–47. Advance online publication. doi:10.1109/MC.2012.106

Bates, P., Biere, M., Weideranders, R., Meyer, A., & Wong, B. (2009). *New Intelligence for a Smarter Planet*. Academic Press.

Bates, A. W. (2016). *Teaching in a digital age: guidelines for designing teaching and learning*. Opentextbook.

Batterlink, M. (2009). *Profiting from external knowledge: How companies use different knowledge acquisition strategies to improve their innovation performance* (PhD thesis). Wageningen University.

Beach, R. (2012). Constructing digital learning commons in the literacy classroom. *Journal of Adolescent & Adult Literacy*, *55*(5), 448–451. doi:10.1002/JAAL.00054

Belk, R. (2014). You are what you can access: Sharing and collaborative consumption online. *Journal of Business Research*, *67*(8), 1595–1600. doi:10.1016/j.jbusres.2013.10.001

Bell, K. (2018). *15 Collaborative Tools for Your Classroom That Are NOT Google*. Retrieved from https://shakeuplearning.com/blog/15-collaborative-tools-for-your-classroom-that-are-not-google/

Bell, A. (2014). Schema Theory, Hypertext Fiction and Links. *Style (Fayetteville)*, *48*(2), 140–161.

Bellair, P. E. (2000). Informal Surveillance and Street Crime: A Complex Relationship*. *Criminology*, *38*(1), 137–170. doi:10.1111/j.1745-9125.2000.tb00886.x

Bell, M., & Pavitt, K. L. R. (1993). Technological accumulation and industrial growth: Contrasts between developed and developing countries. *Industrial and Corporate Change*, *2*(2), 157–210. doi:10.1093/icc/2.2.157

Bell, S., & Shank, J. (2007). *Academic Librarianship by Design: A Blended Librarian's Guide to the Tools and Techniques*. ALA.

Belussi, F., Sammarra, A., & Sedita, S. R. (2010). Learning at the boundaries in an "Open Regional Innovation System": A focus on firms' innovation strategies in the Emilia Romagna life science industry. *Research Policy*, *39*(6), 710–721. doi:10.1016/j.respol.2010.01.014

Benedikt, M. L. (1979). To take hold of space: Isovist and fields. *Environment & Planning B*, *6*(1), 47–65. doi:10.1068/b060047

Benkler, Y. (2004). 'Sharing Nicely': On Shareable Goods and the Emergence of Sharing as a Modality of Economic Production. *The Yale Law Journal*, *114*(2), 273–358. doi:10.2307/4135731

Benner, C. (2015). *Crowdwork - zurück in die Zukunft? Perspektiven digitaler Arbeit*. BUND Verlag.

Benton, T. H. (2009). A Laboratory of Collaborative Learning. *The Chronicle of Higher Education*, *2009*(August). https://www.chronicle.com/article/A-Laboratory-of-Collaborative/47518/

Bernardino, J., & Madeira, H. (2001). Experimental evaluation of a new distributed partitioning technique for data warehouses. *Proceedings of the International Database Engineering and Applications Symposium, IDEAS*, 312-321. 10.1109/IDEAS.2001.938099

Bernardino, J., & Neves, P. C. (2019). Decision-Making with Big Data Using Open Source Business Intelligence Systems. In Web Services: Concepts, Methodologies, Tools, and Applications (pp. 431-458). Hershey, PA: IGI Global. doi:10.4018/978-1-5225-7501-6.ch025

Bernardino, J., Furtado, P., & Madeira, H. (2002). DWS-AQA: a cost effective approach for very large data warehouses. *Proceedings of the International Database Engineering & Applications Symposium, IDEAS'02*, 233-242. 10.1109/IDEAS.2002.1029676

Bernardino, J., & Neves, P. C. (2016). Decision-Making with Big Data Using Open Source Business Intelligence Systems. In H. Rahman (Ed.), *Human Development and Interaction in the Age of Ubiquitous Technology* (pp. 120–147). IGI Global. doi:10.4018/978-1-5225-0556-3.ch006

Berry, D. (2012). *Introduction: understanding the digital humanities. In Understanding Digital Humanities*. Palgrave Macmillan. doi:10.1057/9780230371934

Berthon, H., & Webb, C. (2000). *The Moving Frontier: Archiving, Preservation and Tomorrow's Digital Heritage*. Paper presented at VALA 2000 - 10th VALA Biennial Conference and Exhibition, Melbourne. Retrieved October 27, 2009, from http://www.nla.gov.au/nla/staffpapers/hberthon2.html

Bervoets, E., Van Ham, T., & Ferwerda, H. (2016). *Samen signaleren Burgerparticipatie bij sociale veiligheid*. Platform 31. http://www.beke.nl/doc/2016/PL31-samen%20signaleren.pdf

Bervoets, E. (2014). *Een oogje in het zeil... Buurtpreventie in Ede: Onderzoek en een gefundeerd advies*. Bureau Bervoets. https://www.bureaubervoets.nl//files/pagecontent/OnderzoekEde%20BuurtpreventieDEFDEF.pdf

Bessant, J., & Rush, H. (1995). Building bridges for innovation: The role of consultants in technology transfer. *Research Policy*, *24*(1), 97–114. doi:10.1016/0048-7333(93)00751-E

Bessant, J., & Tidd, J. (2007). *Innovation and Entrepreneurship*. Wiley.

Bessen, J. (2006). Open Source Software: Free Provision of Complex Public Goods. In J. Bitzer & P. J. H. Schröder (Eds.), *The Economics of Open Source Software Development*. Elsevier B.V. doi:10.1016/B978-044452769-1/50003-2

Bianchi, M., Campodall'Orto, S., Frattini, F., & Vercesi, P. (2010). Enabling open innovation in small- and medium-sized enterprises: How to find alternative applications for your technologies. *R & D Management*, *40*(4), 414–431. doi:10.1111/j.1467-9310.2010.00613.x

Bigliardi, B., & Galati, F. (2016). Which factors hinder the adoption of open innovation in SMEs? *Technology Analysis and Strategic Management*, *28*(8), 869–885. doi:10.1080/09537325.2016.1180353

Bilby, M. (2014). *Collaborative Student Research with Google Drive: Advantages and Challenges*. Presentation at the annual conference of the American Theological Library Association, New Orleans, LA. Retrieved October 27, 2015, from http://lanyrd.com/sctzbt

Birt. (2019). *BIRT BI*. Retrieved December 31, 2019, from https://www.eclipse.org/birt/

Björk, B.-C., Welling, P., Laakso, M., Majlender, P., & Hedlund, T. (2010). Open Access to the Scientific Journal Literature. *PLoS ONE, 5*(6). Retrieved January 14, 2012, from www.plosone.org/article/info:doi/10.1371/journal.pone.0011273

Bjork, O. (2012). Digital Humanities and the First-Year Writing Course. In *Digital Humanities Pedagogy: Practices, Principles and Politics* (pp. 97–119). Open Book Publishers. doi:10.2307/j.ctt5vjtt3.9

Blumberg, R., & Atre, S. (2003). The problem with unstructured data. *DM Review*, *13*(2), 42–49.

BOAI: Budapest Open Access Initiative. (2011). *Frequently Asked Questions*. Retrieved December 19, 2012, from http://www.earlham.edu/~peters/fos/boaifaq.htm

Bobish, G. (2011). Participation and Pedagogy: Connecting the Social Web to ACRL Learning Outcomes. *Journal of Academic Librarianship*, *37*(1), 54–63. doi:10.1016/j.acalib.2010.10.007

Boer, H., Hill, M., & Krabbendam, K. (1990). FMS implementation management: Promise and performance. *International Journal of Operations & Production Management*, *10*(1), 5–20. doi:10.1108/01443579010004994

Boiugrain, F., & Haudeville, B. (2002). Innovation, collaboration and SMEs internal research capacities. *Research Policy*, *31*(5), 735–747. doi:10.1016/S0048-7333(01)00144-5

Boltanski, L., & Chiapello, È. (2007). *The New Spirit of Capitalism*. Verso.

Bolter, J. D. (2001). *Writing Space: Computers, Hypertext, and the Remediation of Print* (2nd ed.). Lawrence Erlbaum Associate Publishers. doi:10.4324/9781410600110

Bonner, J., & Walker, O. (2004). Selecting influential business-to-business customers in new product development: Relational embeddedness and knowledge heterogeneity considerations. *Journal of Product Innovation Management*, *21*(3), 155–169. doi:10.1111/j.0737-6782.2004.00067.x

Borsuk, A. (2018). *The book*. MIT Press. doi:10.7551/mitpress/10631.001.0001

Boschma, R. A. (2005). Proximity and innovation: A critical assessment. *Regional Studies*, *39*(1), 61–74. doi:10.1080/0034340052000320887

Box, S. (2009). *OECD Work on Innovation- A Stocktaking of Existing Work*. OECD Science, Technology & Industry Working Papers, 2009/2. OECD Publishing.

Brandenburger, A., & Nalebuff, B. (1996). *Co-Opetition*. Doubleday.

Brockhoff, K. (2003). Customers' perspectives of involvement in new product development. *International Journal of Technology Management*, *26*(5/6), 464–481. doi:10.1504/IJTM.2003.003418

Brogren, C. (2009). *Comments from VINNOVA: Public consultation on Community Innovation Policy, VINNOVA*. Reg. No.

Brown, J. S., & Duguid, P. (2000). *The Social Life of Information*. Harvard Business School Press.

Brown, J. S., & Duguid, P. (2001). Knowledge and Organization: A Social-Practice Perspective. *Organization Science*, *12*(2), 198–213. doi:10.1287/orsc.12.2.198.10116

Brunswicker, S., Bertino, E., & Matei, S. (2015). Big data for open digital innovation–a research roadmap. *Big Data Research, 2*(2), 53-58.

Brunswicker, S., & Vanhaverbeke, W. (2015). Open innovation in small and medium-sized enterprises (SMEs): External knowledge sourcing strategies and internal organizational facilitators. *Journal of Small Business Management*, *53*(4), 1241–1263. doi:10.1111/jsbm.12120

Bukht, R., & Heeks, R. (2018). *Digital Economy Policy: The case example of Thailand*. Development Implications of Digital Economies.

Burstein, F., & Holsapple, C. W. (2008). *Handbook on Decision Support Systems*. Springer Berlin Heidelberg.

Caetano, M., & Amaral, D. C. (2011, July). Original Research Article. *Technovation*, *31*(7), 320–335. doi:10.1016/j.technovation.2011.01.005

Camm, J., Cochran, J., Fry, M., Ohlmann, J., & Anderson, D. (2020). Business Analytics (3rd ed.). Cengage Learning, Inc.

Can, I. (2011). *In-Between Space and Social Interaction: A Case Study of Three Neighborhoods in Izmir* (PhD Thesis). University of Nottingham.

Carayannis, E. G., Popescu, D., Sipp, C., & Steward, M. (2006). Technological learning for entrepreneurial development (TL4ED) in the knowledge economy (KE): Case studies and lessons learned. *Technovation*, *26*(4), 419–443. doi:10.1016/j.technovation.2005.04.003

Cash, D. W. (2001). In order to aid in diffusion useful and practical information: Agricultural extension and boundary organizations. *Science, Technology & Human Values*, *26*(4), 431–453. doi:10.1177/016224390102600403

Center for Teaching Excellence. Carnegie Mellon University. (2009). *Collaboration tools*. Teaching with Technology White Paper. Carnegie Mellon university. Retrieved October 27, 2015, from https://www.cmu.edu/teaching/technology/whitepapers/CollaborationTools_Jan09.pdf

Chai, H. (2012). *Making ''Invisible Architecture Visible'': A Comparative Study Nursing Unit Typologies in The United States and China* (PhD Thesis). Georgia Institute of Technology.

Chamoni, P., & Gluchowski, P. (2004). Integration trends in business intelligence systems-An empirical study based on the business intelligence maturity model. *Wirtschaftsinformatik*, *46*(2), 119–128. doi:10.1007/BF03250931

Chandra, R. (2015). Collaborative Learning for Educational Achievement. *IOSR Journal of Research & Method in Education, 5*(3), 4-7. Retrieved October 3, 2015, from http://www.iosrjournals.org/iosr-jrme/papers/Vol-5%20Issue-3/Version-1/B05310407.pdf

Charmaz, K. (2014). Constructing Grounded Theory. *Sage (Atlanta, Ga.).*

Chattopadhyay, B., Lin, L., Liu, W., Mittal, S., Aragonda, P., Lychagina, V., Kwon, Y., & Wong, M. (2011). Tenzing a SQL Implementation on the MapReduce Framework. *Proceedings of the VLDB Endowment International Conference on Very Large Data Bases*, *4*(12), 1318–1327. doi:10.14778/3402755.3402765

Cheng, R., Zhang, X. (2018). *Business Intelligence & Analytics (BI&A) Training: The Guidelines for Improving the Effectiveness of BI&A Systems Training for Business Users*. Academic Press.

Chesbrough, H. & Crowther, A. K. (2006). Beyond high tech: early adopters of open innovation in other industries. R*&D Management, 36*(3), 229-236.

Chesbrough, H. (2003). Open Innovation: How Companies Actually Do It. *Harvard Business Review, 81*(7), 12-14.

Chesbrough, H. (2006a). Open Innovation: A New Paradigm for Understanding Industrial Innovation. In Open Innovation: Researching a New Paradigm. Oxford University Press.

Chesbrough, H. W., & Bogers, M. (2014). Chapter 1: Explicating Open Innovation: Clarifying an Emerging Paradigm for Understanding Innovation. In New Frontiers in Open Innovation. Oxford University Press.

Chesbrough, H., Vanhaverbeke, W., & West, J. (Eds.). (2006). Open innovation: Researching a new paradigm. Oxford University Press.

Chesbrough, H. (2003a). *Open Innovation: The New Imperative for Creating and Profiting from Technology*. Harvard Business School Press.

Chesbrough, H. (2003b). The era of open innovation. *MIT Sloan Management Review*, *44*(3), 35–41.

Chesbrough, H. (2006b). *Open Business Models: How to Thrive in the New Innovation Landscape*. Harvard Business School Press.

Chesbrough, H. (2007). The market for innovation: Implications for corporate strategy. *California Management Review*, *49*(3), 45–66. doi:10.1177/000812560704900301

Chesbrough, H. W. (2003). *Open innovation: The new imperative for creating and profiting from technology*. Harvard Business School Press.

Chesbrough, H. W. (2003a). The era of open innovation. *MIT Sloan Management Review*, *44*(3), 35–41.

Chesbrough, H. W., Vanhaverbeke, W., & West, J. (2006). *Open Innovation: Researching a new paradigm.* Oxford University Press.

Chesbrough, H., & Crowther, A. K. (2006). Beyond high tech: Early adopters of open innovation in other industries. *R & D Management, 36*(3), 229–236. doi:10.1111/j.1467-9310.2006.00428.x

Christensen, J. F., Olesen, M. H., & Kjær, J. S. (2005). The Industrial Dynamics of Open Innovation: Evidence from the transformation of consumer electronics. *Research Policy, 34*(10), 1533–1549. doi:10.1016/j.respol.2005.07.002

Church, K., & de Oliveira, R. (2013). What's Up with Whatsapp?: Comparing Mobile Instant Messaging Behaviors with Traditional SMS. *Proceedings of the 15th International Conference on Human-Computer Interaction with Mobile Devices and Services*, 352–361. 10.1145/2493190.2493225

Chutimaskul, W. (2002). E-government Analysis and Modeling. *Knowledge Management in e-Government KMGov-2002, 7*, 113-123.

Ciborra, C. U. (1992). From thinking to tinkering: The grassroots of strategic information systems. *The Information Society, 8*(4), 297–309. doi:10.1080/01972243.1992.9960124

Clarke, I. (2007). *The role of intermediaries in promoting knowledge flows within global value chains.* Paper no. 12, PhD Doctoral Symposium, Brunel University, UK.

Coase, R. (1937). The Nature of the Firm. *Economica, 4*(16), 386–405. doi:10.1111/j.1468-0335.1937.tb00002.x

Cofis, E., & Marsili, O. (2003). *Survivor: The role of innovation in firm's survival, No. 03-18. WPT. Koopmans Institute, USE.* Utrecht University.

Cohen, B., & Kietzmann, J. (2014). Ride On! Mobility Business Models for the Sharing Economy. *Organization & Environment, 27*(3), 279–296. doi:10.1177/1086026614546199

Cohen, B., & Munoz, P. (2015). Sharing cities and sustainable consumption and production: Towards and integrated framework. *Journal of Cleaner Production, 134*, 87–97. doi:10.1016/j.jclepro.2015.07.133

Collaborative writing. (n.d.). *EdTechTeacher.* Retrieved October 27, 2015, from http://tewt.org/collaborative-writing/

Cooke, P. (2005). Regionally asymmetric knowledge capabilities and open innovation: Exploring 'Globalisation 2'—A new model of industry organisation. *Research Policy, 34*(8), 1128–1149. doi:10.1016/j.respol.2004.12.005

Cooper, B. B. (2015, November 18). *The Science of Collaboration: How to Optimize Working Together* [Web log post]. Retrieved October 27, 2015, from https://thenextweb.com/entrepreneur/2014/07/15/science-collaboration-optimize-working-together/

Corbin, J., & Strauss, A. (2007). *Basics of Qualitative Research: Techniques and Procedures for Developing Grounded Theory* (3rd ed.). SAGE Publications, Inc.

Coughlan, P., Harbison, A., Dromgoole, T., & Duff, D. (2001). Continuous improvement through collaborative action learning. *International Journal of Technology Management, 22*(4), 285–301. doi:10.1504/IJTM.2001.002965

Crawford, C. M. (1983). New Products Management, Richard D. Irwin, Inc.

Crawford, W. (2011). *Open access: what you need to know now.* ALA Publishing.

CSR Europe. (2008a). The European Alliance for CSR Progress Review 2007: Making Europe a Pole of Excellence on CSR. European Commission.

CSR Europe. (2008b). *R&D Open Innovation: Networks with SME.* Open Innovation Network.

Current Publications. (2001). *The copyright act 1957: with short notes*. Current Publications.

Dahlander, L., & Magnusson, M. G. (2005). Relationships between open source software companies and communities: Observations from Nordic firms. *Research Policy, 34*(4), –493.

Dahlander, L., & Gann, D. M. (2010). How open is innovation? *Research Policy*.

Damaskopoulos, P., & Evgeniou, T. (2003). Adoption of New Economy Practices by SMEs in Eastern Europe. *European Management Journal, 21*(2), 133–145. doi:10.1016/S0263-2373(03)00009-4

Darcy, J., Kraemer-Eis, H., Guellec, D., & Debande, O. (2009). *Financing Technology Transfer*. Working paper 2009/002, EIF Research & Market Analysis, European Investment Fund, Luxemburg.

Darsø, L. (2001). *Innovation in the Making*. Samfundslitteratur.

Datta, P. (2007). An agent-mediated knowledge-in-motion model. *Journal of the Association for Information Systems, 8*(5), 288–311. doi:10.17705/1jais.00130

Daunoriene, A., Draksaite, A., Snieska, V., & Valodkiene, G. (2015). Evaluating Sustainability of Sharing Economy Business Models. *Procedia: Social and Behavioral Sciences, 213*, 836–841. doi:10.1016/j.sbspro.2015.11.486

Davenport, T. H. (1993b). Process Innovation: reengineering work through information technology, Ernst & Young, Harvard Business School Press.

Davenport, T. H. (1993). *Process innovation: reengineering work through information technology*. Ernst & Young.

Davenport, T. H. (1993a). *Process Innovation*. Harvard Business School Press.

Davenport, T. H. (1994). Managing in the New World of Process. *Public Productivity & Management Review, 18*(2), 133–147. doi:10.2307/3380643

Davenport, T. H., & Harris, J. G. (2007). *Competing on analytics: The new science of winning*. Harvard Business School Review Press.

Davidsson, J. (2006). *Small Business Innovation Program: Business development and entrepreneurial training with intellectual property in developing countries*. B-Open Nordic AB.

Davis, M. (2011). *Fortress Los Angeles, The Militarization of Urban Space, Cultural Criminology: Theories of Crime*. Academic Press.

Day M. (1998). Electronic Access: Archives in the New Millennium. Reports on a conference held at the Public Record Office, Kew on 3-4 June 1998. *Ariadane,* (16).

De Backer, K., & Cervantes, M. (2008). *Open innovation in global networks*. OECD.

De Jong, J. P. J., Vanhaverbeke, W., Kalvet, T., & Chesbrough, H. (2008). Policies for Open Innovation: Theory, Framework and Cases. Research project funded by VISION Era-Net.

De Jong, J.P.J., Vermeulen, P.A.M. & O'Shaughnessy, K.C. (2004) Effects of Innovation in Small Firms. *M & O, 58*(1), 21-38.

De Jong, J. P. J. (2006). *Open Innovation: Practice, Trends, Motives and bottlenecks in the SMEs [Meer Open Innovatie: Praktijk, Ontwikkelingen, Motieven en Knelpunten in het MKB]*. EIM.

De Jong, J. P. J., & Marsili, O. (2006). The fruit flies of innovations: A taxonomy of innovative small firms. *Research Policy, 35*(2), 213–229. doi:10.1016/j.respol.2005.09.007

De Jong, J. P. J., Vanhaverbeke, W., Kalvet, T., & Chesbrough, H. (2008). *Policies for Open Innovation: Theory, Framework and Cases*. VISION Era-Net.

De Jong, J. P. J., Vanhaverbeke, W., Van de Vrande, V., & De Rochemont, M. (2007). Open innovation in SMEs: trends, motives and management challenges. *Proceeding of EURAM Conference*.

De Jong, J. P. J., & von Hippel, E. (2009, September). Transfers of user process innovations to process equipment producers: A study of Dutch high-tech firms. *Research Policy*, *38*(7), 1181–1191. doi:10.1016/j.respol.2009.04.005

de Vries, A. (2016, February 1). BuurtWhatsApp: Goed beheer is complex. *Social Media DNA*. https://socialmediadna.nl/buurtwhatsapp-goed-beheer-is-complex/

Dehoff, K. & Sehgal, V. (2006) Innovators without Borders. *Strategy+Business*.

Dehoff, K. & Sehgal, V. (2008). Beyond Borders: The Global Innovation 1000. *Strategy+Business*.

Del Brío, J. Á., & Junquera, B. (2003, December). A review of the literature on environmental innovation management in SMEs: Implications for public policies. *Technovation*, *23*(12), 939–948. doi:10.1016/S0166-4972(02)00036-6

Delen, D., Demirkan, H. (2013). *Data, information and analytics as services*. Academic Press.

Descotes, R. M., Walliser, B., Holzmüller, H., & Guo, X. (2011, December). Original Research Article. *Journal of Business Research*, *64*(12), 1303–1310.

Desyllas, J., & Duxbury, E. (2001). Axial Maps and Visibility Graph Analysis A comparison of their methodology and use in models of urban A comparison of their methodology and use in models of urban pedestrian movement. *3rd International Space Syntax Symposium*, Atlanta.

Devlin, B., Rogers, S., & Myers, J. (2012). *Big data comes of age. Tech. Rep.* Retrieved August 25, 2014, from http://www.9sight.com/pdfs/Big_Data_Comes_of_Age.pdf

Di Maria, E., & Micelli, S. (2008). *SMEs and Competitive Advantage: A Mix of Innovation, Marketing and ICT. The Case of "Made in Italy"*. Marco Fanno Working Paper No. 70, Università degli Studi di Padova, Padova, Italy.

Dickson, P. H., Weaver, K. M., & Hoy, F. (2006, July). Opportunism in the R&D alliances of SMES: The roles of the institutional environment and SME size. *Journal of Business Venturing*, *21*(4), 487–513. doi:10.1016/j.jbusvent.2005.02.003

Diener, K. & Piller, F. (2009). *The Market for Open Innovation: Increasing the efficiency and effectiveness of the innovation process*. Open Innovation Accelerator Survey 2009, RWTH Aachen University, TIM Group.

Dillahunt, T. A., & Malone, A. R. (2015). The Promise of the Sharing Economy among Disadvantaged Communities. *CHI 2015 - Proceedings of the 33rd Annual CHI Conference on Human Factors in Computing Systems: Crossings*, 2285–2294.

DiMaggio, P. J., & Powell, W. W. (1983). The Iron Cage Revisited: Institutional isomorphism and collective rationality in organizational fields. *American Sociological Review*, *48*(2), 147–160. doi:10.2307/2095101

Diwan, P., Suri, R. K., & Kaushik, S. (2000). *IT Encyclopaedia.com* (Vol. 1-10). Pentagon Press.

Dodgson, M., Gann, D., & Salter, A. (2006). The role of technology in the shift towards open innovation: The case of Procter & Gamble. *R & D Management*, *36*(3), 333–346. doi:10.1111/j.1467-9310.2006.00429.x

Dolata, U. (2015). Volatile Monopole. Konzentration, Konkurrenz und Innovationsstrategien der Internetkonzerne. *Berliner Journal fur Soziologie*, *24*(4), 505–529. doi:10.100711609-014-0261-8

Don Dingsdag, D., Armstrong, B., & Neil, D. (2000). *Electronic Assessment Software for Distance Education Students*. Retrieved January 14, 2012, from http://www. ascilite.org.au/ conferences/coffs00/papers/don_dingsdag.pdf

Donegan, L., & Goldfarb, M. (2017). New Building for A New Way? New School Pedagogic and its building spatial configuration in Paraíba (Brazil). *Proceedings of the 11th Space Syntax Symposium.*

Dörre, K. (2009). Die neue Landnahme. Dynamiken und Grenzen des Finanzmarktkapitalismus. In K. Dörre, S. Lessenich, & H. Rosa (Eds.), *Soziologie, Kapitalismus, Kritik. Eine Debatte* (pp. 21–86). Suhrkamp.

Douglas, M., Praul, M., & Lynch, M. (2008). Mobile learning in higher education: An empirical assessment of a new educational tool. *The Turkish Online Journal of Educational Technology, 7*(3). http://www.tojet.net/articles/732.pdf

Dresner Advisory Services. (2014). *Wisdom of Crowds Business Intelligence Market Study.* From http://businesssintelligence.blogspot.com/2011/05/released-2011-edition-of-wisdom-of.html

Drucker, P. (1998, November-December). The Discipline of Innovation. *Harvard Business Review*, 149. PMID:10187245

Dursun, P. (2007). Space Syntax in Architectural Design. *6th International Space Syntax Symposium*, İstanbul.

Dyussenov, M. (2017). Who Sets the Agenda in Thailand? Identifying key actors in Thai corruption policy, 2012–2016. *Journal of Asian Public Policy, 10*(3), 334–364. doi:10.1080/17516234.2017.1326448

EC. (2008). *SBA Fact Sheet Portugal, European Commission: Enterprise and Industry.* EC.

Economist. (1996). The family connection. *The Economist, 341*(7986), 62.

Edquist, C., & Johnson, B. (1997). Institutions and organizations in systems of innovation. In C. Edquist (Ed.), *Systems of Innovation: Technologies, Institutions and Organizations.* Routledge.

Edwards, R. C. (1979). *Contested Terrain.* Basic Books.

Edwards, T., Delbridge, R., & Munday, M. (2005). Understanding innovation in small and medium-sized enterprises: A process manifest. *Technovation, 25*(10), 1119–1127. doi:10.1016/j.technovation.2004.04.005

Ehrenreich, B. (2001, November). Welcome to cancerland. *Harper's Magazine.* https://harpers.org/archive/2001/11/welcome-to-cancerland/

Ehsan Abshirini, D. K. (2013). Visibility Analysis, Similarity, and Disimilarity in General Trends of Building Layouts and Their Functions. *Ninth International Space Syntax Symposium*, Seoul.

eLearning 101 – concepts, trends, applications. (2014). San Francisco: Epignosis. Retrieved October 27, 2015, from LLC http://www.talentlms.com/elearning/

Eliot, S., & Rose, J. (2009). Introduction. In S. Eliot & J. Rose (Eds.), *A Companion to the History of the Book* (pp. 1–6). Wiley-Blackwell.

Elmore, R. F. (1980). Backward Mapping: Implementation Research and Policy Decisions. *Political Science Quarterly, 94*(4), 601–616. doi:10.2307/2149628

English-Lucek, J. A., Darrah, C. N., & Saveri, A. (2002). Trusting strangers: Work relationships in four high-tech communities. *Information Communication and Society, 5*(1), 90–108. doi:10.1080/13691180110117677

Enkel, E., Kausch, C., & Gassmann, O. (2005). Managing the risk of customer integration. *European Management Journal, 23*(2), 203–213. doi:10.1016/j.emj.2005.02.005

Enyedy, N., & Stevens, R. (2014). Analyzing collaboration. In R. E. K. Sawyer (Ed.), *The Cambridge handbook of the learning sciences* (2nd ed., pp. 191–212). Cambridge University Press., doi:10.1017/CBO9781139519526.013

Ernst, D., Mytelka, L., & Ganiatsos, T. (1994). Technological Capabilities: A Conceptual Framework. UNCTAD.

European Association for Business and Commerce. (2019). Retrieved from: https://www.eabc-thailand.org › download

European Union. (2005). *EU-Summary*. The joint Japan-EU Seminar on R&D and Innovation in Small and Medium size Enterprises (SMEs), Tokyo.

Evangelista, R. (2000). Sectoral Patterns of Technological Change in Services. *Economics of Innovation and New Technology*, *9*(3), 183–221. doi:10.1080/10438590000000008

Evans, J. (2019). *Business Analytics* (3rd ed.). Pearson.

Evans, S., Vladimirova, D., Holgado, M., Van Fossen, K., Yang, M., Silva, E. A., & Barlow, C. Y. (2017). Business model innovation for sustainability: Towards a unified perspective for creation of sustainable business models. *Business Strategy and the Environment*, *26*(5), 597–608. doi:10.1002/bse.1939

Evelson, B., & Norman, N. (2008). *Topic overview: business intelligence*. Forrester Research. Retrieved August 23, 2014, from http://www.forrester.com/Topic+Overview+Business+Intelligence/fulltext/-/E-RES39218

Eyman, D. (2015). Defining and Locating Digital Rhetoric. In Digital Rhetoric: Theory, Method, Practice (pp. 12-60). Open Book Publishers. doi:10.2307/j.ctv65swm2.5

Fang, B., Ye, Q., & Law, R. (2015). Effect of sharing economy on tourism industry employment. *Annals of Tourism Research*, *57*, 264–267. doi:10.1016/j.annals.2015.11.018

Farber, M. (2019). Social annotation in the digital age. *Edutopia,* 22. https://www.edutopia.org/article/social-annotation-digital-age

Ferneley, E., & Bell, F. (2006). Using bricolage to integrate business and information technology innovation in SMEs. *Technovation*, *26*(2), 232–241. doi:10.1016/j.technovation.2005.03.005

Ferrary, M. (2011). Specialized organizations and ambidextrous clusters in the open innovation paradigm. *European Management Journal*, *2011*(29), 181–192. doi:10.1016/j.emj.2010.10.007

Ferreira, T., Pedrosa, I., & Bernardino, J. (2017). Evaluating Open Source Business Intelligence Tools using OSSpal Methodology. *Proceedings of the 9th International Joint Conference on Knowledge Discovery, Knowledge Engineering and Knowledge Management*, 1, 283-288. 10.5220/0006516402830288

Firnkorn, J., & Müller, M. (2011). What will be the environmental effects of new free-floating car-sharing systems? The case of car2go in Ulm. *Ecological Economics*, *70*(8), 1519–1528. doi:10.1016/j.ecolecon.2011.03.014

Fligstein, N., & McAdam, D. (2012). *A Theory of Fields*. University Press. doi:10.1093/acprof:oso/9780199859948.001.0001

Fredberg, T., Elmquist, M., & Ollila, S. (2008). Managing Open Innovation: Present Findings and Future Directions. Report VR 2008:02, VINNOVA - Verket för Innovationssystem/Swedish Governmental Agency for Innovation Systems.

Freedman, A. (1996). *The computer desktop encyclopedia*. AMACOM.

Freel, M., & De Jong, J. P. J. (2009). Market novelty, competence-seeking and innovation networking. *Technovation*, *29*(12), 873–884.

Freel, M., & Robson, P. J. (2017). Appropriation strategies and open innovation in SMEs. *International Small Business Journal*, *35*(5), 578–596.

Friedman, A. (1987). Managementstrategien und Technologie. Auf dem Weg zu einer komplexenTheorie des Arbeitsprozesses. In E. Hildebrandt & R. Seltz (Hrsg.), Managementstrategien und Kontrolle. Eine Einführung in die Labour-Process-Debatte (pp. 99–131). Berlin: Edition sigma.

Fritsch, M., & Lukas, R. (2001). Who cooperates on R&D? *Research Policy*, *30*(2), 297–312. doi:10.1016/S0048-7333(99)00115-8

Fuchs, B. (2014). The Writing is on the Wall: Using Padlet for Whole-Class Engagement. *LOEX Quarterly, 40*(4), 7-9. Retrieved October 27, 2015, from http://uknowledge.uky.edu /libraries_facpub/240

Funilkul, S., Chutimaskul, W., & Chongsuphajaisiddhi, V. (2011, August). E-government Information Quality: A case study of Thailand. In *International Conference on Electronic Government and the Information Systems Perspective* (pp. 227-234). Springer. 10.1007/978-3-642-22961-9_18

Fu, X. (2012, April). How does openness affect the importance of incentives for innovation? Original Research Article. *Research Policy*, *41*(3), 512–523. doi:10.1016/j.respol.2011.12.011

Gadamer, H. G. (2004). *Truth and Method*. Continuum International Publishing Group. (Original work published 1960)

Gantz, J., & Reinsel, D. (2018). *The Digitization of the WorldFrom Edge to Core*. https://www.seagate.com/www-content/our-story/trends/files/idc-seagate-dataage-whitepaper.pdf

Garcia, D. (2020). *Rethinking Ethics Through Hypertext*. Emerald Publishing Limited.

Gargiulo, E., Giannantonio, R., Guercio, E., Borean, C., & Zenezini, G. (2015). Dynamic ride sharing service: Are users ready to adopt it? *Procedia Manufacturing*, *3*, 777–784. doi:10.1016/j.promfg.2015.07.329

Gassmann, O. (2006). Opening up the innovation process: Towards an agenda. *R & D Management*, *36*(3), 223–228. doi:10.1111/j.1467-9310.2006.00437.x

Gassmann, O., Enkel, E., & Chesbrough, H. (2010). The future of open innovation. *R & D Management*, *40*(3), 213–221. doi:10.1111/j.1467-9310.2010.00605.x

Gereffi, G., Humphrey, J., & Sturgeon, T. (2005). The governance of global value chains. *Review of International Political Economy*, *12*(1), 78–104. doi:10.1080/09692290500049805

Geron, T. (2012). Airbnb Had $56 Million Impact On San Francisco: Study. *Forbes*. https://www.forbes.com/forbes/welcome/?toURL=https://www.forbes.com/sites/tomiogeron/2012/11/09/study-airbnb-had-56-million-impact-on-san-francisco/&refURL=https://www.google.de/&referrer=https://www.google.de/

Gitelman, L. (2014). *Paper knowledge: Towards a media history of documents*. Duke University Press.

Glotz-Richter, M. (2012). Car Sharing – 'Car on call' for reclaiming street space. *Procedia: Social and Behavioral Sciences*, *48*, 1454–1463. doi:10.1016/j.sbspro.2012.06.1121

Goggin, M. L., Bowman, A. O., Lester, J., & O'Toole, L. (1990). *Implementation Theory and Practice: Toward a Third Generation*. Harper Collins.

Gold, M. K. (2012). *The digital humanities moment. Debates in the digital humanities*. University of Minnesota Press.

Gomes-Casseres, B. (1997). Alliance strategies of small firms. *Small Business Economics*, *9*(1), 33–44. doi:10.1023/A:1007947629435

Government of India: Ministry of Human Resource Development: Press Information Bureau. (2018). *Several steps have been taken to promote e-Education in the country*. Retrieved August18, 2019, from https://pib.gov.in/newsite/PrintRelease.aspx?relid=186501

Government of India: Ministry of Human Resource Development: Press Information Bureau. (2018). *Union HRD Minister launches 'ShaGun' - a web-portal for Sarva Shiksha Abhiyan*. Retrieved August18, 2019, from https://pib.gov.in/newsite/PrintRelease.aspx?relid=157488

Govt. of Bangladesh. (2005a). *Policy Strategies for Small & Medium Enterprises (SME) Development in Bangladesh*. Ministry of Industries, Government of the People's Republic of Bangladesh.

Govt. of Bangladesh. (2005b). *Bangladesh Industrial Policy 2005*. Ministry of Industries, Government of the People's Republic of Bangladesh.

Green, L. S. (2019). Online Learning Is Here to Stay: Librarians transform into digital instructors. *American Libraries Magazine*. https://americanlibrariesmagazine.org/2019/06/03/online-learning-is-here-to-stay/

Grindley, P. C., & Teece, D. J. (1997). Managing intellectual capital: Licensing and cross-licensing in semiconductors and electronics. *California Management Review*, *39*(2), 1–34. doi:10.2307/41165885

Groen, A. J., & Linton, J. D. (2010). Is open innovation a field of study or a communication barrier to theory development? *Technovation*, *30*(11-12), 554. doi:10.1016/j.technovation.2010.09.002

Grondin, J. (1994). *Introduction to Philosophical Hermeneutics* (J. Weinsheimer, Trans.). Yale University Press. (Original work published 1991)

Grupp, H., & Schubert, T. (2010, February). Review and new evidence on composite innovation indicators for evaluating national performance. *Research Policy*, *39*(1), 67–78. doi:10.1016/j.respol.2009.10.002

Gunawong, P., & Gao, P. (2017). Understanding E-government Failure in the Developing Country Context: A process-oriented study. *Information Technology for Development*, *23*(1), 153–178. doi:10.1080/02681102.2016.1269713

Gündoğdu, M. (2014). Mekan Dizimi Analiz Yöntemi ve Araştırma. *Art-Sanat*, 252-274.

Güney, Y. İ. (2007). *Analyzing Visibility Structures in Turkish Domestic Spaces. 6th International Space Syntax Symposium*, İstanbul.

Gutiérrez, K., & Jurow, A. S. (2016). Social design experiments: Toward equity by design. *Journal of the Learning Sciences*, *25*(4), 565–598. doi:10.1080/10508406.2016.1204548

Haagedorn, J., & Schakenraad, J. (1994). The effect of strategic technology alliances on company performance. *Strategic Management Journal*, *15*(4), 291–309. doi:10.1002mj.4250150404

Hafkesbrink, J., & Scholl, H. (2010). Web 2.0 Learning- A Case Study on Organizational Competences in Open Content Innovation. In Competence Management for Open Innovation- Tools and IT-support to unlock the potential of Open Innovation. Eul Verlag.

Hage, J. T. (1999). Organizational Innovation and Organizational Change. *Annual Review of Sociology*, *25*(1), 597–622. doi:10.1146/annurev.soc.25.1.597

Hall, J., & Krueger, A. (2015). An Analysis of the Labor Market for Uber's Driver-Partners in the United States. *IRS Working Papers, 587.*

Hamari, J., Sjöklint, M., & Ukkonen, A. (2015). The sharing economy: Why people participate in collaborative consumption. *American Society for Information Science and Technology Journal*, *67*, 2047–2059.

Hamawand, Z. (2011). Morphology in English: Word Formation in Cognitive Grammar. *Continuum.*

Hamdani, J., & Wirawan, C. (2012). Open Innovation Implementation to Sustain Indonesian SMEs. *Procedia Economics and Finance*, *4*, 223–233. doi:10.1016/S2212-5671(12)00337-1

Hampton, K. N. (2007). Neighborhoods in the Network Society the e-Neighbors study. *Information Communication and Society*, *10*(5), 714–748. doi:10.1080/13691180701658061

Harper, A. (2018). *The benefits of collaborating with school librarians*. https://www.educationdive.com/news/the-benefits-of-collaborating-with-school-librarians/533247/

Hartl, B., Hofmann, E., & Kirchler, E. (2015). Do we need rules for 'what's mine is yours'? Governance in collaborative consumption communities. *Journal of Business Research*, *69*(8), 2756–2763. doi:10.1016/j.jbusres.2015.11.011

Hasol, D. (1998). *Ansiklopedik Mimarlık Sözlüğü*. YEM.

Hassan, N. (2019, Oct 12). Getting Digital IDs Right in Southeast Asia: Countries in Southeast Asia are racing to implement digital IDs, but governments must proceed with caution. *The Diplomat*. Retrieved from: https://thediplomat.com/2019/10/getting-digital-ids-right-in-southeast-asia/

He, J., & Brandweiner, N. (2014). *The 20 best tools for online collaboration*. Retrieved October 27, 2015, from http://www.creativebloq.com/design/online-collaboration-tools-912855

Hedhoud, A., Foudil, C., & Amina, B. (2014). *Way-finding based on space syntax, Proceedings* (Vol. I). Space Syntax Today.

Heidegger, M. (1962). *Being and Time* (J. Macquarrie & E. Robinson, Trans.). Blackwell Publishing. (Original work published 1927)

Helper, S., MacDuffie, J. P., & Sabel, C. M. (2000). Pragmatic collaboration: Advancing knowledge while controlling opportunism. *Industrial and Corporate Change*, *9*(3), 443–488. doi:10.1093/icc/9.3.443

Henderson, R. M., & Clark, K. B. (1990). Architectural Innovation: The Reconfiguration of Existing Product Technologies and the Failure of Established Firms. *Administrative Science Quarterly*, *35*(1), 9–30. doi:10.2307/2393549

Henkel, J. (2006). Selective revealing in open innovation processes: The case of embedded Linux. *Research Policy*, *35*(7), 953–969. doi:10.1016/j.respol.2006.04.010

Herstad, S.J., Bloch, C., Ebersberger, B. & van de Velde, E. (2008). *Open innovation and globalisation: Theory, evidence and implications*. VISION Era.net.

Hidreth, C. (2001). Accounting for user's inflated assessment of online catalogue search performance and usefulness an experimental study. *Information Research*, *6*(2). Retrieved January 14, 2012, from http://www.informationR.net/ir/6-2/paper101.html

Hiebing, R. G. Jr, & Cooper, S. W. (2003). *The Successful Marketing Plan: A Disciplined and Comprehensive Approach*. The McGraw-Hill Companies, Inc.

Hillier, B. (2005). The art of place and the science of space. World Architecture, 96-102.

Hillier, B. (2007). Space is the machine. Academic Press.

Hillier, B. H. (1983). Space syntax: a different urban perspective. *Architect's Journal, 30*.

Hillier, B., Burdett, R., Peponis, J., & Penn, A. (1987). Creating Life: Or, does architecture determine anything? *Arch.&Comport Arch. Behav.*, *3*(3), 233–250.

Hillier, B. (1993). *Specificially Architectural Theory: A Partial Account of the Ascent from the Building as Cultural Transmission*. The Harvard Architecture Review.

Hillier, B., Grajewski, T., Jones, L., Jianming, X., & Greene, M. (1990). *Broadgate Spaces Life in Public Places, Report of Research into the use of public spaces in the Broadgate Development. Unit for Architectural Studies*. Bartlett School of Architecture and Planning, University College London.

Hillier, B., & Hanson, J. (1984). *The Social Logic of Space*. Cambridge University Press. doi:10.1017/CBO9780511597237

Hindle, G., Kunc, M., Mortensen, M., Oztekin, A., & Vidgene, R. (2018). *Business Analytics: Defining the field and identifying a research agenda*. Retrieved December 30, 2019, from https://www.journals.elsevier.com/european-journal-of-operational-research/call-for-papers/business-analytics-defining-the-field-and-identifying-a-rese

Hippel, V. (2010). Open User Innovation. Handbook of the Economics of Innovation, 1, 411-427.

Hippel, V. (1975). *The Dominant Role of Users in the Scientific Instrument Innovation Process, WP 764-75*. NSF.

Hippel, V. (1978, January). Successful Industrial Products from Customer Ideas. *Eric von Hippel Journal of Marketing*, *42*(1), 39–49.

Hippel, V. (2001a, Summer). Innovation by User Communities: Learning from Open-Source Software. *MIT Sloan Management Review*, *42*(4), 82–86.

Hippel, V. (2001b). Perspective: User toolkits for innovation. *Journal of Product Innovation Management*, *18*(4), 247–257. doi:10.1111/1540-5885.1840247

Hippel, V. (2007, April). 2007 Horizontal Innovation Networks—By and for Users. *Industrial and Corporate Change*, *16*(2), 293–315. doi:10.1093/icc/dtm005

Hirsch-Kreinsen, H. (2015). Einleitung. Digitalisierung industrieller Arbeit. In H. Hirsch-Kreinsen, P. Ittermann, & J. Niehaus (Eds.), *Digitalisierung industrieller Arbeit* (pp. 9–30). Nomos. doi:10.5771/9783845263205-10

Hjalager, A.-M. (2010, February). A review of innovation research in tourism. *Tourism Management*, *31*(1), 1–12. doi:10.1016/j.tourman.2009.08.012

Hoffman, H., Parejo, M., Bessant, J., & Perren, L. (1998). Small firms, R&D, technology and innovation in the UK: A literature review. *Technovation*, *18*(1), 39–55. doi:10.1016/S0166-4972(97)00102-8

Hoffman, W. H., & Schlosser, R. (2001). Success factors of strategic alliances in small and medium-sized enterprises: An empirical study. *Long Range Planning*, *34*(3), 357–381. doi:10.1016/S0024-6301(01)00041-3

Hohendanner, C., & Walwei, U. (2013). Arbeitsmarkteffekte atypischer Beschäftigung. *WSI-Mitteilungen*, *66*(4), 239–246. doi:10.5771/0342-300X-2013-4-239

Horey, J., Begoli, E., Gunasekaran, R., Lim, S.-H., & Nutaro, J. (2012). Big data platforms as a service: challenges and approach. *Proceedings of the 4th USENIX conference on Hot Topics in Cloud Computing, HotCloud'12*, 16–16.

Hossain, M., & Kauranen, I. (2016). Open innovation in SMEs: A systematic literature review. *Journal of Strategy and Management*, *9*(1), 58–73. doi:10.1108/JSMA-08-2014-0072

Hovious, A. (2013). *5 Free Real-Time Collaboration Tools*. Retrieved October 21, 2015, from https://designerlibrarian.wordpress.com/2013/09/15/5-free-real-time-collaboration-tools/

Howard, N. (2009). *The book: The life story of a technology*. Johns Hopkins University Press.

Howells, J. (2006). Intermediation and the role of intermediaries in innovation. *Research Policy, 35*(5), 715–728. doi:10.1016/j.respol.2006.03.005

Howells, J., James, A., & Malek, K. (2003). The sourcing of technological knowledge: Distributed innovation processes and dynamic change. *R & D Management, 33*(4), 395–409. doi:10.1111/1467-9310.00306

Huang, X., Soutar, G.N. & Brown, A. (2004) Review and new evidence on composite innovation indicators for evaluating national performance. *Research Policy, 39*(1), 67-78.

Huang, X., Soutar, G. N., & Brown, A. (2004). Measuring new product success: An empirical investigation of Australian SMEs. *Industrial Marketing Management, 33*(2), 117–123. doi:10.1016/S0019-8501(03)00034-8

Huisregels. (2015, July 9). *WhatsApp Buurtpreventie*. https://wabp.nl/huisregels/

Huizingh, E. K. R. E. (2009). The future of innovation. *The XX International Society for Professional Innovation Management - ISPIM Conference 2009. Proceedings*, 21.

Humphrey, J., & Schmitz, H. (2000). *Governance and Upgrading: Linking industrial cluster and global value chain research.* IDS Working Paper 120.

Huston, L., & Sakkab, N. (2006). Connect and develop: Inside Procter & Gamble's new model for innovation. *Harvard Business Review, 84*, 58–66.

IfM & IBM. (2008). *Succeeding through service innovation: A service perspective for education, research, business and government.* Cambridge, UK: University of Cambridge Institute for Manufacturing.

Indarti, N., & Langenberg, M. (2004). Factors Affecting Business Success Among SMEs: Empericical Evidences from Indonesia. In *Proceedings of the Second Bi-annual European Summer University*. University of Twente.

Informatica. (2020). *IoT Data Management: The Rise of Industrial IoT and Machine Learning*. Retrieved January 12, 2020, from https://www.informatica.com/resources/articles/iot-data-management-and-industrial-iot.html#fbid=GTwRdF2a3Xb

International Telecommunication Union (ITU). (2013). *The world in 2013: ICT facts and figures*. Retrieved September 10, 2014, from https://www.itu.int/en/ITU-D/Statistics/.../facts/ ICTFactsFigures2014-e.pdf

Izushi, H. (2003). Impact of the length of relationships upon the use of research institutes by SMEs. *Research Policy, 32*(5), 771–788. doi:10.1016/S0048-7333(02)00085-9

Jaspersoft. (2019). *Jaspersoft Business Intelligence*. Retrieved December 31, 2019, from https://www.jaspersoft.com/

Jenkins, H. (2006). *Convergence culture: Where old and new media collide*. New York University Press.

Jenkins, H., Ito, M., & Boyd, D. (2015). *Participatory culture in a networked era*. Polity Press.

Jin, S. K., & Sailer, K., (n.d.). Seeing and being seen inside a museum and a department store A comparison study in visibility and co-presence patterns. *Proceedings of the 10th International Space syntax Symposium.*

Johnson, L., Smith, R., Willis, H., Levine, A., & Haywood, K. (2011). *The 2011 Horizon Report*. The New Media Consortium.

Jones, O., & Tilley, F. (Eds.). (2003). *Competitive Advantage in SMEs: organizing for innovation and change*. Wiley.

Kalir, J., & Garcia, A. (2019). *Annotation*. MIT Press. https://bookbook.pubpub.org/annotation

Kalir, J. (2019). Open web annotation as collaborative learning. *First Monday, 24*(6). Advance online publication. doi:10.5210/fm.v24i6.9318

Kang, K-N., & Park, H. (2012). Original Research Article. *Technovation, 32*(1), 68–78.

Kanugo, S. (1999). *Making IT work*. Sage Publications.

Karapanos, E., Teixeira, P., & Gouveia, R. (2016). Need fulfillment and experiences on social media: A case on Facebook and WhatsApp. *Computers in Human Behavior, 55*, 888–897. doi:10.1016/j.chb.2015.10.015

Karimi, K. (1997). The Spatial Logic of Organic Cities in Iran and the United Kingdom. In *Proceedings of the First International Symposium on Space Syntax*. London: University College London.

Kasekende, L. (2001). Financing SMEs: Uganda's Experience. In *Improving the Competitiveness of SMEs in Developing Countries: The Role of Finance to Enhance Enterprise Development* (pp. 97–107). United Nations.

Katz, R., & Allen, T. J. (1982). Investigating the not-invented-here (NIH)- syndrome: A look at performance, tenure and communication patterns of 50 R&D project groups. *R & D Management, 12*, 7–19.

Kaufmann, A., & Tödtling, F. (2002, March). How effective is innovation support for SMEs? An analysis of the region of Upper Austria. *Technovation, 22*(3), 147–159. doi:10.1016/S0166-4972(00)00081-X

Kawtrakul, A., Pusittigul, A., Ujjin, S., Lertsuchatavanich, U., & Andres, F. (2013, June). Driving connected government implementation with marketing strategies and context-aware service design. In *Proceedings of the European Conference on e-Government* (p. 265). Academic Press.

Kawtrakul, A., Mulasastra, I., Khampachua, T., & Ruengittinun, S. (2011). The Challenges of Accelerating Connected Government and Beyond: Thailand perspectives. *Electronic Journal of E-Government, 9*(2), 183.

Kearney, C. (2012). Emerging markets research: Trends, issues and future directions. *Emerging Markets Review, 13*(2), 159–183. doi:10.1016/j.ememar.2012.01.003

Keil, T. (2002). *External Corporate Venturing: Strategic Renewal in Rapidly Changing Industries*. Quorum.

Keller, B., & Seifert, H. (2013). *Atypische Beschäftigung zwischen Prekarität und Normalität. Entwicklung, Strukturen und Bestimmungsgründe im Überblick*. Berlin: edition sigma.

Keretho, S., Lent, B., Suchaiya, S., & Naklada, S. (2015). Evaluation of National e-Government Development Levels in Thailand. *Evaluation*.

Kern, H., & Schumann, M. (1984). *Das Ende der Arbeitsteilung? Rationalisierung in der industriellen Produktion: Bestandsaufnahme, Trendbestimmung*. C.H. Beck.

Khampachua, T., & Wisitpongphan, N. (2015, June). Implementing Successful IT Projects in Thailand Public Sectors: A Case Study. In *European Conference on e-Government* (p. 403). Academic Conferences International Limited.

Khona, Z. (2000, September 18). Copyright: A safeguard against piracy. *Express Computer, 11*(28), 15.

Kim, M., & Choi, J. (2009). Angular VGA and Cellular VGA An Exploratory Study for Spatial Analysis Methodology Based on Human Movement Behavior. *Proceedings of the 7th International Space Syntax Symposium*.

Kirchner, S., & Beyer, J. (2016). Die Plattformlogik als digitale Marktordnung. Wie die Digitalisierung Kopplungen von Unternehmen löst und Märkte transformiert. *Zeitschrift für Soziologie, 45*(5), 324–339. doi:10.1515/zfsoz-2015-1019

Kirkels, Y., & Duysters, G. (2010). Brokerage in SME networks. *Research Policy, 39*(3), 375–385. doi:10.1016/j.respol.2010.01.005

Kirschbaum, R. (2005). Open innovation in practice. *Research Technology Management, 48*(4), 24–28. doi:10.1080/08956308.2005.11657321

Kirschenbaum, M. (2012). What is digital humanities and what is it doing in English departments? In M. Gold (Ed.), *Debates in the digital humanities*. University of Minnesota Press., doi:10.5749/minnesota/9780816677948.003.0001

Kleemann, F., & Voß, G. G. (2010). Arbeit und Subjekt. In F. Böhle, G. G. Voß, & G. Wachtler (Eds.), *Handbuch Arbeitssoziologie* (pp. 415–450). Springer. doi:10.1007/978-3-531-92247-8_14

Klein, L. R., & Gold, M. (2016). Digital humanities: The expanded field. In M. Gold & L. R. Klein (Eds.), *Debates in the digital humanities 2016*. University of Minnesota Press., doi:10.5749/j.cttlcn6thb.3

Klein-Woolthuis, R., Lankhuizen, M., & Gilsing, V. (2005). A system failure framework for innovation policy design. *Technovation*, *25*(6), 609–619. doi:10.1016/j.technovation.2003.11.002

Klerkx, L. W. A. (2008). Establishment and embedding of innovation brokers at different innovation system levels: insights from the Dutch agricultural sector. In *Proceedings of the Conference Transitions towards sustainable agriculture food chains and peri-urban areas*. Wageningen University.

Knöll, M., Li, Y., Neuheuser, K., & Rudolph-Cleff, A. (2015). Using space syntax to analyse stress ratings of open public spaces. *Proceedings of the 10th International Space Syntax Symposium.*

Knowage. (2019). *Knowage Business Intelligence*. Retrieved January 4, 2020 from https://knowage.readthedocs.io/en/latest/

Knudsen, M.P. & Mortensen, T.B. (2010). Some immediate – but negative – effects of openness on product development performance. *Technovation.*

Koehler, A. (2013). Digitizing Craft: Creative Writing Studies and New Media: A Proposal. *College English*, *75*(4), 279–397.

Kohn, S., & Hüsig, S. (2006, August). Potential benefits, current supply, utilization and barriers to adoption: An exploratory study on German SMEs and innovation software. *Technovation*, *26*(8), 988–998. doi:10.1016/j.technovation.2005.08.003

Kohsari, M., Oka, K., Owen, N., & Sugiyama, T. (2019). Natural movement: A space syntax theory linking urban form and function with walking for transport. *Cities (London, England).*

Kolb, D. (1996). Discourse Across Links. In C. Ess (Ed.), *Philosophical Perspectives on ComputerMediated Communication* (pp. 15–26). State University of New York Press.

Kole, E. S. (2001). *Internet Information for African Women's Empowerment*. Paper presented at the seminar 'Women, Internet and the South', organized by the Vereniging Informatie en International Ontwikkeling (Society for Information and International Development, VIIO), 18 January 2001, Amsterdam. Retrieved January 14, 2012, from https://www.xs4all.nl/~ekole/public/endrapafrinh.html

Kovacs, A., Van Looy, B., & Cassiman, B. (2015). Exploring the scope of open innovation: A bibliometric review of a decade of research. *Scientometrics*, *104*(3), 951–983. doi:10.100711192-015-1628-0

Kowalski, S. P. (2009). SMES, Open Innovation and IP Management: Advancing Global Development, A presentation paper on the Theme 2. In *The Challenge of Open Innovation for MSMEs - SMEs, Open Innovation and IP Management - Advancing Global Development*. WIPO-Italy International Convention on Intellectual Property and Competitiveness of Micro, Small and Medium-Sized Enterprises (MSMEs).

Kowalski, S. P. (2009). *SMES, Open Innovation and IP Management: Advancing Global Development.* A presentation at the WIPO-Italy International Convention on Intellectual Property and Competitiveness of Micro, Small and Medium-Sized Enterprises (MSMEs), Rome, Italy.

Koyuncugil, A. S., & Ozgulbas, N. (2009). Risk modeling by CHAID decision tree algorithm. *ICCES*, *11*(2), 39–46.

Krauss, R. (1979). Sculpture in the expanded field. *October, 8*, 30–44.

Kroll, H., & Liefner, I. (2008). Spin-off enterprises as a mean of technology commercialisation in a transforming economy – evidence from three universities in China. *Technovation, 28*(5), 298–313. doi:10.1016/j.technovation.2007.05.002

Krongard, S., & McCormick, J. (2013). *Real Time Visual Analytics to Evaluate Online Collaboration.* Presentation at the annual conference of NERCOMP, Providence, RI, Retrieved October 20, 2015, from http://www.educause.edu / nercomp-conference/2013/2013/real-time-visual-analytics-evaluate-online -collaboration.

Kuban, D. (1998). *Mimarlık Kavramları, Tarihsel Perspektif İçinde Mimarlığın Kuramsal Sözlüğüne Giriş*. İstanbul: YEM Yayın.

Kumar, A. (2010). *An exploratory study of unsupervised mobile learning in rural India.* Retrieved November 18, 2012, from http://www.cs.cmu.edu/~anujk1/ CHI2010.pdf

Kumar, A. (1999). *Mass Media*. Anmol.

Kutucu, S., & Yılmaz, E. (2011). *Architectural Competitions as A Method of Building Production Process in Turkey. In 23rd International Building & Life Congress: The Milieu of the Architectural Profession* (p. 477). Bursa: Bursa Chamber of Architects Bursa Branch.

Laforet, S. (2008). Size, strategic, and market orientation affects on innovation. *Technovation, 25*(10), 1119-1127.

Lakhani, K. R., & von Hippel, E. (2003). How open source software works: “Free” user-to-user assistance. *Research Policy, 32*(6), 923–943. doi:10.1016/S0048-7333(02)00095-1

Landow, G. P. (1992). *Hypertext 2.0. The Convergence of Contemporary Critical Theory and Technology*. The John Hopkins University Press.

Landow, G. P. (1994). *Hyper/Text/Theory*. The Johns Hopkins University Press.

Lapa, J., Bernardino, J., & Figueiredo, A. (2014). A comparative analysis of open source business intelligence platforms. *International Conference on Information Systems and Design of Communication, ISDOC 2014. ACM International Conference Proceeding Series*, 86-92. 10.1145/2618168.2618182

Lark, J. (n.d.). Collaboration Tools in Online Learning Environments. *ALN Magazine.* Retrieved October 27, 2015, fromhttp://www.nspnvt.org/jim/aln-colab.pdf

Larson, M. S. (1994). Architectural Competitions as Discursive Events. *Theory and Society, 23*(4), 469–504. doi:10.1007/BF00992825

Larsson, S. (2017). A First Line of Defence? Vigilant Surveillance, Participatory Policing, and the Reporting of ‘Suspicious’ Activity. *Surveillance & Society, 15*(1), 94–107.

Latour, B. (1992). Where Are the Missing Masses? The Sociology of a Few Mundane Artifacts. In W. E. Bijker & J. Law (Eds.), *Shaping Technology / Building Society: Studies in sociotechnical change* (pp. 225–258). MIT Press. https://mitpress.mit.edu/books/shaping-technology-building-society

Latour, B. (1994). On technical mediation. *Common Knowledge, 3*(21), 29–64.

Laursen, K., & Salter, A. (2006). Open for innovation: The role of openness in explaining innovation performance among UK manufacturing firms. *Strategic Management Journal, 27*(2), 131–150. doi:10.1002mj.507

Lawson, C. P., Longhurst, P. J., & Ivey, P. C. (2006). The application of a new research and development project selection model in SMEs. *Technovation, 26*(2), 242–250. doi:10.1016/j.technovation.2004.07.017

Lazzarotti, V., Manzini, R., & Pizzurno, E. (2008). Managing innovation networks of SMEs: a case study. *Proceeding of the International Engineering Management Conference: managing engineering, technology and innovation for growth (IEMC Europe 2008)*, 521-525. 10.1109/IEMCE.2008.4618024

Leavitt, N. (2010). Will NoSQL databases live up to their promise? *IEEE Computer*, *43*(2), 12–14. doi:10.1109/MC.2010.58

Lecocq, X., & Demil, B. (2006). Strategizing industry structure: The case of open systems in low-tech industry. *Strategic Management Journal*, *27*(9), 891–898. doi:10.1002mj.544

Lee, R. (2004). *Scalability report on triple store applications.* Retrieved August 25, 2014 from http://simile.mit.edu/reports/stores/

Lee, E. W. J., Ho, S. S., & Lwin, M. O. (2017). Explicating problematic social network sites use: A review of concepts, theoretical frameworks, and future directions for communication theorizing. *New Media & Society*, *19*(2), 308–326. doi:10.1177/1461444816671891

Lee, M. R., & Lan, Y.-C. (2011, January). Toward a unified knowledge management model for SMEs. *Expert Systems with Applications*, *38*(1), 729–735. doi:10.1016/j.eswa.2010.07.025

Lee, S., Park, G., Yoon, B., & Park, J. (2010). Open innovation in SMEs- An intermediated network model. *Research Policy*, *39*(2), 290–300. doi:10.1016/j.respol.2009.12.009

Leimeister, J. M., Durward, D., & Zogaj, S. (2016). *Crowd Worker in Deutschland: Eine empirische Studie zum Arbeitsumfeld auf externen Crowdsourcing-Plattformen*. Hans-Böckler-Stiftung.

Leiponen, A., & Byma, J. (2009, November). If you cannot block, you better run: Small firms, cooperative innovation, and appropriation strategies. *Research Policy*, *38*(9), 1478–1488. doi:10.1016/j.respol.2009.06.003

Lemola, T. & Lievonen, J. (2008). *The role of innovation policy in fostering open innovation activities among companies*. Vision ERAnet.

Lemola, T., & Lievonen, J. (2008). The role of Innovation Policy in Fostering Open Innovation Activities Among Companies. *Proceedings from "European Perspectives on Innovation and Policy" Programme for the VISION Era-Net Workshop.*

Lenzerini, M. (2002). Data integration: A theoretical perspective. *Proceedings of the 21st ACM SIGMOD-SIGACT-SIGART Symposium on Principles of Database Systems*, 233–246.

Lerman, Y., Rofé, Y., & Omer, I. (2014). Using Space Syntax to Model Pedestrian Movement in Urban Transportation Planning. *Geographical Analysis*, *46*(4), 392–410. doi:10.1111/gean.12063

Lesk, M. (1995). *Keynote address: preserving digital objects: recurrent needs and challenges.* Papers from the National Preservation Office Annual Conference - 1995 Multimedia Preservation: Capturing The Rainbow. Retrieved October 27, 2009, from http://community.bellcore.com/lesk/auspres/aus.html

Levy, M., Powell, P., & Galliers, R. (1998). Assessing information systems strategy development frameworks in SMEs. *Information & Management*, *36*(5), 247–261. doi:10.1016/S0378-7206(99)00020-8

Lichtenthaler, U. (2008a). Open Innovation in Practice: An Analysis of Strategic Approaches to Technology Transactions. *Engineering Management, IEEE Transactions on, 55*(1), 148 – 157.

Lichtenthaler, U. (2005). External commercialization of knowledge: Review and research agenda. *International Journal of Management Reviews*, *7*(4), 231–255. doi:10.1111/j.1468-2370.2005.00115.x

Lichtenthaler, U. (2006). Technology exploitation strategies in the context of open innovation, International Journal of Technology Intelligence and Planning, Volume 2. *Number*, *1/2006*, 1–21.

Lichtenthaler, U. (2007). The drivers of technology licensing: An industry comparison. *California Management Review, 49*(4), 67–89. doi:10.2307/41166406

Lichtenthaler, U. (2007a). Hierarchical strategies and strategic fit in the keep-or-sell decision. *Management Decision, 45*(3), 340–359. doi:10.1108/00251740710744990

Lichtenthaler, U. (2008). Open innovation in practice: An analysis of strategic approaches to technology transactions. *IEEE Transactions on Engineering Management, 55*(1), 148–157. doi:10.1109/TEM.2007.912932

Lichtenthaler, U. (2008b, May/June). Integrated roadmaps for open innovation. *Research Technology Management, 51*(3), 45–49. doi:10.1080/08956308.2008.11657504

Lichtenthaler, U. (2008c, September). Relative capacity: Retaining knowledge outside a firm's boundaries. *Journal of Engineering and Technology Management, 25*(3), 200–212. doi:10.1016/j.jengtecman.2008.07.001

Lichtenthaler, U. (2008d, April). Leveraging technology assets in the presence of markets for knowledge. *European Management Journal, 26*(2), 122–134. doi:10.1016/j.emj.2007.09.002

Lichtenthaler, U. (2008e, July). Externally commercializing technology assets: An examination of different process stages. *Journal of Business Venturing, 23*(4), 445–464. doi:10.1016/j.jbusvent.2007.06.002

Lichtenthaler, U. (2009b, August). Absorptive capacity, environmental turbulence, and the complementarity of organizational learning processes. *Academy of Management Journal, 52*(4), 822–846. doi:10.5465/amj.2009.43670902

Lichtenthaler, U. (2009c, September). Outbound open innovation and its effect on firm performance: Examining environmental influences. *R & D Management, 39*(4), 317–330. doi:10.1111/j.1467-9310.2009.00561.x

Lichtenthaler, U. (2010b, November). Organizing for external technology exploitation in diversified firms. *Journal of Business Research, 63*(11), 1245–1253. doi:10.1016/j.jbusres.2009.11.005

Lichtenthaler, U. (2010c, July-August). Technology exploitation in the context of open innovation: Finding the right 'job' for your technology. *Technovation, 30*(7-8), 429–435. doi:10.1016/j.technovation.2010.04.001

Lichtenthaler, U. (2011a, February–March). 'Is open innovation a field of study or a communication barrier to theory development?' A contribution to the current debate. *Technovation, 31*(2–3), 138–139. doi:10.1016/j.technovation.2010.12.001

Lichtenthaler, U. (2011b, February). Open Innovation: Past Research, Current Debates, and Future Directions. *The Academy of Management Perspectives, 25*(1), 75–93.

Lichtenthaler, U., & Ernst, H. (2006). Attitudes to externally organizing knowledge management tasks: A review, reconsideration and extension of the NIH syndrome. *R & D Management, 36*(4), 367–386. doi:10.1111/j.1467-9310.2006.00443.x

Lichtenthaler, U., & Ernst, H. (2009). Opening up the Innovation Process: The Role of Technology Aggressiveness. *R & D Management, 39*(1), 38–54. doi:10.1111/j.1467-9310.2008.00522.x

Light, A., & Miskelly, C. (2015). Sharing Economy vs. Sharing Cultures? Designing for social, economic and environmental good. *Interaction Design and Architecture(s). Journal, 24*, 49–62.

Li, J., & Kozhikode, R. K. (2009). Developing new innovation models: Shifts in the innovation landscapes in emerging economies and implications for global R&D management. *Journal of International Management, 15*(3), 328–339. doi:10.1016/j.intman.2008.12.005

Lilien, G. L., Morrison, P. D., Searls, K., Sonnack, M., & von Hippel, E. (2002). Performance assessment of the lead user idea-generation process for new product development. *Management Science, 48*(8), 1042–1059. doi:10.1287/mnsc.48.8.1042.171

Lim, K., Sia, C., Lee, M., & Benbasat, I. (2006). Do I Trust You Online, and If So, Will I Buy? An Empirical Study of Two Trust-Building Strategies. *Journal of Management Information Systems*, *23*(2), 233–266. doi:10.2753/MIS0742-1222230210

Lincoln, Y. S., & Guba, E. G. (1985). *Naturalistic Inquiry* (1st ed.). SAGE Publications. doi:10.1016/0147-1767(85)90062-8

Linder, J. C., Jarvenpaa, S., & Davenport, T. H. (2003). Toward an Innovation Sourcing Strategy. *MIT Sloan Management Review. Reprint*, *4447*, 43–49.

Lindermann, N., Valcareel, S., Schaarschmidt, M., & Von Kortzfleisch, H. (2009). SME 2.0: Roadmap towards Web 2.0- Based Open Innovation in SME-Network- A Case Study Based Research Framework. CreativeSME2009, IFIP International Federation for Information Processing, 28-41.

Linge, D. E. (Ed.). (2008). Hans-Georg Gadamer Philosophical Hermeneutics (D. E. Linge, Trans.). Berkeley, CA: University of California Press. (Original work published in 1976)

Lin, Y. H., Tsai, K. M., Shiang, W. J., Kuo, T. C., & Tsai, C. H. (2009). Research on using ANP to establish a performance assessment model for business intelligence systems. *Expert Systems with Applications*, *36*(2), 4135–4146. doi:10.1016/j.eswa.2008.03.004 PMID:32288333

Lipsky, M. (1971). Street-level Bureaucracy and the Analysis of Urban Reform. *Urban Affairs Quarterly*, *6*(4), 391–409. doi:10.1177/107808747100600401

Lipsky, M. (2010). *Street-level Bureaucracy: Dilemmas of the individual in public service*. Russell Sage Foundation.

Liu, X. (2008). *China's Development Model: An Alternative Strategy for Technological Catch-Up*. SLPTMD Working Paper Series No. 020, University of Oxford.

Livieratos, A. D., & Papoulias, D. B. (2009). *Towards an Open Innovation Growth Strategy for New, Technology-Based Firms*. National Technical University of Athens. Retrieved October 30, 2010 from http://www.ltp.ntua.gr/uploads/GJ/eK/GJeKw8Jf5RqjCO9CWapm4w/Growth.pdf

Lopes, C. M., Scavarda, A., Hofmeister, L. F., Thomé, A. M. T., & Vaccaro, G. L. R. (2017). An analysis of the interplay between organizational sustainability, knowledge management, and open innovation. *Journal of Cleaner Production*, *142*, 476–488. doi:10.1016/j.jclepro.2016.10.083

López, A. (2009). Innovation and Appropriability, Empirical Evidence and Research Agenda. In The Economics of Intellectual Property: Suggestions for Further Research in Developing Countries and Countries with Economies in Transition. World Intellectual Property Organization.

Lopez-vega, H. (2009, January 22-24). *How demand-driven technological systems of innovation work?: The role of intermediary organizations*. A paper from the DRUD-DIME Academy Winter 2009 PhD Conference on Economics and Management of Innovation Technology and Organizational Change, Aalborg, Denmark.

Lord, M. D., Mandel, S. W., & Wager, J. D. (2002). Spinning out a star. *Harvard Business Review*, *80*, 115–121. PMID:12048993

Lorsuwannarat, T. (2006, November). Lessons Learned from E-government in Thailand. *Proceeding of Seminar on 'Modernising the Civil Service in Alignment with National Development Goals', Eastern Regional Organization for Public Administration (EROPA).*

Loshin, D. (2011). *The Analytics Revolution 2011: Optimizing Reporting and Analytics to Optimizing Reporting and Analytics to Make Actionable Intelligence Pervasive*. Knowledge Integrity, Inc.

Love, J.H. & Ganotakis, P. (2012). Original Research Article. *International Business Review.*

Lub, V., & De Leeuw, T. (2019). *Politie en actief burgerschap: Een veilig verbond? Een onderzoek naar samenwerking, controle en (neven)effecten* (No. 108; Politiewetenschap). Politie & Wetenschap. https://www.politieenwetenschap.nl/publicatie/politiewetenschap/2019/politie-en-actief-burgerschap-een-veilig-verbond-322/

Lub, V. (2018). Watch Group 2: Countering Burglars. In M. Gill (Ed.), *Neighbourhood Watch in a Digital Age* (pp. 57–72). Palgrave Macmillan., doi:10.1007/978-3-319-67747-7_5

Lub, V., & De Leeuw, T. (2017). Perceptions of Neighbourhood Safety and Policy Response: A Qualitative Approach. *European Journal on Criminal Policy and Research, 23*(3), 425–440. doi:10.100710610-016-9331-0

Luhn, H. P. (1958). A Business Intelligence System. *IBM Journal of Research and Development, 2*(4), 314–319. doi:10.1147/rd.24.0314

Lundvall, B. (1995). *National systems of innovation: Towards a theory of innovation and interactive learning*. Biddles Ltd.

Lundvall, B. A., Johnson, B., Andersen, E. S., & Dalum, B. (2002). National Systems of production and competence building. *Research Policy, 31*(2), 213–231. doi:10.1016/S0048-7333(01)00137-8

Lynch, C. (2008). Big data: How do your data grow? *Nature, 455*(7209), 28–29. doi:10.1038/455028a PMID:18769419

Lyon, D. (2007). Surveillance studies: An overview. *Polity.*

MacGregor, S., Bianchi, M., Hernandez, J. L., & Mendibil, K. (2007). Towards the tipping point for social innovation. *Proceedings of the 12th International Conference on Towards Sustainable Product Design (Sustainable Innovation 07),* 145-152.

Madanipour, A. (2003). *Public and Private Space of the City*. doi:10.4324/9780203402856

Madanipour, A. (2019). Rethinking public space: Between rhetoric and reality. *URBAN DESIGN International, 24*(1), 38–46. doi:10.105741289-019-00087-5

Maes, J. (2009). *SMEs´ Radical Product Innovation: the Role of the Internal and External Absorptive Capacity Spheres. Job Market Paper July 2009, Katholieke Universiteit Leuven.* Faculty of Business and Economics.

Mahajan, S. L. (2002). Information Communication Technology in distance education in India: A challenge. *University News, 40*(19), 1–9.

Maijers, W., Vokurka, L., van Uffelen, R. & Ravensbergen, P. (2005). Open innovation: symbiotic network, Knowledge circulation and competencies for the benefit of innovation in the Horticulture delta. *Presentation IAMA Chicago 2005.*

Major, E. J., & Cordey-Hayes, M. (2000). Engaging the business support network to give SMEs the benefit of foresight. *Technovation, 20*(11), 589–602. doi:10.1016/S0166-4972(00)00006-7

Mallon, M., & Bernsten, S. (2015). *Collaborative Learning Technologies. Tips and Trends.* ACRL Instruction Section, Instructional Technologies Committee, Winter. Retrieved October 02, 2015, from http://bit.ly/tipsandtrendswi15

Mann, S. (2016). *Surveillance (Oversight), Sousveillance (Undersight), and Metaveillance (Seeing Sight Itself).* https://www.cv-foundation.org/openaccess/content_cvpr_2016_workshops/w29/html/Mann_Surveillance_Oversight_Sousveillance_CVPR_2016_paper.html

Mantel, S. J., & Rosegger, G. (1987). The role of third-parties in the diffusion of innovations: a survey. In R. Rothwell & J. Bessant (Eds.), *Innovation: Adaptation and Growth* (pp. 123–134). Elsevier.

Margetts, H., John, P., Hale, S., & Yasseri, T. (2015). *Political turbulence: How social media shape collective action*. Princeton University Press. doi:10.2307/j.ctvc773c7

Marinheiro, A., Bernardino, J. (2015). Experimental Evaluation of Open Source Business Intelligence Suites using OpenBRR. *IEEE Latin America Transactions, 13*(3), 810-817.

Marisetty, V., Ramachandran, K., & Jha, R. (2008). *Wealth effects of family succession: A case of Indian family business groups*. Working Paper. Indian School of Business.

Marrs, K. (2010). Herrschaft und Kontrolle in der Arbeit. In F. Böhle, G. G. Voß, & G. Wachtler (Eds.), *Handbuch Arbeitssoziologie* (pp. 331–356). Springer. doi:10.1007/978-3-531-92247-8_11

Martin, C. J. (2016). The sharing economy: A pathway to sustainability or a nightmarish form of neoliberal capitalism? *Ecological Economics, 27*, 149–159. doi:10.1016/j.ecolecon.2015.11.027

Martinez, R., Yacef, K., & Kay, J. (2010). *Collaborative concept mapping at the tabletop*. Technical Report 657. University of Sydney. Retrieved October 27, 2015, from https://sydney.edu.au/engineering/it/research/tr/tr657.pdf

Massa, S., & Testa, S. (2008, July). Innovation and SMEs: Misaligned perspectives and goals among entrepreneurs, academics, and policy makers. *Technovation, 28*(7), 393–407. doi:10.1016/j.technovation.2008.01.002

Maurer, A. (2004). Von der Krise, dem Elend und dem Ende der Arbeits- und Industriesoziologie. Einige Anmerkungen zu Erkenntnisprogrammen, Theorietraditionen und Bindestrich-Soziologien. *Soziologie, 33*, 7–19.

Mayring, P. (2008). *Qualitative Inhaltsanalyse: Grundlagen und Techniken*. Beltz.

McArthur, E. (2015). Many-to-many exchange without money: Why people share their resources. *Consumption Markets & Culture, 18*(3), 239–256. doi:10.1080/10253866.2014.987083

McCabe, L. (2010). *What's a Collaboration Suite & Why Should You Care?* Retrieved October 27, 2015, from https://www.smallbusinesscomputing.com/biztools/article.php/3890601/Whats-a-Collaboration-Suite--Why-Should-You-Care.htm

McEvily, S. K., Eisenhardt, K. M. M., & Prescott, J. E. (2004). The global acquisition, leverage, and protection of technological competencies. *Strategic Management Journal, 25*(8/9), 713–722. doi:10.1002mj.425

Mearian, L. (2010). Data growth remains IT's biggest challenge, Gartner says. *Computerworld*. https://www.computerworld.com/s/article/9194283/Data_growth_remains_IT_s_biggest_challenge_Gartner_says

Mehlbaum, S. L., & van Steden, R. (2018). *Doe-het-zelfsurveillance. Een onderzoek naar de werking en effecten van Whatsapp-buurtgroepen* (No. 95; Politiekunde). Politie & Wetenschap. https://research.vu.nl/ws/portalfiles/portal/73988502/Mehlbaum_Van_Steden_2018.pdf

Mehra, K. (2009). Role of Intermediary Organisations in Innovation Systems- A Case from India. *Proceedings from the 6th Asialics International Conference*.

Mention, A-L. (2010). Co-operation and co-opetition as open innovation practices in the service sector: Which influence on innovation novelty? *Technovation*.

Miles, I. (2005). Innovation in Services. In The Oxford Handbook of Innovation. Oxford University Press.

Millar, C. C. J. M., & Choi, C. J. (2003). Advertising and knowledge intermediaries: Managing the ethical challenges of intangibles. *Journal of Business Ethics, 48*(3), 267–277. doi:10.1023/B:BUSI.0000005788.90079.5d

Minssen, H. (2006). Arbeits- und Industriesoziologie. New York: Campus.

Mirchandani, D. A., Johnson, J. H. Jr, & Joshi, K. (2008). Perspectives of citizens towards e-government in Thailand and Indonesia: A multigroup analysis. *Information Systems Frontiers*, *10*(4), 483–497. doi:10.100710796-008-9102-7

Mirra, N. (2018). Pursuing a commitment to public scholarship through the practice of annotation. *The Assembly: A Journal for Public Scholarship on Education, 1*(1). https://scholar.colorado.edu/concern/articles/xw42n8680

Möhlmann, M. (2015). Collaborative consumption: Determinants of satisfaction and the likelihood of using a aharing economy option again. *Journal of Consumer Behaviour*, *14*(3), 193–207. doi:10.1002/cb.1512

Mohr, J., & Spekman, R. (1994). Characteristics of partnership success: Partnership attributes, communication behavior and conflict resolution techniques. *Strategic Management Journal*, *15*(2), 135–152. doi:10.1002mj.4250150205

Mols, A., & Pridmore, J. (2019). When Citizens Are "Actually Doing Police Work": The Blurring of Boundaries in WhatsApp Neighbourhood Crime Prevention Groups in The Netherlands. *Surveillance & Society*, *17*(3/4), 272–287. doi:10.24908s.v17i3/4.8664

Moore, G. A. (1991). *Crossing the Chasm*. HarperBusiness.

Moore, G. A. (2006). *Dealing with Darwin: How Great Companies Innovate at Every Phase of Their Evolution*. UK Capstone Publishing.

Mortara, L., & Minshall, T. (2011). How do large multinational companies implement open innovation? *Technovation*, *31*(10-11), 586–597. doi:10.1016/j.technovation.2011.05.002

Motlik, S. (2008). Mobile Learning in Developing Nations. *The International Review of Research in Open and Distance Learning, 9*(2). Retrieved November 18, 2012, from http://www.irrodl.orgindex.php/irrodl/article/view/564

Murugan, S. (2013). User Education: Academic Libraries. *International Journal of Information Technology and Library Science Research, 1*(1), 1-6. Retrieved October 27, 2015, from http://acascipub.com/Journals.php

Muuro, M., Wagacha, W., Kihoro, R., & Oboko, J. (2014). Students' Perceived Challenges in an Online Collaborative Learning Environment: A Case of Higher Learning Institutions in Nairobi, Kenya. *The International Review of Research in Open and Distributed Learning*, *15*(6). Advance online publication. doi:10.19173/irrodl.v15i6.1768

Nagaoka, S., & Kwon, H. U. (2006). The incidence of cross-licensing: A theory and new evidence on the firm and contract level determinants. *Research Policy*, *35*(9), 1347–1361. doi:10.1016/j.respol.2006.05.007

Nahlinder, J. (2005). *Innovation and Employment in Services: The Case of Knowledge Intensive Business Services in Sweden* (Doctoral Thesis). Department of technology and Social Change, Linköping University, Sweden.

Nählinder, J. (2005). *Innovation and Employment in Services: The case of Knowledge Intensive Business Services in Sweden* (PhD Thesis). Department of Technology and Social Change, Linköping University.

Nalebuff, B. J., & Brandeburger, A. M. (1996). *Co-opetition*. Harper Collins.

Napitupulu, D., & Sensuse, D. I. (2014, October). Toward Maturity Model of e-Government Implementation based on Success Factors. In *2014 International Conference on Advanced Computer Science and Information System* (pp. 108-112). IEEE. 10.1109/ICACSIS.2014.7065887

Narain, S. (2001). Development Financial Institutions' and Commercial banks' Innovation Schemes for Assisting SMEs in India. In *Improving the Competitiveness of SMEs in Developing Countries: The Role of Finance to Enhance Enterprise Development*. United Nations.

Narula, R. (2004). R&D collaboration by SMEs: New opportunities and limitations in the face of globalization. *Technovation*, *24*(2), 153–161. doi:10.1016/S0166-4972(02)00045-7

Ndonzuau, F. N., Pirnay, F., & Surlemont, B. (2002). A stage model of academic spin-off creation. *Technovation, 22*(5), 281–289. doi:10.1016/S0166-4972(01)00019-0

Niculescu Dinca, V. (2016). *Policing Matter(s)*. Universitaire pers Maastricht. https://cris.maastrichtuniversity.nl/portal/en/publications/policing-matters(b911f31c-f8e8-44e9-999c-5c8edcafd7b7).html

Nine Tools for Collaboratively Creating Mind Maps. (2010). Retrieved from http://www.freetech4teachers.com/2010/03/nine-tools-for-collaboratively-creating.html#.VubuNPl97IV

Novelli, M., Schmitz, B., & Spencer, T. (2006, December). Networks, clusters and innovation in tourism: A UK experience. *Tourism Management*, *27*(6), 1141–1152. doi:10.1016/j.tourman.2005.11.011

Nunes,P.M., Serrasqueiro, Z. & Leitão, J. (2010). Is there a linear relationship between R&D intensity and growth? Empirical evidence of non-high-tech vs. high-tech SMEs. *Research Policy, 41*(1), 36-53.

O'Regan, N., & Ghobadian, A. (2005). Innovation in SMEs: The impact of strategic orientation and environmental perceptions. *International Journal of Productivity and Performance Management*, *54*(2), 81–97. doi:10.1108/17410400510576595

O'Regan, N., Ghobadian, A., & Sims, M. (2006, February). Fast tracking innovation in manufacturing SMEs. *Technovation*, *26*(2), 251–261. doi:10.1016/j.technovation.2005.01.003

Odin, J. K. (2010). *Hypertext and the Female Imaginary*. University of Minnesota Press.

OECD. (1992). Oslo Manual. DSTI, Organization for Economic Co-operation and Development (OECD).

OECD. (1996). Oslo Manual (2nd ed.). DSTI, OECD.

OECD. (2005). Oslo Manual (3rd ed.). DSTI, OECD.

OECD. (2005). *Oslo Manual: Guidelines for Collecting and Interpreting Innovation Data* (3rd ed.). Organization for Economic Co-operation and Development.

OECD. (2006). *The Athens Action Plans for Removing Barriers to SME Access to International Markets*, Adopted at the OECD-APEC Global Conference in Athens.

OECD. (2007). *Innovation and Growth: Rationale for an Innovation Strategy, OECD Report*. OECD.

OECD. (2008). *Open Innovation in Global Networks, Policy Brief, OECD Observer*. OECD.

OECD/ADB. (2019). *Government at a Glance Southeast Asia 2019*. OECD Publishing., doi:10.1787/9789264305915-

Olson, E., & Bakke, G. (2001). Implementing the lead user method in a high technology firm: A longitudinal study of intentions versus actions. *Journal of Product Innovation Management*, *18*(2), 388–395. doi:10.1111/1540-5885.1860388

OpenText. (2019). Retrieved December 31, 2019, from https://www.opentext.com/products-and-solutions/products/opentext-product-offerings-catalog/rebranded-products/actuate

Othman, F., Mohd Yusoff, Z., & Abdul Rasam, A. R. (2019). Isovist and Visibility Graph Analysis (VGA): Strategies to evaluate visibility along movement pattern for safe space. *IOP Conf. Ser.: Earth Environ. Sci., 385*.

Padmadinata, F. Z. S. (2007). Quality Management System and Product Certification Process and Practice for SME in Indonesia. *Proceedings of the National Workshop on Subnational Innovation Systems and Technology Capaicity Building Policies to Enhance Competitiveness of SMEs*. UN-ESCAP and Indonesian Institute of Science (LIPI).

Pandey, A.P. & Shivesh. (2007). *Indian SMEs and their uniqueness in the country*. Munich Personal RePEc Archive, MPRA Paper No. 6086. Available online at https://mpra.ub.uni-muenchen.de/6086/

Papsdorf, C. (2009). Wie Surfen zu Arbeit wird: Crowdsourcing im Web 2.0. New York: Campus.

Parida, V., Westerberg, M., & Frishammar, J. (2012). Inbound open innovation activities in high-tech SMEs: The impact on innovation performance. *Journal of Small Business Management*, *50*(2), 283–309. doi:10.1111/j.1540-627X.2012.00354.x

Partanen, J., Möller, K., Westerlund, M., Rajala, R., & Rajala, A. (2008, July). Social capital in the growth of science-and-technology-based SMEs. *Industrial Marketing Management*, *37*(5), 513–522. doi:10.1016/j.indmarman.2007.09.012

Parvin, A., Ye, A. M., & Jia, B. (2007). Multilevel Pedestrian Movement: Does Visibility Make Any Difference? *6th International Space Syntax Symposium*, İstanbul.

Payne, J. E. (2003). *E-Commerce Readiness for SMEs in Developing Countries: a Guide for Development Professionals*. Available at URL: http://learnlink.aed.org/Publications/Concept_Papers/ecommerce_readiness.pdf

Pedersen, M., Sondergaard, H.A., & Esbjerg, L. (2009). Network characteristics and open innovation in SMEs. *Proceedings of The XX ISPIM Conference*.

Pentaho. (2019). *Pentaho Open Source BI*. Retrieved August 31, 2019, from https://www.hitachivantara.com/en-us/products/data-management-analytics/pentaho-platform.html

Peponis, J., & Wineman, J. (2002). Spatial Structure of Environment and Behavior. In R. B. Bechtel & A. Churchman (Eds.), *Handbook of Environmental Psychology*. John Wiley & Sons, Inc.

Pfeiffer, S. (2010). Technisierung von Arbeit. In F. Böhle, G. G. Voß, & G. Wachtler (Eds.), *Handbuch Arbeitssoziologie* (pp. 231–261). Springer. doi:10.1007/978-3-531-92247-8_8

Pietrobelli, C., & Rabellotti, R. (Eds.). (2006). *Upgrading to Compete:Global Value Chains, Clusters, and SMEs in Latin America*. Inter-American Development Bank.

Pisano, G. P. (1990). The R&D boundaries of the firm: An empirical analysis. *Administrative Science Quarterly*, *35*(1), 153–176. doi:10.2307/2393554

Pollock, R. (2009). Innovation, Imitation and Open Source. *International Journal of Open Source Software & Processes*, *1*(2), 114-127.

Porter, M. E., & Bond, G. C. (1999). *The Global Competitiveness Report 1999*. World Economic Forum.

Pourdehnad, J. (2007). *Idealized Design - An "Open Innovation" Process*. Presentation from the annual W. *Edwards Deming Annual Conference*, Purdue University, West Lafayette, IN.

Powell, W. (1990). Neither market nor hierarchy: network forms of organization. In B. Stow & L. L. Cummings (Eds.), *Research in Organizational Behavior*. JAI Press.

Power, D. J. (2007). *A Brief History of Decision Support Systems, version 4.0*. DSSResources.COM. http://DSSResources.COM/history/dsshistory.html

Prencipe, A. (2000). Breadth and depth of technological capabilities in CoPS: The case of the aircraft engine control system. *Research Policy*, *29*(7-8), 895–911. doi:10.1016/S0048-7333(00)00111-6

Pressman, J. L., & Wildavsky, A. (1973). *Implementation. Barkley and Los Angeles*. University of California Press.

Pridmore, J., Mols, A., Wang, Y., & Holleman, F. (2018). Keeping an eye on the neighbours: Police, citizens, and communication within mobile neighbourhood crime prevention groups. *The Police Journal: Theory. Practice and Principles*, *92*(2), 97–120. doi:10.1177/0032258X18768397

Pries, L., Schmidt, R., & Trinczek, R. (1990). *Entwicklungspfade von Industriearbeit. Chancen und Risiken betrieblicher Produktionsmodernisierung*. Westdeutscher Verlag. doi:10.1007/978-3-322-85577-0

Prügl, R., & Schreier, M. (2006). Learning from leading-edge customers at *The Sims*: Opening up the innovation process using toolkits. *R & D Management, 36*(3), 237–250. doi:10.1111/j.1467-9310.2006.00433.x

Pülzl, H., & Treib, O. (2006). Implementing Public Policy. In Handbook of Public Policy Analysis. Theory, Politics and Methods. CRC Press.

Quinn, J. B. (2000). Outsourcing innovation: The new engine of growth. *Sloan Management Review, 41*, 4, 13–28.

Rahman, H., & Ramos, I. (2010). Open Innovation in SMEs: From Closed Boundaries to Networked Paradigm. *Issues in Informing Science and Information Technology, 7*, 471–487. doi:10.28945/1221

Rahman, H., & Ramos, I. (2012). Open Innovation in Entrepreneurships: Agents of Transformation towards the Knowledge-Based Economy. *Proceedings of the Issues in Informing Science and Information Technology Education Conference.*

Rahman, H., & Ramos, I. (2013). Open Innovation Strategies in SMEs: Development of a Business Model. In *Small and Medium Enterprises: Concepts, Methodologies, Tools, and Applications* (Vols. 1–4; pp. 281–293). Information Resources Management Association. doi:10.4018/978-1-4666-3886-0.ch015

Ramos, E., Acedo, F. J., & Gonzalez, M. A. (2011, October–November). Internationalisation speed and technological patterns: A panel data study on Spanish SMEs. *Technovation, 31*(10–11), 560–572. doi:10.1016/j.technovation.2011.06.008

Ranchordás, S. (2015). *Does Sharing Mean Caring? Regulating Innovation in the Sharing Economy*. Tilburg Law School Legal Studies Research Paper Series, 6.

Rastogi, N., & Hendler, J. (2017). WhatsApp Security and Role of Metadata in Preserving Privacy. In A. R. Bryant, J. R. Lopez, & R. F. Mills (Eds.), *Proceedings of the 12th International Conference on Cyber Warfare and Security* (pp. 269–274). Academic Conferences and Publishing International Limited. https://books.google.nl/books?hl=en&lr=&id=iXWQDgAAQBAJ&oi=fnd&pg=PA269&dq=whatsapp+end-to-end+encryption+privacy+users&ots=HVbblQPNqW&sig=_brz7_Y2SHFL6v1TH40Pqx6Pycs#v=onepage&q=whatsapp%20end-to-end%20encryption%20privacy%20users&f=false

Raymond, L., & St-Pierre, J. (2010). R&D as a determinant of innovation in manufacturing SMEs: An attempt at empirical clarification. *Technovation, 30*(1), 48–56. doi:10.1016/j.technovation.2009.05.005

Reckwitz, A. (2002). Toward a Theory of Social Practices. *European Journal of Social Theory, 5*(2), 243–263. doi:10.1177/13684310222225432

Reeves, J. (2012). If You See Something, Say Something: Lateral Surveillance and the Uses of Responsibility. Surveillance & Society, 10(3/4), 235–248.

Rejeb, H. B., Morel-Guimarães, L., Boly, V., & Assiélou, N. G. (2009). Measuring innovation best practices: Improvement of an innovation index integrating threshold and synergy effects. *Technovation, 28*(12), 838–854. doi:10.1016/j.technovation.2008.08.005

Renaud, P. (2008). *Open Innovation at Oseo Innovation: Example of the Passerelle Programme (A tool to support RDI collaboration between innovative SME's and large enterprises)*. Presentation at the OECD Business Symposium on Open Innovation in Global Networks, Copenhagen.

Rhee, J., Park, T., & Lee, D. H. (2010). Drivers of innovativeness and performance for innovative SMEs in South Korea: Mediation of learning orientation. *Technovation, 30*(1), 65–75. doi:10.1016/j.technovation.2009.04.008

Richardson, L. (2015). Performing the sharing economy. *Geoforum, 67*, 121–127. doi:10.1016/j.geoforum.2015.11.004

Robb, A. (2015, January 14). 92 percent of college students prefer reading print books to e-readers. *New Republic*. https://newrepublic.com/article/120765/naomi-

Robson, E. (2009). The Clay Tablet Book in Sumer, Assyria, and Babylonia. In S. Eliot & J. Rose (Eds.), *A Companion to the History of the Book* (pp. 67–83). Wiley-Blackwell.

Roemer, C. (2009). The Papyrus Roll in Egypt, Greece, and Rome. In S. Eliot & J. Rose (Eds.), *A Companion to the History of the Book* (pp. 84–94). Wiley-Blackwell.

Rogers, E. M. (1962). *Diffusion of Innovations*. Free Pres.

Rohloff, I. K., Psarra, S., & Wineman, J. (2009). Experiencing Museum Gallery Layouts through Local and Global Visibility Properties in Morphology an inquiry on the YCBA, the MoMA and the HMA. *7th International Space Syntax Symposium*, Stockholm.

Rolf, A., & Sagawe, A. (2015). *Des Googles Kern und andere Spinnennetze. Die Architektur der digitalen Gesellschaft*. UVK.

Roos, D., Eaton, C., Lapis, G., Zikopoulos, P., & Deutsch, T. (2011). *Understanding Big Data: Analytics for Enterprise Class Hadoop and Streaming Data*. McGraw-Hill Osborne Media.

Roper, S., & Hewitt-Dundas, N. (2004). *Innovation persistence: survey and case - study evidence*. Working Paper, Aston Business School, Birmingham, UK.

Rosenblat, A., & Stark, L. (2016). Algorithmic Labor and Information Asymmetries: A Case Study of Uber's Drivers. *International Journal of Communication, 10*, 3758–3784.

Rotchanakitumnuai, S. (2008). Measuring E-government Service Value with the E-GOVSQUAL-RISK model. *Business Process Management Journal, 14*(5), 724–737. doi:10.1108/14637150810903075

Rothwell, R., & Dodgson, M. M. (1994). The Handbook of Industrial Innovation. Edward Elgar.

Rothwell, R. (1991). External networking and innovation in small and medium-sized manufacturing firms in Europe. *Technovation, 11*(2), 93–112. doi:10.1016/0166-4972(91)90040-B

Rouse, M. (2012). *Definition of big data analytics*. Retrieved August 25, 2014, from https://searchbusinessanalytics.techtarget.com/definition/big-data-analytics

Ruben, T. (2012). *Improving Pedestrian Accessibility to Public Space Through Space Syntax Analysis*. Conference: 8th Space Syntax Symposium, Santiago de Chile.

Ruiz, P. P. (2010).*Technology & Knowledge Transfer Under the Open Innovation Paradigm: a model and tool proposal to understand and enhance collaboration-based innovations integrating C-K Design Theory, TRIZ and Information Technologies* (Master's dissertation). Management School of Management, University of Bath, UK.

Ryan, M.-L. (2004). Multivariant narratives. In S. Schreibman, R. S. Siemens, & J. Unsworth (Eds.), *A Companion to Digital Humanities* (pp. 415–430). Blackwell Publishing. doi:10.1002/9780470999875.ch28

Saarikoski, V. (2006). *The Odyssey of the Mobile Internet- the emergence of a networking attribute in a multidisciplinary study* (Academic dissertation). TIEKE, Helsinki.

Sabatier, P. (1986). *Top-down and Bottom-up Approaches to Implementation Research: A critical analysis and suggested synthesis*. Retrieved from: http://0www.jstor.org.opac.sfsu.edu/stable/pdfplus/3998354.pdf?acceptTC=true

Sabatier, P. (1999). *Theories of the Policy Process*. Westview Press.

Saebi, T., & Foss, N. J. (2015). Business models for open innovation: Matching heterogeneous open innovation strategies with business model dimensions. *European Management Journal*, *33*(3), 201–213. doi:10.1016/j.emj.2014.11.002

Sagarik, D., Chansukree, P., Cho, W., & Berman, E. (2018). E-government 4.0 in Thailand: The role of central agencies. *Information Polity*, *23*(3), 343–353. doi:10.3233/IP-180006

Saini, D. S., & Budhwar, P. S. (2008). Managing the human resource in Indian SMEs: The role of indigenous realities. *Journal of World Business*, *43*(4), 417–434. doi:10.1016/j.jwb.2008.03.004

Samara, E., Georgiadis, P., & Bakouros, I. (2012). The impact of innovation policies on the performance of national innovation systems: A system dynamics analysis. *Technovation*, *32*(11), 624–638. doi:10.1016/j.technovation.2012.06.002

Santisteban, M. A. (2006). Business Systems and Cluster Policies in the Basque Country and Catalonia (1990-2004). *European Urban and Regional Studies*, *13*(1), 25–39. doi:10.1177/0969776406059227

Santos, R. J., Bernardino, J., & Vieira, M. (2011). A survey on data security in data warehousing: Issues, challenges and opportunities. *EUROCON 2011 - International Conference on Computer as a Tool - Joint with Conftele 2011*.

Sauer, D., & Döhl, V. (1997). Die Auflösung des Unternehmens? Entwicklungstendenzen der Unternehmensreorganisation in den 90er Jahren. In Jahrbuch Sozialwissenschaftliche Technikberichterstattung 1996. Schwerpunkt: Reorganisation (pp. 19-76). Berlin: edition sigma.

Sautter, B., & Clar, G. (2008). *Strategic Capacity Building in Clusters to Enhance Future-oriented Open Innovation Processes*. Foresight Brief No. 150, The European Foresight Monitoring Network. www.efmn.info

Savioz, P., & Blum, M. (2002). Strategic forecast tool for SMEs: How the opportunity landscape interacts with business strategy to anticipate technological trends. *Technovation*, *22*(2), 91–100. doi:10.1016/S0166-4972(01)00082-7

Savitskaya, I. (2009). Towards open innovation in regional Innovation system: case St. Petersburg. Research Report 214, Lappeenranta University of Technology, Lappeenranta.

Scaraboto, D. (2015). Selling, Sharing, and Everything In Between: The Hybrid Economies of Collaborative Network. *The Journal of Consumer Research*, *42*(1), 152–176. doi:10.1093/jcr/ucv004

Schaffers, H., Cordoba, M. G., Hongisto, P., Kallai, T., Merz, C., & van Rensburg, J. (2007). *Exploring business models for open innovation in rural living labs*. Paper from the *13th International Conference on Concurrent Enterprising*, Sophia-Antipolis, France.

Schaltegger, S., & Wagner, M. (2011). Sustainable entrepreneurship and sustainability innovation: Categories and interactions. *Business Strategy and the Environment*, *20*(4), 222–237. doi:10.1002/bse.682

Schatzki, T. R. (2002). *The site of the social: A philosophical account of the constitution of social life and change*. The Pennsylvania State University Press. https://scholar.google.com/scholar?cluster=5833415857224674858&hl=en&oi=scholarr

Schumpeter, J. A. (1934). *The Theory of Economic Development*. Harvard University Press.

Schumpeter, J. A. (1942). *Capitalism, Socialism, and Democracy*. Harper & Row.

Schumpeter, J. A. (1982). *The Theory of Economic Development: An Inquiry into Profits, Capital, Credit, Interest, and the Business Cycle (1912/1934)*. Transaction Publishers.

Secretariat, U. N. C. T. A. D. (2001). Best Practices in Financial Innovations for SMEs. In *Improving the Competitiveness of SMEs in Developing Countries: The Role of Finance to Enhance Enterprise Development* (pp. 3–58). United Nations.

SelectHub. (2020). *The Future of Business Intelligence (BI) in 2020.* Retrieved January 12, 2020 from https://www.selecthub.com/business-intelligence/future-of-bi/

Shackelford, L. (2014a). Literary Turns at the Scene of Digital Writing. In Tactics of the Human: Experimental Technics in American Fiction (pp. 28-56). University of Michigan Press, Digitalculturebooks.

Shackelford, L. (2014b). Tracing the Human through Media Difference. In LTactics of the Human: Experimental Technics in American Fiction (pp. 57-94). University of Michigan Press, Digitalculturebooks.

Shaheen, S. A., Mallery, M. A., & Kingsley, K. J. (2012). Personal vehicle sharing services in North America. *Research in Transportation Business & Management*, *3*, 71–81. doi:10.1016/j.rtbm.2012.04.005

Sharda, R., Delen, D., & Turban, E. (2014). *Business Intelligence and Analytics: Systems for Decision Support* (10th ed.). Prentice Hall.

Shorthouse, S. (2008). *Innovation and Technology Transfer*. Presentation from the International Conference DISTRICT 2008. International Conference Centre, Dresden.

Shove, E., Pantzar, M., & Watson, M. (2012). The Dynamics of Social Practice: Everyday Life and how it Changes. *Sage (Atlanta, Ga.)*. Advance online publication. doi:10.4135/9781446250655

Shrader, C., Mulford, C., & Blackburn, V. (1989). Strategic and operational planning, uncertainty and performance in small firms. *Journal of Small Business Management*, *27*(4), 45–60.

Sinclair, B. (2009). The blended librarian in the learning commons: New skills for the blended library. *College & Research Libraries News*, *70*(9), 504–516. doi:10.5860/crln.70.9.8250

Singh, A. (2019). ICT initiatives in School Education of India *Indian Journal of Educational Technology, 1*(1), 38-49.

Singh, S. P., & Passi, A. (2014). Real Time Communication. *International Journal of Recent Development in Engineering and Technology*, *2*(3). Retrieved October 27, 2015, from http://www.ijrdet.com/files/Volume2Issue3/IJRDET_0314_23.pdf

Siu, W. S. (2005). An institutional analysis of marketing practices of small and medium-sized enterprises (SMEs) in China, Hong Kong and Taiwan. *Entrepreneurship and Regional Development*, *17*(1), 65–88. doi:10.1080/0898562052000330306

Skok, M. (2008). *The Future of Open Source: Exploring the Investments, Innovations, Applications, Opportunities and Threats*. North Bridge Venture Partners.

Skyep. (n.d.). Retrieved from Wikipedia: https://en.wikipedia.org/wiki/Skype

Slack, E. (2012). *Storage infrastructures for big data workflows.* Storage Switzerland White Paper. Retrieved August 25, 2014, from https://iq.quantum.com/exLink.asp?8615424OJ73H28I34127712

Slideshare. (n.d.). Retrieved from Wikipedia: https://en.wikipedia.org/wiki/SlideShare

Smallbone, D., North, D., & Vickers, I. (2003). The role and characteristics of SMEs. In *Regional Innovation Policy for Small-Medium Enterprises*. Edward Elgar. doi:10.4337/9781781009659.00010

Smeets, M. E., Schram, K., Elzinga, A., & Zoutendijk, J. (2019). *Alerte burgers, meer veiligheid?: De werking van digitale buurtpreventie in Rotterdam.* https://surfsharekit.nl/publiek/inholland/87bc7b97-632b-43d0-906e-e68575308f56

Smith, B. L., & MacGregor, J. T. (1992). What is collaborative learning? In A. S. Goodsell, M. R. Maher, & V. Tinto (Eds.), *Collaborative Learning: A Sourcebook for Higher Education. National Center on Postsecondary Teaching, Learning, & Assessment*. Syracuse University.

Soja, E. (1996). *Thirdspace: Journeys to Los Angeles and Other Real-and-Imagined Places*. Blackwell.

Sousa, M. C. (2008). Open innovation models and the role of knowledge brokers. *Inside Knowledge Magazine*, *11*(6), 1–5.

Spago, B. I. (2019). Retrieved January 04, 2020 from https://www.spagobi.org/

Specifications. (2015). Retrieved from Arkitera: https://sartname.arkitera.com/Main.php?MagID=63&MagNo=220

Spithoven, A., Clarysse, B., & Knockaert, M. (2010). Building absorptive capacity to organize inbound open innovation in traditional industries. *Technovation*, *30*(2), 130–141. doi:10.1016/j.technovation.2009.08.004

Spithoven, A., Vanhaverbeke, W., & Roijakkers, N. (2013). Open innovation practices in SMEs and large enterprises. *Small Business Economics*, *41*(3), 537–562. doi:10.100711187-012-9453-9

Srivastava, D. K. (2005). Human resource management in Indian mid size operations. In Indian mid-size manufacturing enterprises: Opportunities and challenges in a global economy. Gurgaon: Management Development Institute.

Stankiewics, R. (1995). The role of the science and technology infrastructure in the development and diffusion of industrial automation in Sweden. In B. Carlsson (Ed.), *Technological systems and economic performance: The case of the factory automation* (pp. 165–210). Kluwer Academic Publishers. doi:10.1007/978-94-011-0145-5_6

Statistisches Bundesamt. (2008). *Übersicht über die Gliederung der Klassifikation der Wirtschaftszweige*. Ausgabe 2008 (WZ2008).

Steadman, P. (1983). *Architectural Morphology, An Introduction to the Geometry of Building Plans, London*. Pion.

Stewart, J. Jr, Hedge, D. M., & Lester, J. P. (2007). *Public policy: An evolutionary approach*. Nelson Education.

Stiegler, B. (1998). *Technics and time, 1: The fault of epimetheus* (R. Beardsworth & G. Collins, Trans.). Stanford University Press. (Original work published 1994)

Storey, D. (1994). *Understanding the small business sector*. International Thomson Business Press.

Strigel, C. (2011). *ICT and the Early Grade Reading Assessment: From Testing to Teaching*. Retrieved January 14, 2012, from https://edutechdebate.org/reading-skills-in-primary-schools/ict-and-the-early-grade-reading-assessment-from-testing-to-teaching/

Sukasame, N. (2004). The Development of e-Service in Thai Government. *BU Academic Review*, *3*(1), 17–24.

Sutanonpaiboon, J., & Pearson, A. M. (2006). E-commerce Adoption: Perceptions of managers/owners of small-and medium-sized enterprises (SMEs) in Thailand. *Journal of Internet Commerce*, *5*(3), 53–82. doi:10.1300/J179v05n03_03

Tahar, B., & Brown, F. (2003). The visibility graph: An approach for the analysis of traditional domestic M'zabite spaces. *4th International Space Syntax Symposium*, London.

Takagi, Y., & Czajkowski, A. (2012). WIPO services for access to patent information - Building patent information infrastructure and capacity in LDCs and developing countries. *World Patent Information*, *34*(1), 30–36. doi:10.1016/j.wpi.2011.08.002

Tapscott, D., & Williams, A. D. (2007). *Wikinomics: How mass collaboration changes everything*. Penguin Book.

Task Management. (n.d.). Retrieved from Wikipedia: https://en.wikipedia.org/wiki/Task_management

Taylor, A. (2014). *The people's platform: Taking back power and culture in the digital age*. Metropolitan Books.

TDWI. (2019). *The Data Warehousing Institute*. https://tdwi.org/Home.aspx

TeachThought Staff. (2019). *30 Of The Best Digital Collaboration Tools For Students*. Retrieved from https://www.teachthought.com/technology/12-tech-tools-for-student-to-student-digital-collaboration/

Technologyadvice. (2020). *Business Intelligence Software - Compare over 100 vendors*. Retrieved January 12, 2020 form https://technologyadvice.com/business-intelligence/

Teece, D. J. (2006). Reflections on "Profiting from Innovation". *Research Policy*, *35*(8), 1131–1146. doi:10.1016/j.respol.2006.09.009

Teradata. (2019). *Teradata University Network*. Retrieved December 30, 2019, from https://www.teradatauniversitynetwork.com/

Thampy, A. (2010). *Financing of SME firms in India Interview with Ranjana Kumar, Former CMD, Indian Bank; Vigilance Commissioner, Central Vigilance Commission*. doi:10.1016/j.iimb.2010.04.011

The 451 Group. (2008). *Open Source Is Not a Business Model*. The 451 Commercial Adoption of Open Source (CAOS) Research Service.

These 10 Collaboration Tools for Education are Boosting Student Engagement and Bringing Interactive Fun to the Classroom. (n.d.). Retrieved from https://www.getcleartouch.com/what-are-the-best-online-collaboration-tools-for-students/

Thoben, K. D. (2008) A new wave of innovation in collaborative networks. *Proceedings of the 14th international conference on concurrent enterprising: ICE 2008*, 1091-1100.

Thomson, S. (2014). *6 Online Collaboration Tools and Strategies for Boosting Learning*. Retrieved October 27, 2015, from https://elearningindustry.com/6-online-collaboration-tools-and-strategies-boosting-learning

Thorgren, S., Wincent, J., & Örtqvist, D. (2009). Designing interorganizational networks for innovation: An empirical examination of network configuration, formation and governance. *Journal of Engineering and Technology Management*, *26*(3), 148–166. doi:10.1016/j.jengtecman.2009.06.006

Tidd, J. (2014) Open innovation research, management and practice. Imperial College Press.

Tidd, J., Bessant, J., & Pavitt, K. (2001). *Managing Innovation*. John Wiley.

Tidd, J., Bessant, J., & Pavitt, K. (2005). *Managing Innovation: Integrating Technological, Market and Organizational Change* (3rd ed.). John Wiley & Sons Ltd.

Tikam, M (2013). Impact of ICT on education. *International Journal of Information Communication Technologies and Human Development*, *5*(4), 1-9.

Tikam, M. (2016). ICT Integration in Education Potential and Challenges. In Human Development and Interaction in the Age of Ubiquitous Technology: Advances in Human and Social Aspects of Technology. IGI Global.

Tikam, M., & Lobo, A. (2012). Special Information Center for Visually Impaired Persons – A Case Study. *International Journal of Information Research*, *1*(3), 42–49.

Tilley, F., & Tonge, J. (2003). Introduction. In O. Jones & F. Tilley (Eds.), *Competitive Advantage in SME's: Organising for Innovation and Change*. John Wiley & Sons.

Todoist. (2015). Retrieved from http://www.pcmag.com/article2/0,2817,2408574,00.asp

Tödtling, F., & Trippl, M. (2005). One size fits all? Towards a differentiated regional innovation policy approach. *Research Policy, 34*(8), 1203–1219. doi:10.1016/j.respol.2005.01.018

Toivonen, M. (2004). *Expertise as Business: Long-term development and future prospects of knowledge-intensive business services (KIBS).* Laboratory of Industrial Management, Helsinki University of Technology.

Torkkeli, M., Tiina Kotonen, T., & Pasi Ahonen, P. (2007). Regional open innovation system as a platform for SMEs: A survey. *International Journal of Foresight and Innovation Policy, 3*(4), 2007. doi:10.1504/IJFIP.2007.016456

Torres, R., Sidorova, A., & Jones, M. C. (2018). Enabling firm performance through business intelligence and analytics: A dynamic capabilities perspective. *Information & Management, 2018*(7), 822–839. Advance online publication. doi:10.1016/j.im.2018.03.010

Tourigny, D., & Le, C. D. (2004). Impediments to Innovation Faced by Canadian Manufacturing Firms. *Economics of Innovation and New Technology, 13*(3), 217–250. doi:10.1080/10438590410001628387

Traxler, J. (2007). Defining, discussing and evaluating mobile learning: The moving finger writes and having writ . . .. *International Review of Research in Open and Distance Learning, 8*(2). Advance online publication. doi:10.19173/irrodl.v8i2.346

Trilokekar, N. P. (2000). *A practical guide to information technology act, 2000: Indian cyber law.* Snow White Publications.

Trinczek, R. (2011). Überlegungen zum Wandel von Arbeit. *WSI-Mitteilungen, 11*(11), 606–614. doi:10.5771/0342-300X-2011-11-606

Trottier, D. (2012). Interpersonal Surveillance on Social Media. *Canadian Journal of Communication, 37*(2). Advance online publication. doi:10.22230/cjc.2012v37n2a2536

Trott, P., & Hartmann, D. A. P. (2009). Why 'open innovation' is old wine in new bottles. *International Journal of Innovation Management, 13*(04), 715–736. doi:10.1142/S1363919609002509

Tu, C.-H. (2004). *Online Collaborative Learning Communities: Twenty-One Designs to Building an Online Collaborative Learning Community.* Libraries Unlimited Inc.

Tufekci, Z. (2017). *Twitter and tear gas: The power and fragility of networked protest.* Yale University Press.

Turban, E., King, D., Lee, J. K., & Viehland, D. (2004). *Electronic Commerce 2004: A Managerial Perspective* (3rd ed.). Prentice Hall.

Turner, A. (2004, June). *Depthmap 4 A Researcher's Handbook.* Academic Press.

Turner, A. (2001). Depthmap: A Programme to Perform Visibility Graph Analysis. *Proceedings of the Third International Symposium on Space Syntax*, 31.1.

Turner, A., Doxa, M., O'Sullivan, D., & Penn, A. (2001). From isovists to visibility graphs: A methodology for the analysis of architectural space. *Environment and Planning. B, Planning & Design, 28*(1), 103–121. doi:10.1068/b2684

Turpin, T., Garrett-Jones, S., & Rankin, N. (1996). Bricoleurs and boundary riders: Managing basic research and innovation knowledge networks. *R & D Management, 26*(3), 267–282. doi:10.1111/j.1467-9310.1996.tb00961.x

Tzu, S. (1963). *The Art of War. United Sates of America.* Oxford University Press.

Ulrich Lichtenthaler Lichtenthaler, U. (2009a, April). Retracted: The role of corporate technology strategy and patent portfolios in low-, medium- and high-technology firms. *Research Policy, 38*(3), 559–569. doi:10.1016/j.respol.2008.10.009

UNESCO. (2006). *ICTs in education for people with special needs united nations educational, scientific and cultural organizationunesco institute for information technologies in education specialized training course*. Moscow: UNESCO. Retrieved December 19, 2012, from https://iite.unesco.org/pics/publications/en/files/3214644.pdf

UNESCO. (2009). *Guide to measuring Information and Communication Technologies (ICT) in education*. UNESCO Institute for Statistics.

United Nations. (2006). *Globalization of R&D and Developing Countries, United Nations Conference on Trade and Development*. UNCTAD/ITE/IIA/2005/6, United Nations.

University of Waterloo. (n.d.). *Collaborative online learning: fostering effective discussions*. Retrieved March 21, 2014, from https://uwaterloo.ca/centre-for-teaching-excellence/teaching-resources/teaching-tips/alternatives-lecturing/discussions/collaborative-online-learning

Valk, J., Rashid, A., & Elder, L. (2010). Using Mobile Phones to Improve Educational Outcomes: An Analysis of Evidence from Asia. *International Review of Research in Open and Distance Learning*, *11*(1), 117. doi:10.19173/irrodl.v11i1.794

Van Ark, B., Broersma, L. & Den Hertog, P. (2003). *Service Innovation, Performance and Policy: A Review*. Research Series no. 6, Synthesis report in the framework of the project on Structural Information Provision on Innovation in Services SIID for the Ministry of Economic Affairs of the Netherlands.

Van de Ven, A. H. (1986). Central problems in the management of innovation. *Management Science*, *32*(5), 590–607. doi:10.1287/mnsc.32.5.590

Van de Vrande, V., de Jong, J. P. J., Vanhaverbeke, W., & de Rochemont, M. (2008). Open innovation in SMEs: Trends, motives and management challenges. A report published under the SCALES-initiative (SCientific AnaLysis of Entrepreneurship and SMEs), as part of the 'SMEs and Entrepreneurship programme' financed by the Netherlands Ministry of Economic Affairs, Zoetermeer.

Van de Vrande, V., De Jong, J. P. J., Vanhaverbeke, W., & De Rochemont, M. (2009). Open innovation in SMEs: Trends, motives and management challenges. *Technovation*, *29*(6-7), 423–437. doi:10.1016/j.technovation.2008.10.001

Van de Vrande, V., Vanhaverbeke, W., & Gassman, O. (2010). Broadening the scope of open innovation: Past research, current state and future directions. *International Journal of Technology Management*, *52*(3-4), 221–235. doi:10.1504/IJTM.2010.035974

van der Land, M. (2017). Meer of minder eigenrichting door buurtwachten? In P. Ponsaers, E. Devroe, K. van der Vijver, L. G. Moor, J. Janssen, & B. van Stokkom (Eds.), *Eigenrichting* (pp. 63–72). Maklu.

Van der Land, M., van Stokkom, B., & Boutellier, H. (2014). *Burgers in veiligheid: Een inventarisatie van burgerparticipatie op het domein van de sociale veiligheid*. Vrije Universiteit. https://www.publicspaceinfo.nl/media/bibliotheek/None/LANDSTOKKO%202014%20001.pdf

Van Dijk, C., & Van den Ende, J. (2002). Suggestion systems: Transferring employee creativity into practicable ideas. *R & D Management*, *32*(5), 387–395. doi:10.1111/1467-9310.00270

Van Gils, A., & Zwart, P. (2004). Knowledge Acquisition and Learning in Dutch and Belgian SMEs: The Role of Strategic Alliances. *European Management Journal*, *22*(6), 685–692. doi:10.1016/j.emj.2004.09.031

Van Hemel, C., & Cramer, J. (2002, October). Barriers and stimuli for ecodesign in SMEs. *Journal of Cleaner Production*, *10*(5), 439–453. doi:10.1016/S0959-6526(02)00013-6

Van Hemert, P., & Nijkamp, P. (2010). Knowledge investments, business R&D and innovativeness of countries: A qualitative meta-analytic comparison. *Technological Forecasting and Social Change*, *77*(3), 369–384. doi:10.1016/j.techfore.2009.08.007

Van Zundert, J. (2016). Barely Beyond the Book? In M. J. Driscoll & E. Pierazzo (Eds.), *Digital Scholarly Editing* (pp. 83–106). Open Book Publishers.

Vanhaverbeke, W., Chesbrough, H., & West, J. (2014). Surfing the new wave of open innovation research. *New Frontiers in Open Innovation, 281*, 287-288.

Vanhaverbeke, W. (2006). The Inter-organizational Context of Open Innovation. In H. Chesbrough, W. Vanhaverbeke, & J. West (Eds.), *Open Innovation: Researching a New Paradigm* (pp. 205–219). Oxford University Press.

Vanhaverbeke, W. (2017). *Managing open innovation in SMEs*. Cambridge University Press. doi:10.1017/9781139680981

Vanhaverbeke, W., & Cloodt, M. (2006). Open innovation in value networks. In H. Chesbrough, W. Vanhaverbeke, & J. West (Eds.), *Open Innovation: Researching a New Paradigm*. Oxford University Press.

Varavithya, W., & Esichaikul, V. (2003, September). The Development of Electronic Government: A case study of Thailand. In *International Conference on Electronic Government* (pp. 464-467). Springer. 10.1007/10929179_85

Vathanophas, V., Krittayaphongphun, N., & Klomsiri, C. (2008). Technology Acceptance toward E-government Initiative in Royal Thai Navy. *Transforming Government: People. Process and Policy*, *2*(4), 256–282. doi:10.1108/17506160810917954

Veugelers, M., Bury, J., & Viaene, S. (2010). Linking technology intelligence to open innovation. *Technological Forecasting and Social Change*, *77*(2), 335–343. doi:10.1016/j.techfore.2009.09.003

Von Hippel, E. (1988). The Sources of Innovation. Oxford University Press.

Von Hippel, . (1987). Cooperation between rivals: Informal Know-how trading. *Research Policy*, *16*(6), 291–302. doi:10.1016/0048-7333(87)90015-1

Von Hippel. (1994). “Sticky Information” and the Locus of Problem Solving: Implications for Innovation. Management Science, 40(4), 429-439.

Von Hippel. (1998). Economics of product development by users: The impact of `sticky’ local information. *Management Science, 44*(5), 629-644.

Von Hippel, E. (1986). Lead users: A source of novel product concepts. *Management Science*, *32*(7), 791–805. doi:10.1287/mnsc.32.7.791

Von Hippel, E. (1998). Economics of product development by users: The impact of ‘Sticky’ local information. *Management Science*, *44*(5), 629–644. doi:10.1287/mnsc.44.5.629

Von Hippel, E. (2005). *Democratizing Innovation*. MIT Press. doi:10.7551/mitpress/2333.001.0001

von Hippel, E. (2005). Democratizing innovation: The evolving phenomenon of user innovation. *Journal für Betriebswirtschaft*, *55*(1), 63–78. doi:10.100711301-004-0002-8

Von Hippel, E., & Von Krogh, G. (2003). Open source software and the “private – collective” innovation model: Issues for organization science. *Organization Science*, *14*(2), 209–223. doi:10.1287/orsc.14.2.209.14992

Voß, G. G. (2010). Arbeit als Grundlage menschlicher Existenz: Was ist Arbeit? Zum Problem eines allgemeinen Arbeitsbegriffs. In F. Böhle, G. G. Voß, & G. Wachtler (Eds.), *Handbuch Arbeitssoziologie* (pp. 23–80). Springer VS. doi:10.1007/978-3-531-92247-8_2

Vujovic, S., & Ulhøi, J. P. (2008). Opening up the Innovation Process: Different Organizational Strategies. Information and Organization Design Series, Vol. 7, Springer US. doi:10.1007/978-0-387-77776-4_8

Vygotsky, L. S. (1978). *Mind in society: The development of higher psychological processes*. Harvard University Press.

Walther, J., & D'Addario, K. (2001). The Impacts of Emoticons on Message Interpretation in Computer-Mediated Communication. *Social Science Computer Review, Volume*, *19*(3), 324–347. doi:10.1177/089443930101900307

Walton, R. E. (1989). *Up and Running: Integrating Information Technology and the Organization*. Harvard Business School Press.

Wang, L., Jaring, P., & Wallin, A. (2009) Developing a Conceptual Framework for Business Model Innovation in the Context of Open Innovation. *Proceedings of the Third IEEE International Conference on Digital Ecosystems and Technologies (IEEE DEST 2009)*, 460-465.

Wangpipatwong, S., Chutimaskul, W., & Papasratorn, B. (2008). Understanding Citizen's Continuance Intention to Use e-Government Website: A Composite View of Technology Acceptance Model and Computer Self-Efficacy. *Electronic Journal of E-Government*, *6*(1).

Wangpipatwong, S., Chutimaskul, W., & Papasratorn, B. (2009). Quality Enhancing the Continued Use of E-government Web Sites: Evidence from e-citizens of Thailand. *International Journal of Electronic Government Research*, *5*(1), 19–35. doi:10.4018/jegr.2009092202

Wang, S., Yeoh, W., Richards, G., Wong, S. F., & Chang, Y. (2019). Harnessing business analytics value through organizational absorptive capacity. *Information & Management*, *56*(7), 103–152. doi:10.1016/j.im.2019.02.007

Wang, Y., Vanhaverbeke, W., & Roijakkers, N. (2012). Exploring the impact of open innovation on national systems of innovation - A theoretical analysis. *Technological Forecasting and Social Change*, *79*(3), 419–428. doi:10.1016/j.techfore.2011.08.009

Warner, A. (2001). *Small and Medium Sized Enterprises and Economic Creativity*. A paper presented at UNCTAD's intergovernmental Expert Meeting on "Improving the Competitiveness of SMEs in Developing Countries: the Role of Finance, Including E-finance, to Enhance Enterprise Development".

Warner, A. (2001). *Small and Medium Sized Enterprises and Economic Creativity*. Paper presented at UNCTAD's intergovernmental Expert Meeting on "Improving the Competitiveness of SMEs in Developing Countries: the Role of Finance, Including E-finance, to Enhance Enterprise Development", Geneva.

Wasserman, A. I. (2017). OSSpal: Finding and Evaluating Open Source Software. In *Springer International Publishing* (pp. 193–203). doi:10.1007/978-3-319-57735-7_18

Waterloo, S. F., Baumgartner, S. E., Peter, J., & Valkenburg, P. M. (2017). Norms of online expressions of emotion: Comparing Facebook, Twitter, Instagram, and WhatsApp. *New Media & Society*, *1461444817707349*. Advance online publication. doi:10.1177/1461444817707349 PMID:30581358

Watkins, D., & Horley, G. (1986). Transferring technology from large to small firms: the role of intermediaries. In T. Webb, T. Quince, & D. Watkins (Eds.), *Small Business Research* (pp. 215–251). Gower.

Watson, H. J., Wixom, B. H., & Goodhue, D. L. (2004). Data Warehousing: The 3M Experience. In H. R. Nemati & C. D. Barko (Eds.), *Org. Data Mining: Leveraging Enterprise Data Resources for Optimal Performance* (pp. 202–216). Idea Group Publishing. doi:10.4018/978-1-59140-134-6.ch014

Watts, R., & Flanagan, C. (2007). Pushing the envelope on civic engagement: A developmental and liberation psychology perspective. *Journal of Community Psychology*, *35*(6), 779–792. doi:10.1002/jcop.20178

Weber, T. A. (2014). Intermediation in a Sharing Economy: Insurance, Moral Hazard, and Rent Extraction. *Journal of Management Information Systems*, *31*(3), 35–71. doi:10.1080/07421222.2014.995520

Wenger, E. (1998). *Communities of Practice: Learning, Meaning and Identity*. Cambridge University Press. doi:10.1017/CBO9780511803932

West, J. (2006). Does Appropriability Enable or Retard Open Innovation? In Open Innovation: Researching a New Paradigm. Oxford University Press.

West, J., Salter, A., Vanhaverbeke, W., & Chesbrough, H. (2014). *Open innovation: The next decade*. Academic Press.

West, J., Vanhaverbeke, W., & Chesbrough, H. (2006). Open Innovation: A Research Agenda. In Open Innovation: Researching a New Paradigm. Oxford University Press.

Westergren, U. H., & Holmström, J. (2012). Exploring preconditions for open innovation: Value networks in industrial firms. *Information and Organization*, *22*(4), 209–226. doi:10.1016/j.infoandorg.2012.05.001

West, J., & Bogers, M. (2014). Leveraging external sources of innovation: A review of research on open innovation. *Journal of Product Innovation Management*, *31*(4), 814–831. doi:10.1111/jpim.12125

West, J., & Gallagher, S. (2006). Challenges of open innovation: The paradox of firm investment in open-source software. *R & D Management*, *36*(3), 319–331. doi:10.1111/j.1467-9310.2006.00436.x

West, J., & Gallagher, S. (2006). Patterns of Open Innovation in Open Source Software. In H. Chesbrough, W. Vanhaverbeke, & J. West (Eds.), *Open Innovation: Researching a New Paradigm* (pp. 82–106). Oxford University Press.

West, J., & Lakhani, K. (2008). Getting Clear About the Role of Communities in Open Innovation. *Industry and Innovation*, *15*(2), 223–231. doi:10.1080/13662710802033734

Whyte, W. H. (2000). The Social Life Of Small Urban Spaces, Common Ground? Readings and Reflections on Public Space. Routledge.

Wijffels, H. (2009). Leadership, sustainability and levels of consciousness. A speech at the conference on the Leadership for a Sustainable World, Den Haag, The Netherlands.

Wiki. (n.d.). Retrieved from Wikipedia: http://wiki.org/wiki.cgi?WhatIsWiki

Williams, A. M., & Shaw, G. (2011). Internationalization and innovation in Tourism. *Annals of Tourism Research*, *38*(1), 27–51. doi:10.1016/j.annals.2010.09.006

Winch, G. M., & Courtney, R. (2007). The organization of innovation brokers: An international review. *Technology Analysis and Strategic Management*, *19*(6), 747–763. doi:10.1080/09537320701711223

Wit, B., & Meyer, R. (2010). *Strategy: Process, Content, Context* (4th ed.). Cengage Learning EMEA.

Wolf, M. (2018). *Reader, come home: The reading brain in a digital world*. Harper Collins.

Woodhams, C., & Lupton, B. (2009, June). Analysing gender-based diversity in SMEs. *Scandinavian Journal of Management*, *25*(2), 203–213. doi:10.1016/j.scaman.2009.02.006

Woodward, A. (2009). *Nihilism in Postmodernity*. The Davies Group, Publishers.

World Bank. (2002). *Information and Communication Technologies: a World Bank group strategy*. World Bank.

World Business Council. (2007). Promoting Small and Medium Enterprises for Sustainable Development. World Business Council for Sustainable Development.

Wright, L. (2018, January 1). The dark bounty of Texas oil. *The New Yorker.* https://www.newyorker.com/magazine/2018/01/01/the-dark-bounty-of-texas-oil

Wynarczyk, P. (2013). Open innovation in SMEs: A dynamic approach to modern entrepreneurship in the twenty-first century. *Journal of Small Business and Enterprise Development, 20*(2), 258–278. doi:10.1108/14626001311326725

Xiaobao, P., Wei, S., & Yuzhen, D. (2013). Framework of open innovation in SMEs in an emerging economy: firm characteristics, network openness, and network information. *International Journal of Technology Management, 62*(2/3/4), 223-250.

Yadava, J. S., & Mathur, P. (1998). *Mass communication: the basic concepts* (Vol. 1 & 2). Kanishka.

Yen, Yuyun, Shameen, & Rajen. (2013). *SMU Libraries' Role in Supporting SMU's Blended Learning Initiatives.* Retrieved October 27, 2015, from https://library.smu.edu.sg/sites/default/files/library/pdf/Librarys_Role_in_Blended_Learning.pdf

Zahra, A. Z., & George, G. (2002). Absorptive capacity: A review, reconceptualization, and extension. *Academy of Management Review, 27*(2), 185–203. doi:10.5465/amr.2002.6587995

Zeng, S. X., Xie, X. M., & Tam, C. M. (2010). Relationship between cooperation networks and innovation performance of SMEs. *Technovation, 30*(3), 181–194. doi:10.1016/j.technovation.2009.08.003

Zhang, J., & Zhang, Y. (2009). Research on the Process Model of Open Innovation Based on Enterprise Sustainable Growth. *Proceeding of the International Conference on Electronic Commerce and Business Intelligence*, 318-322.

Zuboff, S. (1988). *In the Age of the Smart Machine: The Future of Work and Power.* Basic Books.

About the Contributors

Hakikur Rahman is an academic over 27 years has served leading education institutes and established various ICT4D projects funded by ADB, UNDP and World Bank in Bangladesh. He is currently serving as a Post Doctoral Researcher at the University of Minho, Portugal under the Centro Algoritmi. He has written and edited over 25 books, more than 50 book chapters and contributed over 100 articles on computer education, ICTs, knowledge management, open innovation, data mining and e-government research in newspapers, journals and conference proceedings. Graduating from the Bangladesh University of Engineering and Technology in 1981, he has done his Master's of Engineering from the American University of Beirut in 1986 and completed his PhD in Computer Engineering from the Ansted University, BVI, UK in 2001.

* * *

Lia Almeida has been full Professor at the Federal University of Tocantins, Brazil, since February 2013. She teaches courses in undergraduate courses of Administration and Graduate Program of Regional Development. Dr. Almeida currently researches on policy analysis, studying theories of the policy process. In the Public administration field, her major focus is on the policy capacity of provincial and local governments, while her empirical focus is on environmental policies.

Jorge Bernardino is a Coordinator Professor of Polytechnic of Coimbra - ISEC, Portugal. He received the PhD degree from the University of Coimbra. He has authored over 100 publications in refereed conferences and journals. He was President of ISEC from 2005 to 2010. During 2014 he was Visiting Professor at CMU. Currently, he is Director of the Institute for Applied Research (i2A) at Polytechnic of Coimbra.

Mergen Dyussenov has been appointed Advisor to the Minister of culture and sports (Kazakhstan, May 2020) overseeing anti-corruption policy formulation. Prior to that, Mergen taught at the Academy of Kazakhstan, and served as the Deputy Director. In 2019 Mergen defended a PhD thesis. Until recently, he served as the Deputy Director at the Institute of Management of the Academy and in 2019 defended a PhD thesis at LKY School of Public Policy, NUS (Singapore). His primary interest lies in actor-centered agenda-setting processes, and policy evaluation. Policy areas of interest are corruption, ICT & e-government in Global South, specifically across Central and Southeast Asia.

Markus Hertwig holds the Professorship of Sociology with Specialization in Work and Organization at Chemnitz University of Technology.

Andrew Klobucar, Associate Professor of English at New Jersey Institute of Technology in Newark, New Jersey is a literary theorist and teacher, specializing in internet research, electronic literature, and semantic technologies. His writings on experimental literary forms and genres continue to analyze the increasingly important role technology plays in contemporary cultural practices in both print and screen formats. He has worked on developing software for writing instruction and written on the use of programmable media in classroom instruction.

Anouk Mols is a PhD candidate at the Department of Media & Communication of the Erasmus University Rotterdam. She is currently involved in the 'Mapping Privacy and Surveillance Dynamics in Emerging Mobile Ecosystems' project and her research revolves around everyday privacy and surveillance practices in the context of local communities, workplaces, messaging apps, families, and smart technologies.

Pedro Neves has enrolled in the bachelor degree at Instituto Superior De engenharia de Coimbra (ISEC) in 2008, from which he graduated in 2012. Later that year he enrolled for the Master degree in Computer Science at the same institution. Since the very beginning of his Master, Pedro always carried out research, being enrolled in two different fellowships in which he was introduced to data analytics and machine learning. In 2015, even before graduating from his master (completed in February that year), he was hired by Accenture as data analyst while researching and preparing the research project he carried out under the Carnegie Mellon University - Portugal program later that year, in which his goal was to identify the existent techniques and methodologies to scale big data systems on the cloud. After the CMU-Portugal research fellowship, in 2016, Pedro returned to Accenture for about two years where he work as a data analyst. He left Accenture to join Aaxians in 2018 where he worked as a data scientist for about one more year. In 2019, Pedro worked at Nokia as a Big Data Engineer where he had the chance to work with break-through big data technology. In 2020, Pedro started working for Talkdesk as a Data Engineer/Data Scientist where he is developping methods for anomaly detection. He has recently enrolled in the PhD program at the University of Coimbra.

Christian Papsdorf holds the Junior Professorship in Sociology of Technology with Specialization in Internet and New Media at Chemnitz University of Technology.

Madhuri V. Tikam is working as the Chief Librarian of a well reputed Commerce College located in Mumbai - H R College of Commerce & Economics, Churchgate since 1997. She received Best Teacher Award for Outstanding Contribution to Teaching & Education (Library Education) from Higher Education Forum (HEF) in September 2015. She also received Nomination from SAARC Documentation Center for Workshop on "Essential skills for new age professionals" at Maldives in November 2014 and Scholarship from American Library Association (ALA) for ALCTS Online Course on Fundamentals of Collection Assessment in 2014. She got selected for University System to System Exchange Programme of HSNC Board and University system of Georgia, USA in 2003. She is Chief Editor of "International Journal of Information Resources and Knowledge Management (IJIRKM)" and Guest Editor of International Journal of Information Communication Technologies and Human Development (IJICTHD) Special issue on "The Impact of Information Communication Technologies (ICT) in Education". She authored a book and many journal articles. She offered information literacy programmes and consultancy to many institutions.

Index

F

G

H

I

J

K

L

M

N

O

P

R

S

T

V

W

Printed in the United States
By Bookmasters